Culture, Body, and Language

Applications of Cognitive Linguistics
7

Editors
Gitte Kristiansen
Michel Achard
René Dirven
Francisco J. Ruiz de Mendoza Ibáñez

Mouton de Gruyter
Berlin · New York

Culture, Body, and Language

Conceptualizations of Internal Body Organs across Cultures and Languages

Edited by
Farzad Sharifian
René Dirven
Ning Yu
Susanne Niemeier

Mouton de Gruyter
Berlin · New York

Mouton de Gruyter (formerly Mouton, The Hague)
is a Division of Walter de Gruyter GmbH & Co. KG, Berlin

⊗ Printed on acid-free paper
which falls within
the guidelines of the ANSI
to ensure permanence and durability.

Library of Congress Cataloging-in-Publication Data

Culture, body, and language : conceptualizations of internal body organs across cultures and languages / edited by Farzad Sharifian ... [et al.].
p. cm. – (Applications of cognitive linguistics ; 7)
Includes bibliographical references and index.
ISBN 978-3-11-019622-1 (hardcover : alk. paper)
1. Language and culture. 2. Body, Human, and language. I. Sharifian, Farzad.
P35.C83 2008
401–dc22
2008030011

Bibliographic information published by the Deutsche Nationalbibliothek

The Deutsche Nationalbibliothek lists this publication in the Deutsche Nationalbibliografie; detailed bibliographic data are available in the Internet at http://dnb.d-nb.de.

ISBN 978-3-11-019622-1
ISSN 1861-4078

Printed in Germany

Acknowledgements

This volume grew out of a theme session with the title of "Looking for the Mind inside the Body", which was part of the 9th International Cognitive Linguistics Conference held at Yonsei University, Seoul, Korea, July 17th–22nd, 2005. The editors wish to thank the organizers of the conference for their help and support.

We also wish to thank the authors of the chapters in this volume not only for their valuable contributions but also for their help in serving as internal reviewers of other contributors' chapters. The external reviewers, both invited and anonymous, also deserve a special word of thanks for their helpful comments on the earlier versions of the chapters.

Finally, we are grateful to the series editors of Applications of Cognitive Linguistics as well as Birgit Sievert at Mouton de Gruyter for their help and support during the preparation of this volume.

Table of contents

List of contributors

Farzad Sharifian
School of Languages, Cultures, and Linguistics
Monash University, Melbourne, Australia
Email: farzad.sharifian@arts.monash.edu.au

Ning Yu
Department of Modern Languages, Literatures, and Linguistics
School of International and Area Studies
University of Oklahoma, Norman, Oklahoma, USA
Email: ningyu@ou.edu

Susanne Niemeier
English Department
University of Koblenz-Landau, Germany
Email: niemeier@uni-koblenz.de

Alice Gaby
Linguistics Department
University of California, Berkeley, USA
Email: agaby@berkeley.edu

Poppy Siahaan
Faculty of Linguistics and Literature
University of Bremen, Germany
Email: siahaan@uni-bremen.de

Cliff Goddard
School of Behavioural, Cognitive and Social Sciences
University of New England, Armidale, Australia
Email: cgoddard@une.edu.au

Iraide Ibarretxe-Antuñano
Departamento de Lingüística General e Hispánica
Universidad de Zaragoza, Spain
Email: iraide@unizar.es

Yoshihiko Ikegami
Graduate School
Showa Women's University, Tokyo, Japan
Email: ikegami@swu.ac.jp

Debra J. Occhi
School of International Liberal Arts
Faculty of Comparative Culture
Miyazaki International College
Email: docchi@miyazaki-mic.ac.jp

Kyung Joo Yoon
School of Language Studies
Australian National University, Canberra, Australia
Email: Kyung-Joo.Yoon@anu.edu.au

Daniel P. Wolk
University of Chicago, USA
Email: dannowolk@gmail.com

Dirk Geeraerts
Research Unit Quantitative Lexicology and Variational Linguistics
University of Leuven, Belgium
Email: dirk.geeraerts@arts.kuleuven.be

Caroline Gevaert
Research Unit Quantitative Lexicology and Variational Linguistics
University of Leuven, Belgium
Email: Caroline.Gevaert@kuleuven-kortrijk.be

Ad Foolen
Department of Linguistics
Radboud University Nijmegen, The Netherlands
Email: a.foolen@let.ru.nl

Zouhair Maalej
Department of European Languages and Translation,
College of Languages & Translation,
King Saud University, Riyadh, Saudi Arabia
Email: zmaalej@gnet.tn

A. Introduction

Culture and language: Looking for the "mind" inside the body

Farzad Sharifian, René Dirven, Ning Yu, and Susanne Niemeier

Abstract

The history of mankind has been characterized by attempts to understand the universe and the human being itself. Humans are so uniquely able to understand because they have a "mind" that distinguishes them from lower species. As the centre of some or all of the faculties of feeling, thinking, and knowing, the mind has traditionally been the defining feature of the "animal rationale", i.e., "reason-possessing or rational animal" that a human being is. Not the least striking of it all is that the understanding of the mind is crucial to the understanding of the human body, the self, and particularly human cognition. Looking back and around, however, we find apparently divergent inroads to the understanding of the mind over history and across cultures, as may become more evident in the studies in this volume about the synchronic variation and the diachronic development of "heart" conceptions in various languages. The central aim of this volume is to contribute to the knowledge about various cultures' conceptualizations of the heart and other internal body organs, and in particular about how feeling, thinking and knowing are related to internal body organs in different cultures, as they are reflected in the respective languages.

Keywords: culture and language, conceptualization, mind-body, metaphor, heart.

1. Cultures' search for the locus of the mind

In this chapter we use the word "mind" as a shorthand term for "ways of knowing, thinking, and feeling" of which different cultures, or different periods of the same culture, may have different understandings, as the chapters in this volume will demonstrate. The topic of this cognitive search for the conceptualization of the mind and body organs is not a new one (see also Jahoda 1992; Palmer, Goddard, and Lee 2003; Shweder and Levine 1984; Shweder 1991; Wierzbicka 1989); what is new is the search for the impact of metaphorical thought, if present, and of cultural models that may have channeled these conceptualizations. Both cultural models of the mind and more scientific approaches in philosophy and/or medicine have in various cultures invoked central parts of the human body as the locus of the mind. The major loci have been the abdomen region, the heart region and the head region or, more particularly, the brain region. These three types of conceptualizations can be labeled "abdominocentrism", "cardiocentrism", and "cerebrocentrism" (or "cephalocentrism"), respectively. These three labels only intend to capture the idea that the region in question is the main centre, which does not exclude a similar role for body parts in other regions. This three-fold orientation in the bodily-based conception of the mind has also served as the basis for the volume's three main sections.

Abdominocentrism is found in various parts of the world. It has been – and still is – the predominant approach in the Southern Asian, Polynesian, and other disparate cultures, including the Basque culture. This is exemplified in the present volume in the chapters on Indonesian (Siahaan), Malay (Goddard), and Kuuk Thaayorre (Gaby), which discuss tendencies of these languages and cultures towards abdominocentrism. Malay and Indonesian reveal conceptualizations of *hati* 'liver' as the seat of both emotional and mental activities, and in Kuuk Thaayorre the belly takes part in expressions that are associated with emotions. This is, of course, not to say that in these languages the head has no conceptual significance nor to say that in cardiocentric languages the liver is not associated with any emotional or mental activities, it is simply a matter of which organs play a central role in these conceptualizations. In both Chinese and Greek medicine and philosophy, the liver is still a very important organ. In traditional Chinese medicine, the liver, as one of the five *zang* organs and as "the organ of the general", has the mental function of making up strategies and storing a sort of "soul" (Yu). According to one view in Greek classics, the liver is one of the three most important organs, only superseded in importance by the heart and the brain. It hosts one part of the three-part soul, which comprises "the animal soul", presiding over rational thought and causing sensation and motion, "the vital soul", governing the emotions and providing the life force, and "the natural soul", controlling nutrition. These are seated in the brain, the heart, and the liver respectively (see Erickson 1997: 4).

Cardiocentrism may have been the traditional view in China, Japan and Korea, the three neighbouring countries in East Asia that share certain cultural roots and traditions. In Chinese culture, for instance, traditional Chinese medicine, guided by the theories of the *yin-yang* and of the five elements or phases in ancient Chinese philosophy, has taken a cardiocentric approach, and ancient philosophical and medical thoughts have laid the foundation for the Chinese cultural model of the mind being fundamentally cardiocentric (Yu). This cardiocentrism has been quite consistent and stable in China, as much in traditional Chinese philosophy as in medical thinking. It is interesting to note that the brain was given little attention in traditional Chinese medicine where it was seen more or less as a reservoir for marrows. Variations of cardiocentrism are found in Japanese culture (Ikegami; Occhi) and Korean culture (Yoon) as well, which have both traditionally localized intellectual and emotional activities in and around the heart. In modern Japanese, however, it is noted that the dichotomy between brain/head and heart is similar to that in English (Ikegami), which may represent a change over time owing to the interaction of cultures. The increasing use of the loanword *haato* 'heart' in Japanese is an obvious case of the influence of English in Japan (Occhi).

As we move our eyes toward the west, we find, in the Greek-based West Asian, European and North African cultures, various forms of dualism, e.g. the dualism between mind and matter, including the body, and the dualism between the head/brain (as the seat of the intellect) and the heart (as the seat of emotions). The second form of dualism actually grew out of the earlier interaction between cardiocentrism and cerebrocentrism in the time of ancient Greek philosophy and medicine (see Foolen). Clarke and O'Malley (1968: 1) summarize the history of this interaction from antiquity to early modernity as follows:

> Throughout antiquity, and to some extent up to the end of the eighteenth century, there was controversy on the question of which organ was the seat of the soul, the intellect, the passions, and the guiding force in the control of the motor and sensory phenomena of the body.

This portion of human history deserves a little more attention here because it presents a complicated situation and traces a zigzag path of development.

In ancient Greece, both heart and brain were listed as critical organs of the body in the theory of the four humors, first attributed to Hippocrates, but further refined by Galen (Geeraerts and Grondelaers 1995).[1] As Clarke and O'Malley (1968: 1) point out, although earlier Western civilizations had selected the heart as the central organ, the Greeks were divided on this matter. The controversial question was whether the heart or the brain was the seat of the soul or the central organ of thought and sensation. This controversy can be found in the writings of the Hippocratic physicians (fifth century BC) known as "Hippocratic Writers" or "Hippocratic Corpus" (see Foolen). For instance, the brain was considered to be the most powerful organ of the body in one writing (Clarke and O'Malley 1968: 5), but the same power was ascribed to the heart as the center of intellectual activity in another (Erickson 1997: 199; see also Jager 2001). Thus, according to the latter position, the "'Hippocratic' heart was a strong muscle with a large hot chamber housing the intelligent ruling principle of the soul" (Erickson 1997: 199). Plato believed that the soul was made up of three parts and its most important part, the rational soul, was assigned to the head or brain. According to him, therefore, the brain was the most important organ of the body (Clarke and O'Malley 1968: 5–7). This cerebrocentric view, however, was "the minority opinion" of the time, and the opposing cardiocentric view was advocated by, among others, Aristotle (fourth century BC), according to whom "the heart was the chief organ of the body and the center for thought and the appreciation of sensation" (Clarke and O'Malley 1968: 7–8). While Aristotle "was the foremost proponent of the cardiocentric theory", Galen (first century AD) "was his most vocal opponent" who "marshaled an imposing array of arguments ... to demolish the ancient belief and to establish his own ... that the seat of intelligence, motion, and sensation is the brain, not the heart" (Clarke and O'Malley 1968: 15).

During the medieval period, the heart ensured its primacy. Avicenna (tenth–eleventh century AD), a Persian physician and philosopher who was active in preserving and even extending Greek medical learning, attempted to reconcile Aristotelianism and Galenic medicine but on the whole sided with Aristotle (Clarke and O'Malley 1968: 21; Stevens 1997: 266; see also Sharifian's contribution). At the same time, the Bible was a preeminent authority representing the heart-centered orientation of Christianity. After the Middle Ages, religious and scientific views began to diverge increasingly. While the Catholic Church stuck to cardiocentrism in that it strengthened the belief in and the symbolism of the "Sacred Heart", sixteenth and seventeenth century medical scientists such as Andreas Vesalius and William Harvey, who were engaged in studies of the "real" function of the heart, discovered its functioning as a "pump" for blood circulation (Foolen; Niemeier). But, although there had been a lingering dualism between conceptions of the rational head (with its intellect) and the emotional heart since Plato, it is not until the rationalism of the mathematician and philosopher René Descartes (1596-1650) that the fully outspoken dualism was born. He is commonly regarded as "the prototypical representative of the shift to dualism" (Foolen). The much discussed Cartesian dualism, characterized by a "disembodied rational mind", is a "mind-brain split" as well as a "mind-body split" (Stevens 1997: 268–269; see also Damasio 1994). The diachronic shift to the dominance of dualism in West Asian and Western thought, involving replacement of heart-expressions by head-expressions in related languages (see Sharifian for Persian; Wolk for Northeastern Neo-Aramaic; Niemeier for English; and Foolen for Dutch), is generally linked to the development in medicine and philosophy in the sixteenth and seventeenth centuries, when the "real" function of the heart was discovered.

This general intellectual climate represented by dualism has persisted in the cultural models and scientific theories in the Western world up to the present day. In the twentieth century, it was seriously questioned in philosophy by phenomenology and in language philosophy and linguistics by experientialism or embodied realism (see e.g., Geeraerts 1985, 1993; Johnson 1987; Lakoff 1987; Lakoff and Johnson 1999).

2. Cognitive linguistics' search for the nature of the mind

One of the central themes in cognitive linguistics is the uniquely human development of some higher potential called the "mind" and, more particularly, the intertwining of body and mind, which has come to be known as *embodiment* (e.g., Gibbs 2006; Johnson 1987; Lakoff 1987; Lakoff and Johnson 1999). The embodiment thesis, in general terms, not only views cognition as mediated by our bodily experiences, as has traditional wisdom since Thomas Aquinas (*Nihil in intellectu quod non fuerat in sensu* [There is nothing in the mind that was not previously in the senses]), it also views the more abstract target domains of cognition, e.g. those of thought, emotion and language, as based on concrete source domains such as the human body and the conceptualizations of the internal body parts. In his recent book *Embodiment and Cognitive Science*, Gibbs (2006: 9) provides the following as the "embodiment premise":

> People's subjective, felt experiences of their bodies in action provide part of the fundamental grounding for language and thought. Cognition is what occurs when the body engages the physical, cultural world and must be studied in terms of the dynamical interactions between people and the environment. Human language and thought emerge from recurring patterns of embodied activity that constrain ongoing intelligent behaviour. We must not assume cognition to be purely internal, symbolic, computational, and disembodied, but seek out the gross and detailed ways that language and thought are inextricably shaped by embodied action.

He then adds, "The key feature of this premise is the idea that understanding the embodied nature of human cognition demands that researchers specifically look for possible mind-body and language-body connections."

As Foolen points out, "The observation that the human body is an important source of semiosis fits in with the more general idea of embodied cognition, which is now considered as the foundation of cognitive linguistic thinking." This view, as Foolen exemplifies, has been supported by many empirical studies within the paradigm of cognitive linguistics in the past two decades. While embodiment is programmatic in cognitive linguistic research, similar ideas were proposed earlier in the twentieth century within the context of philosophy and cultural studies as the "corporeal turn" in twentieth-century thinking (for details see Foolen). In cognitive linguistics, the "corporeal turn" had "as its most direct impetus the felt need for extra-linguistic 'grounding' of (formal) linguistic categories. In a first step, language was linked to conceptualization processes. This step was, in fact, the starting point of the cognitive enterprise in the 1980s. Soon it was realized that cognition is embodied."

The exact relation between the body and human cognition and the interpretations given to the terms "cognition" and "body", however, have varied, from overlapping views to conflicting and contrasting ones (e.g., Violi 2004, 2008; Sharifian 2003; see also Wilson 2002 for six views of embodied cognition).

Lakoff and Johnson (1980, 1999) explore the embodiment of cognition as laid down and reflected in conceptual metaphors, which are prevalent in language and underlie the hundreds of clusters of actual metaphorical linguistic expressions. As the conceptual metaphor THE MIND IS THE BODY may illustrate, for them the more abstract conceptualizations, such as that of the human mind, have only been made possible thanks to the bodily-based conceptual tools of metaphor and metonymy. As heuristic instruments, these automatically take the body domain to explore and understand one of the most abstract domains, i.e., that of the mind. In Lakoff and Johnson's framework metaphor and metonymy are not purely imaginative leaps for the purpose of mere aesthetic sense-creation, they are rather more fundamentally rooted in and motivated by the bodily experiences of humans.

One critique of Lakoff and Johnson's views is that they only see embodiment in relation to individual agents, not in relation to humans as social agents (see Frank et al. 2008). Another point, recognized, but not explored by Lakoff and Johnson, is that the conceptualization of abstract domains can very well come about without metonymy or metaphor, although metonymy and metaphor often enrich the understanding of more abstract concepts (see Haser 2005 for a fundamental critique). In other words, while metonymy and metaphor are very important instruments in abstract conceptualization, they are not the only instruments in the conceptualization of the more abstract world (see especially Geeraerts and Gevaert).

Another crucial question in studies of (embodied) conceptual metaphors, which has occasionally been touched upon by Lakoff and Johnson, but which has, with the exception of Kövecses (e.g., 2005), Yu (e.g., 2003, 2007), and a few others, hardly ever been systematically explored, is the role of culture and its relationship to conceptual metaphors. This question about the role of culture naturally involves cognitive anthropology, and in fact it was turned into a critical question in the volume *Cultural Models in Language and Thought*, edited by two cognitive anthropologists, Dorothy Holland and Naomi Quinn. In the volume's introduction, the central analytical tool is presented as *cultural models*, which Quinn and Holland (1987: 4) define as follows:

> [Cultural models are] presupposed, taken-for-granted models of the world that are widely shared (although not necessarily to the exclusion of other, alternative models) by the members of a society and that play an enormous role in their understanding of that world and their behaviour in it.

On the instigation of Roger Keesing, the notion of *cultural model* was accepted in place of *folk model* in the volume's title and by most of the contributors. This was meant to replace the traditional anthropological myth of the communal knowledge of the "folk", offering instead a more refined combination of knowledge in individual minds and socially distributed knowledge (see Keesing 1987: 373–375). In the same vein, a cultural model is defined by D'Andrade (1987: 112) as "a cognitive schema that is intersubjectively shared by a social group". The use of the term *schema* originated in cognitive psychology (e.g., Bartlett 1932; Rumelhart 1980; Rumelhart et al. 1986), and this definition builds further on the notion of *schema* as a mental set or representation.

A typical linguistic contribution to this exploration of cultural models in Holland and Quinn (1987) is the chapter on "The cognitive model of anger inherent in American English" by Lakoff and Kövecses, who view metaphors such as *You make my blood boil* as elements of the cultural model of the physiological effects of anger. This cultural model is defined by Lakoff and Kövecses (1987: 196) as follows:

- THE PHYSIOLOGICAL EFFECTS OF ANGER ARE INCREASED BODY HEAT, INCREASED INTERNAL PRESSURE (BLOOD PRESSURE, MUSCULAR PRESSURE), AGITATION, AND INTERFERENCE WITH ACCURATE PERCEPTION.
- AS ANGER INCREASES, ITS PHYSIOLOGICAL EFFECTS INCREASE.
- THERE IS A LIMIT BEYOND WHICH THE PHYSIOLOGICAL EFFECTS OF ANGER IMPAIR NORMAL FUNCTIONING.

For Lakoff and Kövecses (1987), the relation between metaphor and cultural model is such that the metaphors *constitute* a cultural model. Quinn (1987: 176) and Quinn and Holland (1987: 24), however, think that metaphors rather *reflect* cultural models. And Quinn (1991) explicitly rejects the claim that metaphors constitute cultural models. More specifically, in her analysis of the cultural model of American marriage, Quinn (1991) argues that metaphors are selected by people because they fit into already existing cultural models, a position also advocated recently by Wee (2006) for the cultural model of modernity and the way it operates in people's search for local identities in Singapore. Kövecses (1999) addresses the problem directly in his paper entitled "Metaphor: Does it constitute or reflect cultural models?" He proposes a wider definition of the marriage model arguing that the model of marriage cannot be detached from the concept of love, which is centered around the metaphorically conceptualized notion of unity, namely the conceptual metaphor MARRIAGE IS THE PHYSICAL UNITY OF TWO COMPLEMENTARY PARTS underlying the conceptualization of various social, legal, psychological, and political unities as well as love. He, therefore, embeds the treatment of the marriage model in a general discussion of the relation between conceptual metaphors and cultural models (see Dirven, Wolf, and Polzenhagen 2007).

Lakoff and Johnson (1999) refer to the role of culture in relation to *primary* and *complex* metaphors. They maintain that "complex, everyday metaphors are built out of primary metaphors plus forms of commonplace knowledge: cultural models, folk theories, or simply knowledge or beliefs that are widely accepted in a culture" (Lakoff and Johnson 1999: 60). This decompositional account of conceptual metaphors in terms of primary and complex metaphors adds greatly to the explanatory power of conceptual metaphor theory regarding the questions of whether and how specific conceptual metaphors are potentially universal, widespread, or culture-specific. However, Lakoff and Johnson do not make a clear distinction between "folk theories" (knowledge based in a so-called collective mind) and "cultural models" (shared knowledge based in individual minds). They seem to use the notions of "folk" and "cultural" interchangeably; parallel to this, they use the terms "model" and "belief" interchangeably. For example, they refer to the idea that "people are supposed to have purposes in life, and that they are supposed to act so as to achieve those purposes" as both a "folk model" and a "cultural belief". This statement is in fact what is known as a *proposition schema* (Quinn 1987: 179). The notion of culture did not become a central theoretical concern in Lakoff's further research, which has increasingly concentrated on the neural basis of language, metaphor and thought and is culminating into his Neural Theory of Language (NTL).

In spite of this later non-culturally-oriented development in Lakoff's work, the priority to be attributed to culture and to the cultural components in the conception of "body" itself (see Violi 2004) or of any of the body parts has challenged linguistics again and again, and from different perspectives. While Lakoff and Kövecses (1987) presented their analysis of the American concept of anger as reflecting a universally valid physiological process, Geeraerts and Grondelaers (1995: 162) show that many linguistic expressions of anger have a specific cultural basis and largely conform to the cultural model furnished by the theory of the four humors, according to which anger is caused by yellow bile or choler, the production of which may be stimulated by certain kinds of food. As stated before, the message to cognitive linguists in general is that we, while focused on metaphorical conceptualizations, should look out for all onomasiological realizations, including the non-metaphorical part of conceptions that come to us as whole packages (see Geeraerts and Gevaert).

In his later work, Kövecses (1995, 2005) has taken up the "cultural" challenge, discussing the controversy that surrounds the relationship between metaphor, body, and culture and making an attempt to resolve the issue by referring to data from four languages: English, Hungarian, Japanese, and Chinese. Noting some differences as well as similarities across these languages in terms of conceptualizations of anger, Kövecses (1995: 70) arrives at five elements, "(possibly universal) actual human physiology, conceptualized physiology (metonymy), metaphor, cultural model (with its schematic basic structure), and the broader cultural context". He maintains that the cultural models of anger across the languages that he selected "are the joint products of metaphor, metonymy, (possibly universal) actual physiology, and cultural context". Here, Kövecses uses the notion of "cultural context" to refer to cultural systems such as the four humors or traditional Chinese medicine (see Kövecses 2005: 234).

The position that we adopt in this volume is that cultural models are complex conceptual systems that act as building blocks of a cultural group's *cultural cognition* (see Sharifian 2008). We maintain that cognitive systems such as traditional Chinese medicine and the theory of the four humors also lend themselves to the notion of cultural model in that they provide the members of a cultural group with "templates" for understanding certain aspects of their lives (see also Yu 2007). The locus of such systems may be a conceptual system developed by one individual, but over time they become part of the cultural cognition of a cultural group to the extent that after several generations, many members of the cultural group may not know about the origin of such conceptual systems anymore. Thus, any divide between the original conceptual systems, such as the theory of the four humors, and the metaphorical/metonymic systems that have arisen from them would be arbitrary.

If we agree to also view deeply internalized systems of conceptualization, such as the theory of the four humors, as *cultural models*, then it can be argued that metaphors can be *rooted* in certain cultural models, rather than just *reflecting* them. It should be added that cultural models that are driven by the ethnomedical traditions, such as the four humors or the five elements, are sometimes the result of conceptual mappings in the sense that they may reveal mappings from one category to another. For example, the theory of the five elements categorizes and conceptualizes the universe and the human body in terms of the five elements (wood, fire, earth, metal, and water). In a sense, in such cases, conceptual mappings may be viewed as fundamental to the construction of cultural models.

The above argument of course does not contradict the view that metaphors are very often grounded in immediate bodily experience, but expands it so as to include the mediating role of cultural models, such as those embodying ethno-medical/religious beliefs. For example, the metaphorical system of anger in English can then be viewed as originating from and being part of the cultural model of the four humors. One might of course undertake to identify a conceptual system, for example of ANGER, that is currently reflected in a particular language and call it "the cultural model of ANGER in language X", but it should be kept in mind that the identified conceptual system may be part of a more complex cultural model that has its own roots in the history of the cultural cognition of the cultural group in question. This observation may act as a precaution in ascribing certain cognitive attributes to the individual speaker of a language. It may also shed light on the intricate relationship between language, body, conceptualization, and culture.

It should be added here that since cultural models originate from various sociocultural traditions within the history of a community, the cultural cognition of the group as a whole may have several cultural models for a particular domain of conceptualization, such as that of emotions (Sharifian 2008). This is reflected in the fact that often one finds sets of linguistic expressions that reflect different, or even conflicting, cognitive models for the same experience. It should also be noted that not all cultural models of human experience are associated with the body. In some cultures, such as many Aboriginal Australian cultures, even the sources of diseases that affect the body are conceptualized in terms of non-bodily cultural models that are rooted in the Aboriginal worldviews.

3. Insights gained in and issues raised by this volume

The observations noted above provided an incentive to bring together a group of scholars from various language backgrounds to make a collective attempt to further explore the relationship between mind, body, language, and culture, by focusing on conceptualizations of the heart and other internal body organs across a number of languages. The general aim of this venture is to explore (a) the ways in which internal body organs have been employed in different languages to conceptualize human experiences such as emotions and/or workings of the mind, and (b) the cultural models that appear to account for the observed similarities as well as differences of the various conceptualizations of internal body organs.

While working on this volume, we found that several general patterns emerged from the studies included here. First of all, all the languages included reveal a tendency to employ the domain of internal body organs as a source of conceptualizations for the human faculties of thought and emotion, for personality traits, for mood, etc. This is not astonishing, in view of the fact that people all over the world, independent of their cultural backgrounds and culture-specific conceptualization of the body itself, do have bodies that are basically similar and therefore necessarily share bodily experiences. Such bodily experiences are immediately noticeable and whenever we find that something is happening in or with our bodies, we try to look for explanations, be they lay views or expert ones. Therefore, explaining the body is arguably a general human need. One step beyond this basic interest in the body, we not only try to explain our bodies themselves, but also other, not immediately body-related aspects of our selves, by referring to our bodies. Even in the age of neuroscience we identify more with our bodies than with the "other entity" – as Goddard calls it – since that "other entity" is too abstract and still too little understood. This is why we tend to metonymize and to metaphorize such experiences by using body-related terms. Still, we want to stress again that we are not exclusively dealing with metaphor here, as such conceptualizations can also come about metonymically or even literally, as was for example the case in Old English (Geeraerts and Gevaert).

Second, the observations made in this volume suggest similarities and variations in different languages and cultures in terms of conceptualizations of internal body organs. The individual papers show that such conceptualizations are mainly due to underlying cultural models which in closely related cultures are often shared. Such observations have led us, as stated in Section 1, to subdivide the volume into three parts. Although each part includes widely differing languages, these nevertheless show certain patterns of similarity among themselves. As pointed out in Section 1, we distinguish between abdominocentrism, cardiocentrism, and cerebrocentrism. The first part of the volume includes languages which are guided by – tentatively speaking – cultural models of "animistic spirituality" and which show abdomen-centering conceptualizations. The second part includes languages which are guided by philosophical and medical cultural models and which show holistic heart-centering conceptualizations, whereas the third part includes languages which are guided by philosophical, medical and/or religious cultural models, showing dualistic heart/head-centering conceptualizations that have grown out of historical interaction between cardiocentrism and cerebrocentrism.

The languages included in the first part tend to locate emotions in the abdomen region.[2] Thus, in Kuuk Thaayorre (Gaby) emotions are mainly located in the belly. Simplifying somewhat, in Kuuk Thaayorre this is due to the Aboriginal worldview, in which food has a significant role in determining spiritual and physical health. The place of the food, i.e. the belly, is also connected to spiritual and physical health. In Indonesian (Siahaan) as well as in Malay (Goddard) – two related languages – the central body-organ for emotion concepts is the liver, which may be due to the old ritual of liver divination in which the liver was seen as the central inner organ by means of which spiritual beings communicate with human beings. Basque

(Ibarretxe-Antuñano) is a borderline case in this first part, as it connects emotions to various inner organs, namely the heart, the liver, the stomach and the intestines, and uncovers different cultural models in order to account for the various emotion conceptualizations. The heart is primarily connected with positive emotions, which the author attributes to the influence of Cartesianism (see Part 3 for full-blown Cartesian cultural models), and the liver is connected with negative feelings (which is explained by the theory of the four humors). The stomach and the intestines are not very productive domains in emotion conceptualizations and the author intimates that we are not dealing with a cultural model here, but with a more universal instantiation of embodiment. While Basque could also have been situated within Part 2, together with the "holistic" group of languages, in our view it is a member of the first group because of the existence of the concept *gogo*, a "primitive thought" and an "intellectual reasoning process based on intuition and emotion". *Gogo* is untranslatable and somewhat hard to understand for non-members of Basque culture and thus, as an "(animistic-)spiritual" cosmovision, warrants the inclusion of this paper in part one. Basque differs strongly from other cultures in Europe, where the heart was once seen as the centre of both the intellect and the emotions (see Section 1 above and the contribution by Geeraerts and Gevaert). Although Basque has obviously been influenced by the dominant religious and philosophical streams in Europe, it has not completely adopted them, but has developed its mixed system of cardio-abdomino-centrism.

The second part of the volume is devoted to Chinese, Japanese, and Korean, which are all East Asian languages that display holistic heart-centering conceptualizations. The "heart" conceptualizations in these languages show a remarkable similarity to each other. Of course this unity is only relative, as apart from the similarities we also detect differences. Still, all the languages in this part see the heart as the central organ of the body, which determines both emotion and thought, although the brain may play a role in thought processing as well (see, especially, Ikegami for Japanese).

The chapter on the Chinese heart by Yu explores the origin of the conceptualization of *xin* 'heart' in ancient Chinese philosophy and traditional Chinese medicine. The heart is traditionally believed to be the central faculty of cognition as well as being the physiological center of the human being – and even in a cosmic view as the "mirror of the universe". The heart is seen as governing the body, including the brain. This cultural model is still widely manifested in the Chinese language today, in spite of newer scientific knowledge. Crossing over to Japan, we have included two studies which both focus on the meaning of heart, but from different perspectives. One study (Ikegami) closely analyzes the different expressions that Japanese possesses for "heart" and points out their origins (a Chinese loanword for the organ itself, a highly polysemous native word for the workings of the heart, and an English loanword for emotions connected with the heart). The other study (Occhi) shows that Japanese has begun to be influenced by the English way of conceptualizing the heart. Occhi's chapter reveals that the conceptualizations of the heart as the seat of emotions did not exist in ancient Japanese but that in contemporary Japanese the term *haato* is used in conceptualizing emotions such as that of romantic love, which appears to be an influence from English. The final language considered in this part of the volume is Korean (Yoon), in which we also find three different concepts roughly corresponding to the English concept of "heart", reflecting the Korean cultural conceptualization of the human faculties related to emotion and cognition. One of these concepts refers to the body organ as such, just as in Japanese; another is a highly polysemous concept referring to the body part, as well as to the psychological faculties of a person, again similar to Japanese; the third concept refers to psychological and mental faculties and can be used as a metonymy and/or metaphor for a person.

The third part of the volume discusses languages which are guided by philosophical and/or religious cultural models, showing dualistic heart/head-centering conceptualizations. Prototypical examples can be found in the major Indo-European languages, which follow the cultural model of the Cartesian split between the mind and the body or, more concretely, between the head and the heart. Cartesianism was a major influence in all of Western Europe, which is why the languages in question can claim reasonably similar conceptualizations of these body organs. What at first sight may appear to be somewhat surprising is that this similarity is even more widespread than in Western Europe alone, since also North African cultural models show dualistic heart/head-centering conceptualizations (Northeastern Neo-Aramaic; Tunisian Arabic; Persian – which, however, is an Indo-European language). As pointed out in Section 1, the remarkable unity of these cultural models of the body may be seen as a continuation of the ideas in ancient Greek-Roman civilization and its empires. Of course, similarity does not mean uniformity, as competing cultural models have also left their imprints in the languages analyzed. Thus Northeastern Neo-Aramaic has been influenced by Ancient Syriac thought on ethnomedicine and by early Christian mythical thought; Tunisian Arabic has been influenced by the

Koran; and Persian has been influenced by Sufism. Still, we consider the languages similar enough to justify having them in one single part of this volume, as they are the only examples in this volume where we find clear reflections of a dualistic conceptualization. This may be seen as proof that cultural models do not align with language families but can transcend boundaries and frontiers. Cultural transfer of scientific ideas, albeit outdated today, has always been an important issue, and its repercussions are still noticeable in the languages in question today. However, we have to be aware of the fact that the differences between these languages are just as strong as their commonalities.

Concerning Persian (Sharifian), we find a correspondence between conceptualizations of *del* 'heart' in spiritual accounts of Sufism, Avicenna's version of the theory of the four humors, and the metaphorical expressions in Persian that refer to the internal body organ "heart". In Northeastern Neo-Aramaic (Wolk), we find a lot of correspondences with English and also traces of the Galenic model. However, we do not find complete similarity concerning the place of the heart in the system of body organs, as the expression *libbā* also encompasses the stomach. Concerning the Galenic humor theory, the organs liver and spleen are largely absent, whereas the differentiation between wet/dry and hot/cold properties – not markedly present in English – has survived. As a noticeable difference from English, Northeastern Neo-Aramaic also entails a cultural model of morality.

Turning to English, it seems that at some earlier stage emotions and reason were both associated with the heart (Geeraerts and Gevaert), similarly to the languages analyzed in the second part of this volume. However, for the sake of diachronic continuity we have decided to place the chapter on Old English in the third section. Geeraerts and Gevaert argue very convincingly that Old English expressions referring to "mood" – rather than those for "heart" – were to a large extent used non-metaphorically. Still, the humoral doctrine on which the later Middle English and Early Modern English view was founded has been superseded by newer models and increasingly the mind has come to be associated with the head. Even emotions are being increasingly linked to the head, thanks to the advancements in neurobiological medical research. This of course begs the question of why such a change can come about at all. Language change is always a very slow process, and today's language generally mirrors a worldview that is no longer completely present. But we are not only talking about a change in lexis or syntax; what we are talking about is a change on the conceptual level which is then mirrored by the changing use of internal body-part expressions.

Niemeier discusses the nearly absolute dichotomy between head and heart expressions in present-day English, which is mainly due to the overwhelming influence of Cartesian philosophy. Similar results have been found for Dutch (Foolen). And also for Dutch, we find a similar development as for English, as Middle Dutch still maintained a cardiocentric view which is lost in today's Dutch. This reflects the fact that in language change, expressions may increasingly become used non-literally, which in the end makes them seem to be metaphors, at least for the average language user who has no way to, and no motivation to, trace them back etymologically (cf. Niemeier 2000, who argues that all English "heart" expressions ultimately rely on metonymies).

Finally, Tunisian Arabic (Maalej) also shows a Cartesian split, but at the same time it has been influenced by a cultural model founded on a religious tradition, namely that of the Koran. The Koran teaches a conflated model of body and mind, but, owing to translations of Greek philosophy into Arabic, this view has progressively been replaced by a view of body and mind as separate, so that today the heart is no longer associated with mental faculties.

The significance of the internal body organs in human beings' conceptualization is partly physiologically motivated, an issue that is pointed out in several chapters in the volume. For example, the physiological effect of an emotional experience on an internal body organ may lead to conceptualizing the affected organ as the seat of the emotion. Thus, in romantic situations, the increased heart beat that is often experienced leads to the conceptualization of THE HEART AS THE SEAT OF LOVE. In this sense, the conceptualization may be viewed as a metonymically-based metaphor. But it should be stressed that physiological motivation does not seem to play a major role in conceptualizations of internal body organs, otherwise there would be many more similarities across different languages and cultures. As Gaby puts it, "although biological events may in some cases inspire the conceptualization of emotion (and other non-corporeal experiences), this conceptualization is in no way predetermined by the biological events, which may be differently interpreted and described according to the cultural models resulting from the major preoccupations of the people and communities concerned".

Overall, the chapters in this volume suggest a strong role for internal body organs in the human conceptualization of internal and external experiences. They also establish significant links between conceptualizations of internal body organs, linguistic expressions, and cultural models of various kinds that have prevailed at different stages in the history of different speech communities. Further research, as Foolen suggests, "should thus continue to explore three levels of analysis: the language-specific details, the culture-specific generalities ('Western cultures', 'Chinese culture', etc.) and the universal constraints that play a role in the embodied conceptualization of inner life".

Finally, it should be emphasized again that English terms such as *emotion* and *intellect* refer to faculties or experiences that may not be conceptualized similarly across different cultures (see for example Harkins and Wierzbicka 2001; Wierzbicka 1999); after all, these terms capture human construals of experience rather than objective entities in the world. This is the price that research on culture and language pays by publishing in English, an inevitable choice if one wants to reach a wide, international audience.

Acknowledgements

The authors wish to thank Dirk Geeraerts and Cliff Goddard as well as Ad Foolen for valuable hints on the wider cognitive research in the field and the relation of cognitive-linguistic thinking to the phenomenologist tradition.

Notes

1. In a nutshell, the four organs and their corresponding fluids (or humors) are: the liver (yellow bile), the heart (blood), the brain (phlegm), and the spleen (black bile). Of these four organs, however, only the heart or the brain were regarded as the primary or central organs of the body, according to either the cardiocentric or the cerebrocentric view (see also Yu for a profound discussion of the theory of the four humors in comparison with the Chinese theory of the five elements in traditional Chinese culture).
2. Since the theme session at ICLC 9, from which this volume has grown, originally focused on the heart and other internal body organs as the seat of emotion, the papers in this first part have mainly concentrated on emotion, not on both emotion and thought like the two other parts, which for obvious reasons had to do so. Of course, this is not to say that the languages in the first part have not conceptualized the domain of thought as well, only it was not in focus in these papers. The domain of thought is never absent, though. Thus, for example, thought is associated with the ear in Kuuk Thaayorre (Gaby) and with the liver in Malay (Goddard) and Indonesian (Siahaan).

References

Bartlett, Frederic Charles
1932 *Remembering: A Study in Experimental and Social Psychology.* Cambridge: Cambridge University Press.

Clarke, Edwin, and Charles Donald O'Malley
1968 *The Human Brain and Spinal Cord: A Historical Study Illustrated by Writings from Antiquity to the Twentieth Century.* Berkeley/Los Angeles: University of California Press.

Damasio, Antonio
1994 *Descartes' Error: Emotion, Reason, and the Human Brain.* New York: Putnam.

D'Andrade, Roy
1987 A folk model of the mind. In *Cultural Models in Language and Thought*, Dorothy Holland, and Naomi Quinn (eds.), 112–148. New York: Cambridge University Press.

Dirven, René, Hans-Georg Wolf, and Frank Polzenhagen
2007 Cultural studies. In *Handbook of Cognitive Linguistics*, Dirk Geeraerts, and Hubert Cuyckens (eds.). Oxford: Oxford University Press.

Erickson, Robert A.
1997 *The Language of the Heart, 1600–1750.* Philadelphia: University of Pennsylvania Press.

Frank, Roslyn M., Enrique Bernárdez, René Dirven, and Tom Ziemke (eds.)
2008 *Body, Language, and Mind. Vol. 2: Sociocultural Situatedness.* Berlin/New York: Mouton de Gruyter.

Geeraerts, Dirk
1985 *Paradigm and Paradox: Explorations into a Paradigmatic Theory of Meaning and its Epistemological Background.* Leuven: Universitaire Pers.
1993 Cognitive linguistics and the history of philosophical epistemology. In *Conceptualizations and Mental Processing in Language*, Brygida Rudzka-Ostyn, and Richard A. Geiger (eds.), 53–79. Berlin/New York: Mouton de Gruyter.

Geeraerts, Dirk, and Stefan Grondelaers
1995 Looking back at anger. Cultural traditions and metaphorical patterns. In *Language and the Construal of the World*, John R. Taylor and Robert E. MacLaury (eds.), 153–179. Berlin: Mouton de Gruyter.

Gibbs, Raymond W.
2006 *Embodiment and Cognitive Science.* Cambridge: Cambridge University Press.

Harkins, Jean, and Anna Wierzbicka (eds.)
2001 *Emotions in Cross-Linguistic Perspective*. Berlin: Mouton de Gruyter.

Haser, Verena
2005 *Metaphor, Metonymy, and Experientialist Philosophy: Challenging Cognitive Semantics*. Berlin/New York: Mouton de Gruyter.

Holland, Dorothy, and Naomi Quinn (eds.)
1987 *Cultural Models in Language and Thought*. New York: Cambridge University Press.

Jager, Eric
2001 *The Book of the Heart*. Chicago: The University of Chicago Press.

Jahoda, Gustav
1992 *Crossroads between Culture and Mind: Continuities and Change in Theories of Human Nature*. New York: Harvester Wheatsheaf.

Johnson, Mark
1987 *The Body in the Mind: The Bodily Basis of Meaning, Imagination, and Reason*. Chicago: The University of Chicago Press.

Keesing, Roger
1987 Models, 'folk' and 'cultural': Paradigms regained? In *Cultural Models in Language and Thought*, Dorothy Holland, and Naomi Quinn (eds.), 369–393. New York: Cambridge University Press.

Kövecses, Zoltán
1995 Metaphor and the folk understanding of the anger. In *Everyday Conceptions of Emotions: An Introduction to the Psychology, Anthropology, and Linguistics of Emotion*, James A. Russell, José-Miguel Fernández-Dols, Anthony S. R. Manstead, and Jane C. J. Wellenkamp (eds.), 49–71. Dordrecht: Kluwer.
1999 Metaphor: Does it constitute or reflect cultural models? In *Metaphor in Cognitive Linguistics*, Raymond W. Gibbs Jr., and Gerard J. Steen (eds.), 167–188. Amsterdam: John Benjamins.
2005 *Metaphor in Culture: Universality and Variation*. Cambridge: Cambridge University Press.

Lakoff, George
1987 *Women, Fire, and Dangerous Things. What Categories Reveal about the Mind*. Chicago: The University of Chicago Press.

Lakoff, George, and Mark Johnson
1980 *Metaphors We Live By*. Chicago: The University of Chicago Press.
1999 *Philosophy in the Flesh. The Embodied Mind and its Challenge to Western Thought*. New York: Basic Books.

Lakoff, George, and Zoltán Kövecses
1987 The cognitive model of anger inherent in American English. In *Cultural Models in Language and Thought,* Dorothy Holland, and Naomi Quinn (eds.), 195–221. New York: Cambridge University Press.

Niemeier, Susanne
2000 Straight from the heart: Metonymic and metaphorical explorations. In *Metaphor and Metonymy at the Crossroads*, Antonio Barcelona (ed.), 195–211. Berlin: Mouton de Gruyter.

Palmer, Gary B., Cliff Goddard, and Penny Lee (guest eds.)
2003 *Talking about Thinking across Languages*. Special Issue of *Cognitive Linguistics* 14 (2).

Quinn, Naomi
1987 Convergent evidence for a cultural model of American marriage. In *Cultural Models in Language and Thought,* Dorothy Holland and Naomi Quinn (eds.), 173–192. New York: Cambridge University Press.
1991 The cultural basis of metaphor. In *Beyond Metaphor: The Theory of Tropes in Anthropology*, James W. Fernandez (ed.), 56–93. Stanford: Stanford University Press.

Quinn, Naomi, and Dorothy Holland
1987 Introduction. In *Cultural Models in Language and Thought*, Dorothy Holland and Naomi Quinn (eds.), 3–40. New York: Cambridge University Press.

Rumelhart, David E.
1980 Schemas: The building blocks of cognition. In *Theoretical Issues in Reading Comprehension: Perspectives in Cognitive Psychology, Linguistics, Artificial Intelligence and Education*, Rand J. Spiro, Bertram C. Bruce, and William F. Brewer (eds.), 33–58. Hillsdale, NJ: Lawrence Erlbaum.

Rumelhart, David E., Paul Smolensky, James L. McClelland, and Geoffrey E. Hinton
1986 Schemata and sequential thought processes in PDP models. In *Parallel Distributed Processing: Explorations in the Microstructure of Cognition. Vol. 2: Psychological and Biological Models*, David E. Rumelhart, and James L. McClelland (eds.), 7–57. Cambridge, MA: The MIT Press.

Sharifian, Farzad
2003 On cultural conceptualisations. *Journal of Cognition and Culture* 3: 187–207.
2008 Distributed, emergent cultural cognition, conceptualisation, and language. In *Body, Language, and Mind. Vol. 2: Sociocultural Situatedness*, Roslyn M. Frank, René Dirven, Tom Ziemke, and Enrique Bernárdez (eds.), 109–136. Berlin/New York: Mouton de Gruyter.

Shweder, Richard A.
1991 *Thinking Through Cultures*. Cambridge: Harvard University Press.

Shweder, Richard A., and Robert A. Levine (eds.)
1984 *Culture Theory: Essays on Mind, Self, and Emotion*. New York: Cambridge University Press.

Stevens, Scott Manning
1997 Sacred heart and secular brain. In *The Body in Parts: Fantasies of Corporeality in Early Modern Europe*, David Hillman, and Carla Mazzio (eds.), 263–282. New York/London: Routledge.

Violi, Patrizia
2004 Embodiment at the crossroads between cognition and semiosis. *Recherches en Communication* 19: 199–217.
2008 Beyond the body: Towards a full embodied semiosis. In *Body, Language, and Mind. Vol. 2: Sociocultural Situatedness*, Roslyn M. Frank, René Dirven, Tom Ziemke, and Enrique Bernárdez (eds.), 53–76. Berlin/New York: Mouton de Gruyter.

Wee, Lionel
2006 The cultural basis of metaphor revisited. *Pragmatics and Cognition* 14 (1): 111–128.

Wierzbicka, Anna
1989 Soul and mind: Linguistic evidence for ethnopsychology and cultural history. *American Anthropologist* 91 (1): 41–58.
1999 *Emotions Across Languages and Cultures: Diversity and Universals*. Cambridge: Cambridge University Press.

Wilson, Margaret
2002 Six views of embodied cognition. *The Psychonomic Bulletin and Review* 9 (4): 625–636.

Yu, Ning
2003 Metaphor, body, and culture: The Chinese understanding of *gallbladder* and *courage*. *Metaphor and Symbol* 18 (1): 13–31.
2007 The Chinese conceptualization of the heart and its cultural context: Implications for second language learning. In *Applied Cultural Linguistics: Implications for Second Language Learning and Intercultural Communication*, Farzad Sharifian, and Gary B. Palmer (eds.), 65–85. Amsterdam/Philadelphia: John Benjamins.

B. Abdomen-centering conceptualizations

Gut feelings: Locating intellect, emotion and lifeforce in the Thaayorre body

Alice Gaby

Abstract

The present paper is concerned with the conceptualization of mental processes, emotions, and life force in Kuuk Thaayorre, a Paman language of Cape York Peninsula. Body part terms are ubiquitous in Thaayorre descriptions of emotional, mental and spiritual activity. I argue that this both reflects and constructs the conceptualization of these experiences as embodied. The purposes of this paper are threefold. The first is to present these various metaphorical and metonymic associations. The second is to propose some overarching conceptual metaphors that account for the regularity with which a particular body part is linked to a particular experience. The third goal of this paper is to outline the cultural models (the ethnomedical system, cosmovision and broader linguistic context) that underlie such associations. Having done this, I argue that the *ngeengk* 'belly' is conceptualized as the primary site of emotion, spirit and life force (although other internal organs, such as the *thip* 'liver', man 'throat' and *man-ngeengk* 'heart', are also associated with these), whilst the intellect is metaphorically situated in the *kaal* 'ear' (though also secondarily in the *meer* 'eye' and *paant-thuur* 'brain'). I conclude that both cultural traditions and physiological responses to internal experiences inspire these conceptual metaphors.

Keywords: Kuuk Thaayorre, emotion, conceptualization, metaphor, Australian, body, Pama-Nyungan, metonymy.

1. Introduction

1.1. Locating intellect, emotion and life force in the human body

Around the world, the body is repeatedly invoked in the conceptualization and description of mental processes, emotions, and life force (hereafter, all subsumed under the term *internal experiences*). Why should this be so? These internal experiences frequently give rise to physiological symptoms; as abundantly illustrated in Ekman and Davidson (1994), we may infer that a person is happy if we see them smile, or that they are thinking if their brow is furrowed. Similarly, there are a range of physiological characteristics by which we judge whether someone is alive or dead (e.g. heart beat, breathing, response to stimuli). Since we cannot access other people's internal states directly, these physical indexes play an important role in our understanding of internal experiences. It has also been well documented that the human body and associated experiences help shape conceptualization more generally (Johnson 1987, Lakoff and Johnson 1999). Similarly, the tendency for languages to express more abstract, intangible concepts in terms of the more concrete (cf. Sweetser 1990) gives rise to the body's frequent use as a metaphorical source domain across the languages of the world, the body being arguably the most concrete and familiar object in our personal universes (see, e.g., Kövecses 2002).

But although the human physical form remains basically constant across all human communities, it is less clear that the same associations between physiological symptoms and internal states are made in culture after culture. So whilst some researchers (e.g. Kövecses 1995) stress the "universal" patterns in how internal experiences are conceptualized in terms of the body, others (e.g. Van Geert 1995, Enfield and Wierzbicka 2002, Kövecses 2002, and Yu 2002) emphasize the cultural and cognitive forces that mediate the interpretation of physiological and internal experiences. For instance, even within a single speech community, one emotional state may be associated with a number of different physical organs and physiological symptoms. So if a person is in love, they might experience a palpitating heart, sweaty palms, flushed cheeks, light-headedness or a fluttering stomach, to name just a few symptoms to which English-speakers attend. But the converse is also true: a single physiological experience may be symptomatic of a number of different emotions. A palpitating heart might signify that a person is in love, but it could also be a symptom of nervousness, anger, enthusiastic excitement, competitiveness, or may simply follow from the fact that they climbed the stairs too fast. A speech community, then, can pick out any one of these emotion-and-symptom pairings to create an idiom.

Numerous researchers have stressed the role of culture in determining which physiological symptoms are invoked in the description of emotion. As Hupka et al. (1996: 246) state, "although the genetic and physiological bases of emotions may be similar in all human beings, talk about emotions may vary because of cultural scripting". This cultural scripting of emotion and other internal experiences has been the subject of a number of interesting studies. Ameka (2002), for instance, describes the cultural construction of emotion in reference to the Ewe linguistic community. Sharifian (2003, this volume) describes how *cultural conceptualisations* arise from cultural interactions and become internalised by members of that cultural group. The literal or figurative status of the linguistic expressions codifying these conceptualizations has also been debated. Ameka argues that the body part terms invoked in the description of emotion in Ewe are not figurative, but must instead be taken literally. Enfield (2002), however, argues that many Lao body part terms are polysemous, and that it is their non-physical senses (e.g. "locus of intellect" or "locus of emotion") that are involved in expressions describing internal states. Accordingly, Enfield (2002) argues, Lao speakers do not necessarily conceptualize their internal experiences in terms of internal organs at all. These debates will be reviewed in section 3 below.

1.2. The language and its speakers

Kuuk Thaayorre is a Southwest Paman language spoken on the west coast of Cape York Peninsula. Traditionally hunter-gatherers, many of the Thaayorre people moved to the Anglican mission established at Edward River in 1937. Nearly seventy years later, Kuuk Thaayorre is still spoken by close to 300 people, most of whom are resident in the community of Pormpuraaw, situated on the former mission site. The traditional lands of the Thaayorre include the community of Pormpuraaw and areas extending to the south and east, and most Kuuk Thaayorre speakers maintain strong links to their traditional lands, spending at least some of the year at outstations thereon.

2. Embodiment of internal experiences

2.1. The ear as main locus of the intellect

In Kuuk Thaayorre, the ear is conceptualized as the locus of the intellect. The dominant metaphor in this domain equates thought with audition. This is evidenced by the polysemous verb *ngeey* (which means both 'hear' and 'think') and dozens of expressions relating to cogitation that make reference to the *kaal* 'ear' (e.g., *kaal-aath-* 'believe'). These terms – *ngeey* 'hear, think' and *kaal* 'ear, locus of intellect' – are clearly polysemous, and expressions of cogitation involving these terms (e.g. *kaal-aath-* 'believe') should not be taken to refer to the ear (or hearing) literally (cf. Enfield

2002). But speakers nevertheless appear to associate such expressions with the ear as an organ, invoking the ear itself in novel descriptions of the thought process, and touching or otherwise indicating their ear when talking about cogitation. The same is also true for many of the body parts – internal experience associations described below. This points to a metaphorically-structured polysemy (Sweetser 1990) founded in the regular and unidirectional semantic shift from the concrete domain of the body to the more abstract realm of the intellect. For clarity, I will consistently gloss these polysemous terms with their body part senses, although in many cases they are used without literal reference to the body part in question. Nevertheless, this polysemy will be shown to be motivated by a number of cultural conceptualizations and pervasive metaphors; diachronically if not synchronically. Further, there is less danger of ethnocentric bias in interpreting body part glosses since in most cases there is close alignment between the denotata of the corresponding Thaayorre and English terms (cf. Gaby 2006).

Wilkins and Evans (2000) have documented a strong association between knowing and hearing in numerous Australian languages. Beyond the simple linkage of ear and mind, an overarching schema that KNOWLEDGE IS AN OBJECT [RECEIVED BY AND STORED IN THE EAR] finds expression in numerous Thaayorre lexemes and idioms. This conceptualization of the ear as a receptacle and container for the knowledge object is based upon several submetaphors. The hierarchical relationship between these can be represented as in Figure 1:

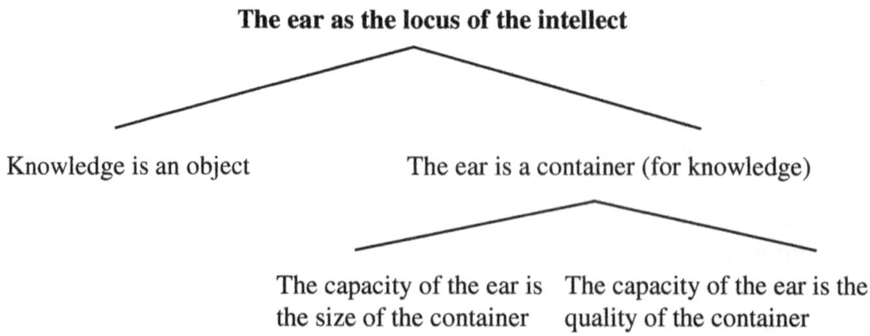

Figure 1. Locating the intellect in the ear

The metaphor KNOWLEDGE IS AN OBJECT is cross-linguistically commonplace. Note that I am using the term *knowledge* very broadly, denoting information, memories, and thoughts. This metaphor is related to the conceptualization of the ear as a vessel (i.e. THE EAR IS A CONTAINER). The conceptualization of the ear as a vessel used to store the received knowledge object, is evidenced by Thaayorre expressions such as *kaalkaalu kal* (ear:RDP^{2}-ERG carry) 'be mindful, remember' (literally 'carry with/in the ear'), or *kaal-purng-m* ('ear-block-VBR') 'forget'. The latter perhaps suggests that the ear vessel must be open if information is to be retrieved. For one's ear to become blocked, then, is to lose access to previously acquired and stored knowledge (i.e. to forget). Conversely, the ear vessel must be large and strong in order to effectively receive and retrieve information; THE CAPACITY OF THE INTELLECT IS THE SIZE/QUALITY OF THE EAR. Accordingly, young children are often exhorted to *kaal piinth* (ear grow) 'pay attention', or *kaal piinth ngeey* (ear strong hear) 'listen carefully'. Although it may appear that these expressions are more straightforwardly related to audition than cognition, it is clear in context that a request is being made for comprehension, not the simple perception of sound waves. It should also be noted that the verb *piinth* 'grow' and the adjective *piinth* 'strong' are both homophonous with the noun *piinth* 'bone'.

Where *kaal piinth* (ear grow) 'pay attention' links a large ear vessel to the effective reception of the knowledge object, the size of the ear is also linked to knowledge retrieval, as evidenced by (1):

(1)	*nhul*	*kaal*	*pii<pi>nth-r*
	3sg(NOM)	ear(NOM)	grow<RDP>-P.PFV
	'he remembered again'		(lit. 'he grew earwise / in the ear')

A large ear presumably makes for a better knowledge-vessel, with a larger capacity. The verb *piinth* 'grow' may also be associated with development or maturation, however. Thus an ear that grows might be expected to be a more sophisticated, highly developed knowledge archive.

The pervasiveness of the cultural belief that knowledge is received by and stored in the ear can be seen in the following excerpt from a conversation I had with Thaayorre elder Alfred Charlie. In it, Charlie is excusing his forgetfulness:

(2) a. *ngay* *pam* *nhangkn* *nhamp* *ngay*
1sg(NOM) man 2sgPOS(ACC) name(ACC) 1sg(NOM)
pamngoongkom
be.ignorant[3]
'I've forgotten your man's name'

b. *ngay* *kaal* *waarr=kaak*
1sg(NOM) ear bad=PROP
'I have a bad ear'

c. *ngay* *kaal<kal>-u* *kaarp kal<al>-ø,*
1sg(ERG) ear<RDP>-ERG NEGcarry<RDP>-NPST
ngay *pam* *thuump*
1sg(NOM) man old
'I can't carry with/in the ear, I'm an old man'

Note that line (2b) contains a further exponent of the idea that the ear must be strong to receive and retrieve information. Where the ear is weak (or 'bad'), retrieval of the knowledge object is difficult, hence Charlie employs the phrase *ngay kaal waarr=kaak* 'I have a bad ear' to describe his failing memory. Giving an English translation of what he had said in Kuuk Thaayorre, Charlie said "I got no good mind, you know, to remember". Though losing the metonymic motivation, this paraphrase makes explicit the parallelism between the Thaayorre conceptualization of the ear, and the English concept of *mind*. Further to this, the conceptual metaphors identified here for Kuuk Thaayorre are very similar to metaphors identified by Jäkel (1995) for English, with the exception that where English speakers conceptualize the mind as the locus of the intellect, for Thaayorre speakers it is the ear. Compare, for example, Jäkel's "understanding is taking idea objects into the mind container" (1995: 204) and "the valuables of knowledge are kept in the memory store for further use" (1995: 213) with the conceptual metaphors outlined here. We might connect the conceptualization of the ear as the metonymical basis for the mind with the dominance of oracy in cultures like the Thaayorre. The fact that knowledge about history, cosmography and other important cultural information is primarily transmitted through the ear (in the form of sound waves) is a clear motivation for privileging this organ.

In addition to the conceptualization of the mind as the locus of thought, there is a metaphorical association between sight and knowledge in English and most European languages. Although this is rarer in Australian languages than the association between hearing and knowledge, Kuuk Thaayorre does possess one term that expresses a KNOWING IS SEEING metaphor, namely; *wal-meer-em* (brow-eye-VBR) 'remember'. Several other Thaayorre terms and phrases express a related EYE FOR ATTENTION metonymy. For example, the act of showing (i.e. to draw something to somebody's attention) is described as *meer-en* (literally, eye-V^). Similarly, the eye is invoked in a request for attentiveness in the following:

(3)	*paanth*	*ngathn*	*meer<meer>-e*	*kal-ø!*
	female	1sgPOS(ACC)	eye<RDP>-ERG	carry-IMP
	'look after my girls!' (lit. 'carry my girls with/in your eyes')			

Finally, Foote and Hall (1992: 96) document two Thaayorre expressions that locate the intellect in the *paant-thuur* (head-marrow) 'brain': (1) *pam paant-thuur waarr* (man brain bad) 'fool'; and (2) *pam paant-thuur menethurr* (man brain MENETHURR) 'genius'. Although it seems that THE BRAIN AS THE LOCUS OF INTELLECT competes with THE EAR AS THE LOCUS OF INTELLECT, it should be noted that these two expressions are nowhere attested in my own data, whereas expressions linking the intellect to the ear/audition are extremely frequent.

2.2. The belly as the locus of the spirit and intuition

The *ngeengk* 'belly' is conceptualized not only as a part of the physical body, but also as a person's spirit. Accordingly, a person's soul or spirit is often termed *ngeengk-rithrr* (belly-fat), the same term also being used for the concept of 'Holy Spirit' in the Thaayorre Bible translation. Similarly, the spirit of a *wangath* 'doctor (a person with spiritual powers)' is labelled *pam ngeengk* (man belly). It is believed that this spirit can travel large distances to visit the infirm, or be projected into the body of a sick person in order to detect the cause of their illness (Taylor 1977: 426). When the *wangath* puts his spirit inside the patient, it is his *pam ngeengk* (man belly) 'spirit' that enters them.

Related to this spiritual sense, the belly is symbolically invoked in expressions of intuition and premonition, such as *ngeengkman rumparr* 'signify, prophesy'. The *man* 'throat' is also conceived of as a locus of intuitive apprehension. For example, a Thaayorre man who was worried about his brother's safety during a rodeo, said:

(4) *ngay* *man-u* *wa<a>k-r*
1sg(ERG) throat-ERG follow<RDP>-NPST
yarrman *ith=okun* *wont-nhan*
horse(NOM) dem:dist=DUB fall-GO&:NPST
'I'm worried that the horse might fall'
(lit. 'my throat follows [in case] the horse might fall')

The spirit is also described as located within the belly in a number of other Australian languages (see, e.g., Turpin 2002, Myers 1986, Valiquette 1993).

2.3. The belly, and especially the liver, as loci of emotion and character

We can say that the *ngeengk* 'belly' is also conceptualized as the locus of emotion and character[4] in Kuuk Thaayorre. Such an association is cross-linguistically common, both within the Australian context (as documented by, e.g., Turpin 2002 for Kaytetye, Goddard 1996 for Yankunytjatjara and Myers 1986 for the Pintupi), and outside (see, e.g., Bauer 1974 for Nigerian English, Matsuki 1995 and Hasada 2002 for Japanese, or Sharifian, this volume, for Persian). There are several more specific conceptual metaphors, however, that subcategorise Thaayorre expressions in this domain. Firstly that THE *NGEENGK* 'BELLY' IS A CONTAINER (OF FEELINGS, ETC.), analogous to the HEART AS A CONTAINER metaphor Niemeier (1997: 93) identifies in English. This is evidenced by the following line from a contemporary song composed by Gilbert Jack:

(5) *raak* *waarr* *ngathnma* *mong* *minc,* *kar*
thing bad 1sgABL many very like
ngeengk-an *wanyc*
belly-LOC pain
'I have experienced many bad things, [this is] like a pain in the belly'

Here, the pain caused by life experiences is explicitly located by Jack within his *ngeengk* 'belly'. Similarly, worries may be 'held' in the belly. A feeling of anxiety for an absent loved one can thus be described as follows:

(6) *ngay* *ngeengk-an* *piit-ø*
1sg(NOM) belly-DAT hold-NPST
'I hold (worries) in my belly'

When asked under what circumstances he might utter (6), Jack said: "If [my brother] got sick and I'm worried about him [this is] my feeling in my heart, whether he might live". Revealingly, Jack employs the English HEART AS CONTAINER metaphor in order to translate the Thaayorre BELLY AS CONTAINER metaphor. Finally, the belly-container might rupture under emotional stress. Thus regret is described as *ngeengk thaariic* 'belly tearing', revealing the EMOTIONAL PERSON IS A PRESSURISED CONTAINER metaphor analysed by Kövecses (2000).

As well as containing feelings, it could more broadly be said that in Kuuk Thaayorre THE *NGEENGK* 'BELLY' IS THE SEAT OF EMOTION/CHARACTER. This association between emotion and character is not unusual, given that sustained emotional states are, in various theories, e.g., the four-humour theory (see Geeraerts & Grondelaers 1995, Yu this volume) reflective and/or formative of one's character – someone who is habitually happy (emotion) is generally perceived to be a happy (character) person[5]. In Kuuk Thaayorre, then, a compassionate person is described as *pam ngeengk pork* (man belly large) 'a big-bellied man'. The belly is associated equally with positive and negative emotions and characteristics. This can be seen in expressions ranging from surprise to love to anger. For instance, to say that somebody is 'with belly' is to describe them as 'amazed' (*ngeengk=kak* [belly=COM] 'amazed'). This is also semantically close to the idiom *ngeengk miinng* (belly occur) 'be astonished'. Love is lexicalised by the verb *ngeengkmam* (belly:VBR), whilst someone who is 'heartbroken' can be described as *ngeengk watp* (belly dead). If somebody is scared, they *ngeengk ngeerngor=aak* (belly twitch=PROP) 'have a twitching belly'. If they are 'deeply troubled', they are *ngeengk waat* (belly wrongly[6]) (Foote and Hall 1992: 86). Finally, 'anger' is termed *ngeengk-kul* (belly-lap[7]).

The metaphorical and metonymic abstractions based on the *ngeengk* 'belly' in Kuuk Thaayorre, oftentimes correspond with patterns of metaphor in English based on the "heart". Compare, for instance, *ngeengk watp* (belly dead) with its translation 'heartbroken', or *ngeengk pork* (belly big) 'kind' with the English *big-hearted* (see Niemeier 1997: 101). Indeed, many of the English expressions Niemeier groups together under "heart as a metonymy for the person" – such as *big-hearted* generosity and *stout-hearted* bravery – might alternatively be analysed as based on the "heart as a metaphor for emotion/character". If the heart were metonymically standing for the whole person, we would expect the expression *big-hearted* to literally refer to a big person. However, a large stature does not necessarily imply generosity. Similarly, if somebody is 'broken-hearted', or *ngeengk watp* (dead-bellied), it is their love and/or happiness that has been destroyed.

There is also a strong conceptual link between the *thip* 'liver' and emotion. Significantly, the *thip* 'liver' is conceptualized as both within and part of the *ngeengk* 'belly', and accordingly may be referred to equally as *thip* 'liver' or *ngeengk-thip* 'belly-liver'. The *thip* 'liver', then, is singled out as a focal point for emotion/character within the *ngeengk* 'belly'. So, for example, if somebody is brave they might be referred to as *ngeengk-thip thaarrn* (belly-liver strong) 'strong-livered'. Here the liver acts as a metonym for good character. Although it might be tempting to view the *thip* 'liver' as metonymically standing for the whole person, the same idiom may also be used to describe somebody who is not greedy (and therefore a good provider). Clearly there is no necessary correlation between physical strength and altruism. Positive character is also attributed to the liver by the expression *ngeengk-thip thono* (belly-liver one) 'faithful, dependable' (lit. 'one-livered') (Foote and Hall 1992: 86). Not all emotions associated with the liver are positive, however. For instance, a worried person may be described as *thip waarramnam* (liver bad-ABL) 'worried' (Foote and Hall 1992: 77). Finally, drawing on a submetaphor ANGER IS HEAT, the following expression locates hot anger in the liver:

(7)	*ngay*	*ngeengk-thip*	*paapath*
	1sg(NOM)	belly-liver	hot
	'I'm furious'		

Finally, a number of other body parts may also be invoked in the description of emotion. The fixed expression *pancr rirk* 'body hair rises', for instance, describes shame. Romantic love is associated with the eyes in expressions such as *meer nhiin* 'fall in love' (eye be) (Foote and Hall 1992: 67), or the following example:

(8)	*pul*	*paanth*	*meer*	*wowurr-ø*
	3du(NOM)	woman(NOM)	eye(NOM)	meet-NPST
	'the lovers are meeting together'			
	(lit. 'they two, including the woman, meet eyewise/in the eye)			

Anger is referred to by terms invoking the lap, the belly and the throat, such as *ngeengk-kul* (belly-lap) 'angry' and *man-kul* (throat-lap) in (9):

(9)

nhul	*nganh*	*man-kul*	*thongk<ongk>a-n-r*
3sg(ERG)	1sgACC	throat-lap	arrive<RDP>-V^-NPST

'he's making me angry' (lit. 'he's causing my throat-lap to arrive')

The *kaal* 'ear' is associated with worry (e.g. *kaal waarram,* lit. ear bad:VBR) and with awe (*kaal weneth,* lit. ear fear). The compound *meer-kun-waarr* (eye-bottom-bad) is used to describe empathetic sorrow.

2.4. Belly of chest as locus of life force

Many Thaayorre metaphorical expressions, based primarily on the *ngeengk* 'belly', have been shown to play a role comparable to that of English metaphors based on the heart. The heart itself, however, plays only a very minor symbolic role in Kuuk Thaayorre. Interestingly, it is labelled by a term compounded from the lexemes *man* 'chest'[8] and *ngeengk* 'belly' *(man-ngeengk,* lit. 'belly of the chest'). This can most plausibly be explained by the fact that the belly is viewed as the core of the person overall (analogous to the English *heart*), while the heart is characterised as the core of the chest cavity (due to its central importance if not location). It is perhaps significant that while the physiological manifestation of the pulse is described with reference to the heart (10), the heartbeat as symbolic of life force is described in terms of the belly (11):

(10)

man-ngeengk	*thanp<anp>a-rr*
heart(NOM)	jump<RDP>-P.PFV

'[his] heart is beating'

(11)

nhul	*kaar*	*ngeengk*	*ngeerngor=aak*
3sg(NOM)	NEG	belly	twitch=PROP

'he is dead / has no pulse'

Similarly, many vital physiological processes are described in Kuuk Thaayorre in terms of the *ngeengk* 'belly'; e.g. *ngeengk wuunp-* (belly lay) 'breath' and *ngeengk thongk* (belly arrive) 'short of breath'.

3. Roots of cultural conceptualizations

3.1. Ethnomedicine and cosmovision

In order to explain the regularity with which particular internal organs (e.g. the *ngeengk* 'belly', the *thip* 'liver') are associated with mental, emotional, and spiritual experiences, we might seek to find some broader cultural beliefs and practices that encourage such cultural conceptualizations. Perhaps most promising are the ethnomedical beliefs and practices that attempt to rationalize bodily symptoms and occurrences. Unlike many ethnomedical systems from both Western and Eastern cultures, which account for disease and discomfort in terms of forces internal to the body (e.g. the European "four humours" [see Geeraerts and Grondelaers 1995] or Chinese *qi* [see Yu this volume, Kövecses 1995]), for the Thaayorre "death and disease were explained largely in terms of supernatural agents" (Taylor 1977: 425). In particular, most illnesses were seen to result either from *puripuri* 'magic' or *yuk may* (stick/thing vegetable) 'sorcery' being practiced against the invalid by some enemy, or from the invalid's having broken some sacred taboo. Most of these taboos involved the proscription of particular foods to people in various conditions. Young men, for example, would abstain from many key foods in the Thaayorre diet prior to their initiation. Pregnant women and young parents, too, were expected to avoid particular foods considered dangerous to the foetus/infant.

The relationship between food taboos and the present discussion is twofold. Firstly, we saw in §2.2., §2.3. and §2.4. that the *ngeengk* 'belly' is conceptualized as central to the spirit, emotions and life force of Thaayorre people. It is perhaps significant, then, that when asked to define the *ngeengk* 'belly', Thaayorre speakers often respond:

(12)	*ngeengk*	*may-i*
	belly	food-DAT
	'the belly is for food'	

The *ngeengk* 'belly'is where food is digested – the point at which foods become a part of the body. Where foods play such a significant role in determining both spiritual and physical health, then, it is not surprising that the *ngeengk* 'belly' is also viewed as connected to spiritual and emotional health.

Secondly, food taboos are largely anthropomorphic. The proscription of many foods is explained by drawing analogies between the particular food and the human body part that might be harmed by its consumption. For example,

> A pregnant woman could not eat carpet snake. It was thought that when she came to labour, the birth would be impeded as the snake wrapped itself around her abdomen. Similarly, pregnant women could not dig open the nesting mounds of the scrub turkey for this was likened to ripping open a woman's belly and the foetus might be harmed thereby. Parents had to avoid eating scrub turkey eggs while their first-born was still a baby lest the child develop large egg shaped lumps. (Taylor 1977: 424)

This anthopomorphism fits within a much broader connection between the landscape, its inhabitants and the supernatural world. In the Thaayorre cosmology, these three are inextricably linked, and the Thaayorre language is filled with metaphors that describe the natural world in terms of the human body. The *pungk* 'knee' is thus invoked in the description of other angular projections, such as *ngok pungk* (water knee) 'waves', or may have spiritual significance, as in *raak pungk* (place knee) 'tribal lands'. The branches of a tree are described as *punth* 'arms', with twigs being *wuurr* 'digits' (see Gaby 2004, 2006 for further examples). The description of mental, emotional and spiritual experiences in terms of body parts fits within this broader understanding of the external world in corporeal terms.

3.2. Physiological basis for conceptualization

An alternative source of explanation may lie within the human body itself. Although what we conceive of as *thought, emotion*, and *life force* are non-corporeal, each of these does usually have some physiological manifestations. As Enfield and Wierzbicka (2002: 5) assert, "biological events associated with emotion are more or less the same everywhere ... but it is well established that the interpretation and linguistic description of such responses, and the cognitive and cultural routines associated with these, are not universal or species-determined at all". That is to say, although biological events may in some cases inspire the conceptualization of emotion (and other non-corporeal experiences), this conceptualization is in no way predetermined by the biological events, which may be differently interpreted and described according to the cultural models resulting from the major preoccupations of the people and communities concerned. As Van Geert (1995: 261) puts it[9] "It is very possible that the number of physiological patterns [of experiencing emotion] is limited and universal, but there is no universality in the corresponding subjective experience". Accordingly, although many of the Thaayorre conceptual metaphors appear inspired by physiological effects (e.g. the description of one's *ngeengk ranc* 'belly jumping' when one gets a fright), only some conform to cross-linguistically common patterns. Geeraerts and Grondelaers (1995), arguing that the ethnomedical theory of the four humours mediates between bodily experiences of emotion and the English descriptions thereof, contrast the physiological symptom of heat with metaphors based on heat. They point out that, although shame is often manifest in a flushed face and a physical sensation of heat, English 'heat metaphors' are restricted to the semantic domains of anger and love. As they point out; "it would make no sense, for instance, to say that one's blood boils with shame, or that someone is fuming with shame" (1995: 167). Interestingly, in Kuuk Thaayorre the experience of shame is described as *pancr rirk* 'body hair rises'. The Thaayorre thus focus on quite a different physiological symptom of shame to the one privileged by English speakers, namely the *flushed face*. Indeed, an English speaker is much more likely to associate their body hair standing on end with terror or premonition. Thus a single physiological symptom can be differently interpreted and conceptualized by speakers of different languages, leading to alternative patterns of association.

4. Conclusion

Cultural models associating the intellect, emotions and life force with bodily loci are evidenced by languages spoken in every corner of the globe. But just how similarly do speakers of different languages conceptualize emotion, reason and vitality? This paper has presented an array of embodied metaphors and metonymies across these domains in Kuuk Thaayorre, and has proposed some overarching conceptual metaphors and metonymies to account for them. Notably, the *ngeengk* 'belly' was found to be central in the embodiment of emotion, spirit and life force, and the *kaal* 'ear' in the embodiment of intellect. Whilst some of these conceptual metaphors and metonymies have analogues in English and other languages around the world, many do not. The data presented in this paper thus show the associations between internal experiences (associated with emotions, the spirit and intellect) and particular internal organs to vary between linguistic communities.

Acknowledgements

The research reported here was initiated during my candidature at the University of Melbourne and the Max Planck Institute for Psycholinguistics, and funded in large part by these organisations and by the Pormpuraaw Community Council. My thanks to the editors and anonymous reviewers for their extremely helpful suggestions, and in particular to my Kuuk Thaayorre language teachers who with enthusiasm and good humour taught me all I know about embodiment in Kuuk Thaayorre. In particular, Alfred Charlie, Gilbert Jack and Elizabeth Norman contributed data and analyses cited herein.

Notes

1. The acronyms/abbreviations employed in this chapter are as follows: ABL ablative case; ACC accusative case; GO& associated motion; COM comitative; DAT dative case; dem:dist distal demonstrative; DUB dubitative; ERG ergative case; GEN genitive case; IMP imperative mood; LOC locative case; NEG negative; NOM nominative case; NPST nonpast tense; P.PFV past perfective tense/aspect; PROP proprietive; RDP reduplication; V^ valence increasing suffix; VBR verbaliser (derivational suffix).
2. *Pamngoongkom* 'be ignorant' is a bivalent predicate nominal.
3. In my consideration of human emotions, it should be noted, I will also look at what might more properly be considered human "character".
4. My thanks to the anonymous reviewer who pointed out this link between a person's habitual emotional state and what is perceived as their character.
5. The adverb *waat* 'wrongly' is usually coupled with a verb to indicate that the action described by the verb was carried out without the desired, intended, or culturally 'correct' effect. For example; *waat ke'err* (wrongly speared) 'threw a spear (at something) and missed'. It is apparently combined with *ngeengk* 'belly' here as part of a fixed expression, of which the meaning is non-compositional.
6. It is unclear exactly how anger relates to the physiology of the *kul* 'lap', which describes the supportive area formed by the legs of a seated (usually cross-legged) person.
7. The principal extension of *man* is to refer to the 'throat', but it is also found in compounds referring to body parts located anywhere in the throat / chest area

(n.b. the word for 'chest' is itself *man-ruuw*, *ruuw* also meaning 'front', though this could well be derived from 'chest', rather than the reverse analysis of *man-ruuw* as the 'front of the *man*').

8. Sweetser (1990: 30) similarly points out "the Mind-As-Body metaphor is very probably *motivated* by correlations between our external experience and our internal emotional and cognitive states, but the correlations alone will not explain the observed patterns of polysemy and semantic change".

References

Ameka, Felix

2002 Cultural scripting of body parts for emotions: On 'jealousy' and related emotions in Ewe. In *Pragmatics and Cognition*, special issue on *The body in description of emotion*,Nick Enfield and Anna Wierzbicka (eds.). 10 (1-2): 1–25.

Bauer, Anton

1974 *Das melanesische und chinesische Pidginenglisch.* Regensburg: Carl.

Ekman, Paul and Richard J. Davidson (eds.)

1994 *The Nature of Emotions: Fundamental Questions.* Oxford: Oxford University Press.

Enfield, Nick and Anna Wierzbicka

2002 Introduction: The body in description of emotion. In *Pragmatics and Cognition*, special issue on *The body in description of emotion*, Nick Enfield and Anna Wierzbicka (eds). 10 (1-2): 1–24.

Enfield, Nick

2002 Semantic analysis of body parts in emotion terminology: Avoiding the exoticisms of 'obstinate monosemy' and 'online extension'. In *Pragmatics and Cognition*, special issue on *The body in description of emotion*, Nick Enfield and Anna Wierzbicka (eds). 10 (1-2): 81–102.

Foote, Tom and Allen H. Hall

1992 *Kuuk Thaayorre dictionary: Thaayorre / English.* Toowoomba, Jolen Press.

Gaby, Alice

2004 Extended uses of Thaayorre body part terminology. *Melbourne Papers in Linguistics and Applied Linguistics.* 4(1): 21–34.

2006 The Thaayorre 'true man': lexicon of the human body in an Australian language. *Language Sciences.* 28 (2): 201–220.

Geeraerts Dirk and Stefan Grondelaers
1995 Looking back at anger: cultural traditions and metaphorical patterns. In John R. Taylor and Robert E. MacLaury (eds.). *Language and the cognitive construal of the world*. Berlin: Mouton de Gruyter.

Goddard, Clifford
1996 Cross-linguistic research on metaphor. *Language and Communication*. 16(2): 145–151.

Hasada, Rie
2002 'Body part' terms and emotion in Japanese. *Pragmatics and Cognition* 10: 107–128.

Hupka, Ralph B., Zbigniew Zaleski, Jürgen Otto, Lucy Reidl, and Nadia V. Tarabrina
1996 Anger, envy, fear, and jealousy as felt in the body: A five-nation study. *Cross-Cultural Research*, 30, 243–264.

Jäkel, Olaf
1995 The metaphorical conception of mind: 'mental activity is manipulation'. In John Taylor and Robert MacLaury (eds). *Language and the cognitive construal of the world*. Berlin: Mouton de Gruyter.

Johnson, Mark
1987 *The body in the mind: the bodily basis of meaning, imagination, and reason*. Chicago: The University of Chicago Press.

Kövecses, Zoltán
1995 Anger: Its language, conceptualization and physiology in the light of cross-cultural evidence. In J. Taylor and R. MacLaury (eds). *Language and the cognitive construal of the world*. Berlin: Mouton de Gruyter.
2000 *Metaphor and Emotion. Language, Culture and Body in Human Feeling*. Cambridge: Cambridge University Press.
2002 *Metaphor: a practical introduction*. New York: Oxford University Press.

Lakoff, George and Mark Johnson
1999 *Philosophy in the flesh: The embodied mind and its challenge to Western thought*. New York: Basic Books.

Matsuki, Keiko
1995 Metaphors of anger in Japanese. In John Taylor and Robert MacLaury (eds). *Language and the cognitive construal of the world*. Berlin: Mouton de Gruyter.

Myers, Fred R.
1986 *Pintupi Country, Pintupi Self: Sentiment, Place and Politics among Western Desert Aborigines*. Berkeley: University of California Press.

Niemeier, Susanne
1997 To have one's heart in the right place – Metonymic and metaphorical evidence for the folk model of the heart as the site of emotions in English. In Birgit Smieja and Meike Tasch (eds.) *Human Contact through Language and Linguistics.* Frankfurt: Peter Lang. 87–106

Sharifian, Farzad
2003 On cultural conceptualisations. *Journal of Cognition and Culture,* 3 (3): 187–207.

Sweetser, Eve
1990 *From Etymology to Pragmatics: The Mind-as-Body Metaphor in Semantic Structure and Semantic Change*. Cambridge: Cambridge University Press.

Taylor, John
1977 A pre-contact Aboriginal medical system on Cape York Peninsula. *Journal of Human Evolution*, 6: 419–432.
1984 Of acts and axes. Unpublished PhD Thesis. Townsville: James Cook University.

Turpin, Myfany
2002 Body part terms in Kaytetye feeling expressions. *Pragmatics and Cognition*, 10 (1-2): 271–305.

Valiquette, Hilaire (ed.).
1993 *A Basic Kukatja to English Dictionary*. Wirrimanu: Luurnpa Catholic School.

Van Geert, Paul
1995 Green, red and happiness: towards a framework for understanding emotion in universals. *Culture and Psychology* 1: 259–268.

Wilkins, David and Nick Evans
2000 In the mind's ear: The semantic extensions of perception verbs in Australian languages. *Language* 76: 546–592.

Yu, Ning
2002 Body and emotion: Body parts in Chinese expression of emotion. *Pragmatics and Cognition,* 10(1/2): 341–367.

Did he break your *heart* or your *liver*? A contrastive study on metaphorical concepts from the source domain ORGAN in English and in Indonesian

Poppy Siahaan

Abstract

In cognitive linguistics there is a growing interest in pursuing the question of how metaphors are used in cultural models (see, e.g., Quinn 1987). It is in this context that the present paper studies cultural models of the Indonesian concept of LIVER as the centre of both emotional and mental activities. In Indonesian, the word *hati* denotes primarily the liver organ but also means both HEART and MIND as it is understood in English. There are two aspects of the cultural model that I would like to present in this paper as the motivation for this conceptualization. First, there is the old ritual of liver divination in Indonesia, according to which the liver is regarded as the seat of life. In this tradition, humans' characters and fates are believed to be written on the livers by gods or divine spirits. Second, the ethnic religious beliefs in Indonesia regard the liver as the part of the body in which the living soul is located. The LIVING SOUL, a concept different from the concept of BODY and SOUL in English, is understood according to the ethnic religious beliefs to be a person within a person, with his or her own will or wish independent from the human's. A brief historical look at the LIVER concept in antiquity reveals that the conceptualization of the LIVER as the seat of life and soul was also very common among the ancient Babylonians, Assyrians, Etruscans, and, later, among the Hebrews, Greeks and Romans. By examining linguistic expressions in the present-day Indonesian language, known as Bahasa Indonesia, I demonstrate that these two almost forgotten and hardly practised cultural models of the LIVER as the centre of both emotional and mental activities are still reflected in the present-day Indonesian language. A comparison with the English HEART and MIND shows differences from and similarities with the Indonesian conceptualization of the LIVER.

Keywords: Indonesian, metaphor, metonymy, liver, liver divination, living soul.

1. Introduction

The conceptualization of various human body parts or organs as seats of emotional and mental activities in different languages reveals different cultural models that have been developed and shared by societies. In present-day English, for example, the conceptualization of the HEART organ as an abstract EMOTIONAL ACTIVITY has undergone the influence of several cultural models over time that have dominated English societies, so that it now differs from its conceptualization in older English. In Indonesian, the association of the liver organ with emotional and mental activities also derives from cultural tradition and ethnic religious beliefs that have been shared by the Indonesian societies. By exploring this cultural model, I wish to shed light on this specific worldview of Indonesian speakers and to contribute to a greater understanding of different cultural universes.

In Indonesian the word *hati* primarily denotes the liver, but it also means what are referred to as *heart* and *mind* in English, so that both these English glosses are currently used (see Goddard 2001). In Indonesian the LIVER is therefore understood as the seat of both emotional and mental activities. Consider the following linguistic evidence:

(1) a.	*Patah*	*hati*				
	broken	liver				
	'broken heart (lit. broken liver)'					
b.	*...direnungkan*	*dalam*	*lubuk*	*hati*	*nurani*	*yang*
	thought.of	in	deep.pool	liver	lighted	COMP
	paling	*dalam*				
	most	deep				
	'...(it should be) considered in mind (lit. in the depths of one's lighted liver)'[1]					

The first example shows that in Indonesian the LIVER could be regarded as the equivalent of the English HEART in the sense that both organs are associated with the abstract concept of LOVE. In the second example the conceptualization of HATI is more similar to the concept of MIND in present-day English. The fact that Indonesians regard the LIVER as the seat of emotional and mental activities has become one of the more challenging topics of discussion with which an Indonesian speaker is often confronted. In such discussions, the following questions typically arise: "How does the understanding of the liver as the seat of emotions arise?" or "What is the function of the liver in our body that has caused it to be regarded as the seat of emotions?"

In cognitive linguistics, there are two directions that explore the conceptualization of human body parts as representations of abstract notions. First, the notion "embodiment" emphasizes the role of the human body in the experience of the self and the environment and the way the human body functions in the perception and conceptualization of the world, thus providing the basis for our understanding of most abstract concepts (Lakoff and Johnson 1999: 19). From modern medicine we know that the liver is the largest organ in our body which processes blood and helps to clean unwanted substances out of it. One could then argue that the anatomical or configurational aspect of the organ provides the conceptual foundation for the mental understanding of the organ, i.e. as something which comes *from inside our body*. Yet the configurational aspect does not rule out the possibility of other organs being source domains of conceptual mappings. This ahistorical analysis seems to be too limited in scope, since it cannot provide any explanation for why different inner organs are used to refer to as abstract concepts in different cultures.

Second, as the newer approach to situated and cultural embodiment just referred to indicates, there is a growing interest in exploring the historical and cultural basis of human cognition taking into account cultural traditions and models that are commonly shared in societies and from which our conceptualizations of abstract notions derive. For the domain of emotions, Geeraerts and Grondelaers (1995) have convincingly shown that one should consider the historical and cultural background that provides the basis for the emotional concept of ANGER. Yu (this volume) has also demonstrated that the conceptualization of the Chinese HEART as the seat of emotional and mental activities is based on ancient Chinese philosophy and traditional Chinese medicine.

In the following sections of the present paper, I shall present two phenomena that are manifestations of animistic belief and serve as the basis for the conceptualization of the LIVER in Indonesian as the representation of abstract concepts of EMOTIONAL and MENTAL ACTIVITIES. First, I will trace back the ritual of animal sacrifice, and especially the interpretation of the liver organ, known as "liver divination", which was already practised in Indonesia long before the emergence of world religions, and can still be found in the more remote parts of Indonesia (Schefold 1980; Hoskins 1993; Howell 1996). According to the tradition of liver divination, the liver of a sacrificed animal is its most important organ because messages from gods or divine spirits to human beings in this world are written on it. Moreover, human characters and their fates are also believed to be written on the liver organs of sacrificed animals. Second, I examine the concept of the LIVING SOUL[2] which, according to ethnic religions in Indonesia, is located in the liver. Finally, I examine linguistic evidence from the present-day Indonesian language and compare it with English.

2. Cosmology and cultural models in ancient Indonesian societies

2.1. Liver divinations in ethnic religions

The conceptualization of the LIVER as emotional and mental activities – as it appears in metaphorical mapping in the present-day Indonesian language – reflects a cultural model that is widespread among Indonesian people. This aspect of the cultural model has its basis in an animistic belief that the liver is the seat of life. Before the arrival of world religions, animistic belief was very much alive in Indonesia. One of the rituals carried out in accordance with this belief was sacrificing animals and reading their liver organs, a practice known as liver divination (Schefold 1980; Hoskins 1993; Howell 1996). The world religions forbade the practice of reading animals' entrails in Indonesia, so that nowadays it is hardly encountered in modern Indonesian life (Hoskins, personal communication). Nevertheless, there are remote areas, for example in Sumba (Hoskins 1993) or Mentawai (Schefold 1980), in which liver divination is still part of people's lives. The ritual of liver divination reflects the animistic worldview of indigenous Indonesian people who believe that human characters and fates are written on the liver organ by divine spirits or gods. Although liver divination is hardly practised in modern life, and most Indonesian people have converted to major world religions such as Islam, Buddhism, Hinduism and Christianity, the conceptualization of the LIVER as the seat of life is still widely manifested in the present-day Indonesian language. Before discussing the linguistic expressions of these underlying liver metaphors, I will first give a brief account of liver divination.

Divination is an ancient kind of cosmological study in which human beings try to interpret certain signs of things that exist in this world in order to foretell the future. Apart from liver divination there are other kinds of divination, for example divination by dreams, divination by fire, etc. Astrology is also a kind of divination. In the last part of this section it will be shown that liver divination is not an exclusively Indonesian tradition, but that it was also practised by the ancient Babylonians, Assyrians and Etruscans.

The liver divination of sacrificed animals is carried out during ritual feasts in Indonesia. In such a ceremony, domestic animals such as chickens, pigs or buffalo are slaughtered. Their bodies are opened up and their livers are inspected. Liver divination is the interpretation of certain signs of this organ in order to gain knowledge – usually about what will happen in the future – and to communicate with the gods or divine spirits and the world beyond. A priest in Sumba (an island in the eastern part of Indonesia) defines liver divination (as quoted by Hoskins 1993: 169) as a:

> way for spirits of the dead to send letters to the living through animal victims … They are like your own notebooks and scribblings … made of straight lines and carved grooves, creating a message which you can read with your eyes rather than hearing with your ears. They are also like the printed orders that we receive from the regency office, in that they speak with an authority which cannot be refused.

In divination, certain signs of the liver of sacrificial animals, such as the "relative size, colour and orientation" of parts of the liver, are examined (Hoskins 1993: 170). The interpretation of certain configurations of the liver is also reflected in the Indonesian language, so that there are linguistic expressions deriving from it, such as *kecil hati* (lit. small livered; 'coward'), *besar hati* (lit. wide livered; 'generous'). By examining the livers of domestic animals, the priest can gain much knowledge about the life situation or about the character of human beings. In Sumba, for example, the livers of chickens or pigs can tell the priest whether the sacrifice has been accepted by the dead spirits, whether there are problems in or outside the village, and so forth. The liver of a buffalo is exclusively identified with the personality of its owner.

The liver is believed to be the core and spirit of the human being that can only become visible once it is opened up. As Hoskins (1993: 170) points out:

> [t]he liver is the seat of emotion, knowledge, and intentionality in Sumba … The liver is the organ of sincerity, where it is not possible to dissimulate. A man's conscience and self-awareness are found 'inside his liver' (ate dalo), and a speaker's true intentions, veiled in public declarations, reside 'beneath the liver' (pa kambu ate), only slowly finding their way up to the lips. Opening up the bodies of sacrificial animals opens up secrets normally hidden from humans by the invisible spirits, whose intentions only become visible and tangible once an animal life has been offered to provide a medium for their expression.

According to this understanding, the human being has two parts, i.e. the visible (outer) one and the invisible (inner) one. The outer part is the body itself, whereas the inner one belongs to the spiritual world and has its seat in the liver. Obviously, there is no separation between emotion and knowledge according to this ancient belief. In this way, the liver is regarded as the seat of both emotional and mental activities. The sacrificing of animals and the reading of their entrails are manifestations of an ancient belief in which human beings try to understand the spiritual part of life. The conceptualization of an animal's liver as the seat of a human's life is a metaphorical mapping that reflects a pagan worldview which is hardly practised in the more developed areas today and has gradually vanished in the less developed areas of Indonesia. This ancient understanding does not correspond with modern medicine or science. Nor does it correspond with the major world religions that are now widespread in Indonesia. Nevertheless, the manifestation of this ancient belief in the Indonesian language still remains. It is interesting to note how a language can preserve such an ancient cultural model in a constantly changing environment.

2.2. The concept of LIVING SOUL in ethnic religions

In this section, I will give a brief account of the concept of LIVING SOUL according to ethnic religions in Indonesia that corresponds with the conceptualization of the LIVER. Further, more detailed information on the SOUL concept in Indonesia can be found in the following anthropological and religious studies: Warneck (1909), Fischer (1965), Stöhr (1965) and Sibeth (1990). In almost every ethnic group in Indonesia there are two different concepts of the SOUL, so that they have two different words for SOUL, which are referred to here as the LIVING SOUL and the DEAD SOUL. Table 1 shows the different words in five different cultures, which implies that the LIVING SOUL concept is a completely different concept from that of the DEAD SOUL in each ethnic religion.

Both concepts in one culture correspond in general with the two concepts in another culture (Stöhr 1965: 178), although they vary in details. That is, the concept HAMBARUAN corresponds primarily with the concept TONDI, NOSO, AMIRUEH and TANOANA. On the other hand, the concept LIAU corresponds in general with the concept BEGU, BEGHU, ADIAU and ANGGA. Yet the two concepts LIVING SOUL and DEAD SOUL are not perceived as a dichotomy in the sense that one concept refers to and belongs to the other, such as the English concepts of "BODY and SOUL" or "BODY and MIND" (Wierzbicka 1992: 34–63). The LIVING SOUL refers to a living person, whereas the DEAD SOUL refers to the dead.

Table 1. The different words for LIVING SOUL and DEAD SOUL in some ethnic groups

	LIVING SOUL	DEAD SOUL
Ngaju Dayak (South Borneo)	*hambaruan*	*liau*
Maajan-Lawagan (South Borneo)	*amirueh*	*adiau*
Batak (North Sumatra)	*tondi*	*begu*
Nias (off the western shore of Sumatra)	*noso*	*beghu*
Toraja (Central Celebes)	*tanoana*	*angga*

The definition of the LIVING SOUL varies in details from one ethnic religion to another. But in general it is understood as a kind of substance or stuff that comes from the hands of gods. The LIVING SOUL defines the character and fate of human beings. Consequently, the character and fate of human beings are believed to have been formed by gods. As we have seen in the cultural tradition of liver divination, we find again the predestined aspect of human life in the concept of the LIVING SOUL.

The LIVING SOUL is perceived to be a kind of person within a person who has his or her seat in the liver. There are ethnic religions who believe that the LIVING SOUL can also be found in other body parts, such as in the blood, head, heart, hair and nails, or even in the placenta and saliva. This belief comes from the perception that the LIVING SOUL can temporarily leave the human body, for example during sleep, illness or unconsciousness. It can leave and return to the human body, passing through other organs or body parts. When someone is asleep, the LIVING SOUL is experiencing what is happening in that person's dream. Although the LIVING SOUL can be found in other body parts, the seat of the LIVING SOUL is perceived by most ethnic religions to be the liver.

The LIVING SOUL can be understood as human consciousness, because it is endowed with the ability to think and has its own memory, will and desires (Fischer 1965: 247). Being independent from the human being, the LIVING SOUL can impose its own will on the human person, so that it can quite often be in conflict with him or her. In order to please their LIVING SOUL, human beings must take care of it and keep it in a safe place. The Bataks, for example, bury their cut hairs, placentas and nail clippings with care. After death, the LIVING SOUL goes back to the hands of the gods, who mix it again with other stuff for a new human life.

The DEAD SOUL, by contrast, exists only after death. When a person dies, he or she is believed to become a DEAD SOUL and to go to the kingdom of the dead, where the DEAD SOULS of his or her ancestors are also believed to be. The DEAD SOUL is also quite feared because it is believed to chase the LIVING SOULS and cause human beings to die.

By comparison, in English the SOUL was traditionally not understood in terms of a separation of life and death. The English SOUL is perceived as an abstract concept rather than as a kind of concrete person within a person. It was understood as one of two parts of a person, and referred to the spiritual part of a human being as opposed to their physical body (Wierzbicka 1992: 34–63). On the other hand, the Indonesian LIVING SOUL is perceived as a concrete concept of a person within another person. Independently from that human being, the LIVING SOUL is believed not only to influence both the thoughts and emotions of the human, but also to represent other human aspects such as the will, consciousness, character and memory. This understanding of the LIVING SOUL corresponds closely with the LIVER concept in the present-day Indonesian language. We will see later in the linguistic section of the present paper that in the Indonesian language the LIVING SOUL concept – having its seat in the liver – is metonymically identified with the liver organ itself, which in turn is metaphorically mapped to the target domains of human emotional and mental activities. The liver is also understood as the human character, which is believed to be predestined, and it is also the seat of intention and consciousness. In order not to be in conflict with their soul, human beings are advised to take care of their liver. It is interesting to note that this ancient ethnic concept of the soul in Indonesian is still preserved in present-day language use, although the ethnic religions in Indonesia have long since been replaced by the major world religions.

2.3. The conceptualization of the LIVER in oldest antiquity

The conceptualization of the LIVER as the seat of life and the soul is in fact not exclusively a characteristic understanding of Indonesian culture. It was very widespread among the ancient Babylonians, Assyrians, Etruscans, and later among the Hebrews, Greeks and Romans. The physician Sir William Osler (1849–1919) points out that anatomy in ancient Babylonia was connected with the art of divination. In his book, *The evolution of modern medicine*, Osler (1921: 18) points out:

> … [of] all the organs inspected in a sacrificial animal the liver, from its size, position and richness in blood, impressed the early observers as the most important [organ] of the body. Probably on account of the richness in blood it came to be regarded as the seat of life – indeed, the seat of the soul. From this important position the liver was not dislodged for many centuries … Many expressions in literature indicate how persistent was this belief. Among the Babylonians, the word "liver" was used in hymns and other compositions precisely as we use the word "heart" …

Obviously, the conceptualization of the LIVER as the seat of life and the soul is part of a cultural model that used to be common in the Western world too. It is rooted in ancient medical knowledge and the tradition of divination. The earliest study of the liver of sacrificial animals dates back to as early as 3000 B.C., and the earliest known anatomical model, i.e. a clay model of a sheep's liver with the divination text, was dated from about 2000 B.C. (Osler 1921: 19). The ancient belief about the liver was also manifested in the Babylonian language, so that the LIVER was metaphorically mapped into the same abstract domains as the English HEART now is.

A study of the symbolism and interpretation of the liver among societies of the oldest antiquity by Jastrow (as quoted in Osler, 1921) reveals that the understanding of the LIVER as the seat of life and the soul arose from the belief that through the inspection of the liver of sacrificial animals the future could be predicted. When accepted by the gods, the sacrificial animal was perceived to become one with or identical with them. Since the fate of human beings was understood to be in the hands of the gods, it followed that if human beings could enter or read the gods' minds they would be able to see what would happen in the future. As the liver of the sacrificial animal was understood as the centre of their vitality, the life and the soul of the animal were either seen as identical with the minds of the gods or as mirrors in which their minds were reflected. As already discussed, this understanding of the liver organ corresponds closely with the cultural model of the LIVER in the tradition of liver divination in Indonesia. The conceptualization of the ANIMAL'S LIVER as the seat of a human life is a metaphorical mapping which reflects a pagan worldview that also used to be common in antiquity in the Western world.

It is interesting to note that in present-day German (Duden, Deutsches Universalwörterbuch) and English (Shorter Oxford English Dictionary on Historical Principles, hereafter SOED) we still can find certain uses of the words *Leber/liver* that reflect the conceptualization of the LIVER as the seat of some emotions (as well as thoughts):

(2) a. *frisch/frei von der Leber weg sprechen/reden* (lit. to speak freely out of the liver; to speak out frankly).

b. *sich etwas von der Leber reden* (lit. to speak about something out of the liver; to get something off one's chest).

c. *jemandem an der Leber fressen* (lit. to eat someone else's liver; to cause someone a lot of trouble or to bother someone)

(3) a. *liver-hearted* (cowardly)

In the SOED, the *liver* is defined figuratively as "[the] seat of love or other passionate emotion, as anger, bitterness, etc." and also as "the seat of cowardice". In German the word for liver (*Leber*) refers to a type of emotional attitude that is frank and straightforward, ignoring or overlooking one's own shame or someone else's disgrace. This evidence suggests that the conceptualization of the LIVER as EMOTION was not an understanding exclusive to Indonesian culture, but was also in common use in Western cultures. Nevertheless, it is important to note that the abstract concepts of EMOTION associated with the LIVER in German and English in the above examples are different from the Indonesian conceptualization of the LIVER as the seat of emotion and thought. Although we can find some linguistic evidence of the use of the word *liver* as emotion in German and in English, the important position of the LIVER in German and English culture has long since been replaced by another organ – the HEART – whereas the HEAD/BRAIN/MIND has gradually come to be regarded as the seat of thought (Niemeier, this volume). In Indonesian, cultural knowledge of the LIVER has not been subject to any similar changes. It is still regarded as the seat of emotion and thought. The HEAD, by contrast, is hardly used as THOUGHT in Indonesian. It is rather understood as the abstract concept relating to HIERARCHY or LEADERSHIP, e.g. *kepala negara* (lit. head country; 'the president'). This also underlines that – unlike in English – EMOTION and THOUGHT are not perceived as two dichotomous or even distinct concepts in the Indonesian language.

3. The linguistic expressions of *hati* mappings in Indonesian

In this section I present linguistic expressions in which the Indonesian LIVER is conceptualized as the seat of both emotional and mental activities. They are based on samples from a corpus taken from ten electronic daily newspapers.[3] Being a widespread instrument of communication within society, the newspaper reflects the cultural models of the society in question. After the account of liver divinations and of the concept of SOUL in the old ethnic religions in Indonesia in the previous section, I now investigate how much of these cultural aspects still hold today and can be found in the present-day vernacular Indonesian language. I also compare the use of HATI with that of HEART to establish which abstract notions are associated with these two different organs in both languages as they appear in the vernacular.

From the ten English and ten Indonesian newspapers, 3544 instances of the word *heart* and 1954 instances of the word *hati* are retrieved respectively. It is interesting to note that the non-metaphorical use of the word *heart* (90.6% of all instances) occurs a lot more often than that of the word *hati* (9.4% of all instances). The use of HEART as an organ occurs in the English corpus very frequently, for example to describe illnesses relating to this organ (e.g. *heart attack*, *heart disease*) or as a name of an institution dealing with diseases of this organ (e.g. *British Heart Foundation*). In comparison, HATI is not used to describe diseases of this organ in a medical sense. Thus, *sakit hati* (lit. ill liver) means emotional hurt rather than the illness of this organ. When talking about a liver disorder in the medical sense, Indonesian speakers would use the Dutch loan word, *lever* 'liver': *sakit lever*.[4] This underlines that the word *hati* is almost exclusively understood in its conceptual mapping.

In the following subsections I explore the conceptual mappings of HATI onto several abstract notions in the present-day Indonesian language. The mappings show cultural models, which are valid in Indonesia, in which HATI refers to human and also divine qualities. HUMAN CHARACTER, ATTITUDE, EMOTION, THOUGHT and MORAL VALUE are linked with configurational aspects of the LIVER such as PURITY, LIGHT and DEPTH, CHANGEABLE SHAPE, MOVEABLE OBJECT. The LIVER is also conceptualized as seat of emotion, religious belief and inner thought.

3.1. *Hati* as the seat of romantic love and other basic emotions

As shown before in example (1a), the LIVER is understood as the seat of love. The following examples are linguistic expressions in Indonesian, in which HATI is perceived as the seat of love:

(4) a. *Saya jatuh hati pada jejaka yang*
1SG fall liver onto young.man COMP
penampilannya jauh lebih keren
appearence.3SG far more striking
'I fell in love with (lit. fell liver onto) a young man whose look is far more striking ...'

b. *Akhirnya saya jatuh cinta juga padanya.*
eventually 1SG fall love also onto.3SG
'Eventually I fell in love with him too.'

c. *Anak adalah buah hati...*
child be fruit liver
'Children are (your) sweetheart (lit. fruit liver) ...'

The concept FALLING IN LOVE can be expressed in Indonesian as *jatuh hati* (4a) or *jatuh cinta* (lit. *fall love*; *fall in love*) (4b). The concept FALLING IN LOVE implies a sudden change and is based on the conceptual metaphor CHANGE IS MOTION, which is an entailment of the more general metaphor CHANGE OF STATE IS CHANGE OF LOCATION. One could then argue, that FALLING IN LOVE is a near-universal image. It is interesting to note that in Indonesian the concrete image LIVER can be used in terms of FALLING IN LOVE, whereas in English only the abstract concept of LOVE itself can be used. One cannot say in English **I fell in heart with him.* In terms relating to FALLING IN LOVE, the liver stands metonymically for the abstract concept of LOVE and is associated with the concept of LOVE itself. In example (4c), the concept FRUIT is linked with the understanding of LOVE. The *fruit of a liver* is understood as the outcome of a love relationship, that is, the child.

In the following examples, HATI is used to describe several other basic emotions such as HAPPINESS, ANGER, WORRY and SADNESS.

(5) a. *Kehidupan itu saya jalani dengan senang hati...*
life DET 1SG go with happy liver
'I go that (way of) life "contentedly"... (lit. with happy[5] liver)'

b. *Walau hati kita panas tetapi kepala harus tetap*
although liver 1PL hot but head must always
dingin sehingga dapat mengatasi masalah...
cold so.that can overcome problem
'Although we can get angry easily (lit. hot liver), we must stay cool (lit. cold head) to be able to overcome problems...'

c. *Ada sesuatu yang membuatnya tidak enak*
exist something COMP make.3SG NEG tasty
hati, dan pikirannya terganggu.
liver and thought.POSS disturbed
'There was something that worried him (lit. caused his liver unpalatable) and disturbed his thought.'

d. *Dia menceritakan kepedihan hatinya merasakan*
3SG tell smart liver.POSS feel
tingkah laku suaminya yang lebih
behaviour attitude husband.POSS COMP more
mementingkan teman-teman...
consider.important friend-RED
'She told about her heartache (lit. smarting of her liver) that her husband considered his friends to be more important...'

Example (5a) is about happiness, in which HATI stands metonymically for the person himself (PART FOR WHOLE). To do something *with a happy liver* is to do something contentedly. In example (5b), the *hot liver* is contrasted with the *cold head*. Obviously, these two body parts are conceptualized, at first sight, similarly to the Western dichotomy HEAD/HEART (see Niemeier, this volume). That is, the LIVER is conceptualized as the seat of emotion, whereas the HEAD is the seat of thought. The *hot liver* is an instantiation of ANGER IS HEAT. The *cold head*, by contrast, means the RATIONAL MIND. But, as I mentioned before, the conceptualization of Indonesian HEAD as mind occurs very rarely. I am not sure whether the dichotomy LIVER and HEAD in (5b) is an influence from Western (or English-speaking) culture. Nevertheless, it is interesting to note that the Indonesian LIVER still remains the seat of emotion, although in most Western (or English-speaking) cultures, the HEART stands in opposition to HEAD. Example (5c) is an instantiation of the LIVER as the seat of innermost feelings. Interestingly, the bodily experience EATING is linked with EMOTION. However, you can only say that someone has *an unpalatable liver* which means that they are worried or uneasy. You cannot say that someone's **liver tastes good*. Example

(5d) is an instantiation of SADNESS. We have seen in the example (1a) above that *broken liver* means broken heart in the sense of love. Similarly, being an entailment of the concept SADNESS, *smarting liver* in (5d) is the result of *cut liver*.

Furthermore, the Indonesian LIVER is understood as the seat of all feelings, whereas the English HEART is understood as a seat of emotions which are valued as either good or bad (Wierzbicka 1992: 47). Thus, one hardly says in English **His heart was full of surprise/amazement/interest*. Wierzbicka points out that "Surprise, amazement, or interest can be felt, but they cannot be felt 'in one's heart'. The reason is, I believe, that they make no reference to the concepts of 'good' and 'bad'."

3.2. The use of *hati* to describe desire and volition

In the following examples I will discuss another aspect of the LIVER, that is, HUMAN WILL or DESIRE. Interestingly, this aspect is combined with a certain configuration of the LIVER: FORM and WEIGHT. This conceptualization of the LIVER is similar to that of the English HEART, in which the HEART can have CHANGEABLE SIZE (cf. Niemeier, this volume).

(6) a.

Meskipun	*dengan*	*berat*	*hati,*	*Fitria*	*memenuhi*
although	with	heavy	liver	NAME	fulfil
permintaan	*suaminya*				
request	husband.POSS				

'Reluctantly (lit. with heavy liver), Fitria complied with the request of her husband.'

b.

...meminta	*semua*	*unsur*	*parpol*	*untuk*	*dapat*
ask	all	element	political.party	for	can
berbesar	*hati*	*menerima*	*kekalahan*	*dan*	*kemenangan.*
big	liver	accept	defeat	and	victory

'...(he) asked all political parties that they should be encouraged (lit. have a big liver) in accepting defeat as well as victory.'

c.

...berbulat	*hati ...*
round	liver

'...to be determined (lit. to have a round liver)...'

d.

...berjuang	*sepenuh*	*hati...*
fight	full	liver

'...to fight whole-heartedly (lit. with the whole liver) ...'

e.	*...diterima*	*dengan*	*setengah*	*hati ...*
	accepted	with	half	liver

'... it was accepted half-heartedly (lit. with a half liver) ...'

The examples given above are fixed idiomatic expressions in the present-day Indonesian language. Although these expressions refer to certain aspects of the LIVER in the source domain, they conceptualize certain aspects of HUMAN QUALITIES in the target domain. If we consider the account of liver divination, in which the form, the size or the colour of the liver is "read" in order to find out about certain qualities of human beings, we can find this cultural belief manifested in the linguistic instances above that present aspects of the LIVER being understood as HUMAN QUALITIES. In the present-day Indonesian language, to do something with a *heavy liver* (6a) means to do it reluctantly; *to have a big liver* (6b) means to be encouraged; *to have a round liver* (6c) means to be determined. Interestingly, we can find some similarity between the conceptualization of the Indonesian LIVER and the English HEART, when both are combined with the concepts WHOLE and HALF. To do something *with the whole liver* (6d) means to do something whole-heartedly, whereas to do something *with a half liver* (6e) means to do something half-heartedly. Consider the following examples in English:

(7) a. *I didn't vote for them and disagree with their policies, but will give them my whole hearted support in the service of our Iraq.*

b. *It would be nice if the UK could make at least a half hearted attempt to deal with the issue before more children end up dead.*

These examples demonstrate that in terms of its conceptualization as the seat of emotion the Indonesian LIVER is very similar to the conceptualization of the English HEART (cf. Niemeier, this volume).

3.3. The use of *hati* to describe human attitudes

In the following examples the LIVER is conceptualized as a representation of HUMAN ATTITUDES. In this conceptualization, the LIVER – like the HEART in English – can also be understood as A MOVEABLE OBJECT (cf. Niemeier, this volume) and linked with the concepts HIGH, LOW. Another concept linked with the LIVER in this understanding is PRICE. Like the English HEART, the LIVER is here understood as OBJECT OF VALUE (cf.

Niemeier, this volume). But, we can find a specific cultural model valid in Indonesia. Consider the following examples:

(8) a. *...sang* *Bupati* *seolah* *tak* *menyadar* *atau*
honorific.epithet Regent as.if NEG realise or
terlalu *tinggi* *hati* *sehingga* *tak* *mau*
too high liver so.that NEG will
bersikap *mengalah* *demi*
have.certain.attitude give.in for.the.sake.of
penyelesaian *masalah*
solution problem
'...as if the Regent did not realize or was too conceited (lit. to have a high liver), so that he does not want to give in for the sake of solving the problem.'

b. *Keterbukaan* *adalah* *sebuah* *kerendahan* *hati*
frankness be a low liver
untuk *tidak* *selalu* *merasa* *benar...*
for NEG always feel true
'Frankness is modesty (lit. a low liver), (that is) not to always feel self-opinionated...'

c. *Bak* *ilmu* *padi,*
Like knowledge rice
kian *berisi* *kian* *runduk*
the.more have.contents the.more stoop
'The more educated one is, the more humble one should be (lit. like the rice, the more content it has, the deeper it is stooping).'

d. *Memiliki* *kemurahan* *hati,* *senang* *membantu*
have plentiful liver like help
dan *tidak* *mementingkan* *diri* *sendiri*
and NEG consider.important self alone
'To be generous (lit. to have plenty of liver), willing to help and not to consider yourself to be important.'

The above examples demonstrate the association of the LIVER with HUMAN ATTITUDES. Interestingly, there is an added specific cultural value in the linguistic expressions: an attitude that is high is not regarded as positive, and an attitude that is low is not regarded as negative. To the contrary, *a high liver* (8a) is understood as ARROGANCE and *a low liver* (8b) is understood as MODESTY. Indeed, this conceptualization can be seen to correspond with an aphorism (8c) that reflects the cultural model valid in Indonesia. In this particular case, HIGH is understood as NEGATIVE and LOW is understood as POSITIVE in terms of human qualities. Furthermore, *plenty of liver* (8d) is also understood as positive. In terms of human attitude, *kemurahan hati* refers to GENEROSITY.

3.4. The use of *hati* to denote attention and care

In the following examples we find a characteristic of Indonesian language – and of other Austronesian languages – that is, reduplication which has a semantic function. The reduplication of a noun (*anak* 'child') could have a plural meaning (*anak-anak* 'children'), the reduplication of an adjective (*keras* 'loud') can derive an adverb (*keras-keras* 'loudly') (Sneddon 1996: 19). The reduplicated form of the word *hati* is understood as the adverb denoting ATTENTION or CARE.

(9) a. *Tindakan seperti ini harus dicermati secara hati-hati...*
action like DET must observed as liver-RED
'An action like this has to be observed carefully (lit. liver-liver)'

b. *bom bisa meledak bila tidak hati-hati membukanya...*
bomb can explode if NEG liver-RED open.3SG
'the bomb can explode, if we open it without care (lit. without liver-liver)'

In (9) to do something with *hati-hati* (lit. liver-liver) means to do something with full attention or with care. These examples demonstrate that the Indonesian liver is understood as the seat of emotional activities.

3.5. Equation of purity of *hati* with moral qualities

As we have seen in the previous section, according to the ethnic religions the liver is conceptualized as representing human attitudes and behaviour. It is thus also believed to be the seat of moral character. This ancient belief is manifested in the following linguistic examples:

(10) a.

Salah	*satu*	*bentuk*	*kejahatan*	*hati*	*di*	*dalam*
any	one	form	wickedness	liver	in	inside
majelis		*ialah*	*berbicara*	*dengan*	*hati*	*sombong.*
counsel		that.is	speak	with	liver	arrogant

'One of the moral offences (lit. the wickedness of the liver) of the counsel is to speak arrogantly (lit. to speak with arrogant liver).'

b.

Ada	*kalanya*	*kita*	*memang*	*merasa*	*begitu*	*sebal*
exist	time	1PL	really	feel	so	resentful
kepada	*seseorang*		*sehingga*	*hati*	*pun*	*serta-merta*
towards	someone		so.that	liver	also	immediately
kotor						
dirty						

'There are some times when we feel so resentful towards someone else, that we have a dark mind (lit. our liver becomes dirty).'

c.

Bila	*hati*	*kian*	*bersih,*	*pikiran*	*pun*	*selalu*	*jernih,*
if	liver	more	clean	mind	also	always	clear
semangat	*hidup*	*kan*	*gigih,*	*prestasi*	*mudah*	*diraih.*	
zest	life	will	stable	success	easy	achieved	
Tapi	*bila*	*hati*	*busuk,*	*pikiran*	*jahat*	*merasuk...*	
but	if	liver	rotten	mind	wicked	enter	
akhlaqpun		*kan*	*terpuruk,*		*dia*	*jadi*	
moral.also		will	sunk		3SG	become	
makhluk		*terkutuk*					
creature		wretched					

'If we have a good heart (lit. if the liver is clean), we also have a clear mind, we will have a steady zest for life, success will be easy to achieve. But if we are evil-minded (lit. if we have a rotten liver), wicked thoughts will come to us, our morals will sink and we will become wretched creatures.'

d.

... yang	*anti*		*demokrasi*	*pun*	*jika*	*membaca*
COMP	anti		democracy	also	if	read
analisis	*ini*	*akan*	*tertegun*	*hati*	*nuraninya.*	
analysis	DET	will	startled	liver	lighted.3SG	

'... even those who are anti-democracy will be startled (lit. their lighted liver will be startled) when reading this analysis.'

In examples (10) HATI is linked with NEGATIVE or NEUTRAL FEELINGS and ATTITUDES or with MORAL VALUES. In example (10a) *the wickedness of the liver* means inappropriate human moral behaviour, whereas *to speak with arrogant liver* means to speak arrogantly. Both examples reflect the cultural belief common in Indonesia that the LIVER defines HUMAN CHARACTERS and ATTITUDES. In example (10b), the LIVER is combined with metaphors for morality: BEING MORAL IS BEING CLEAN, BEING IMMORAL IS BEING DIRTY. If we feel resentful to someone, our *liver becomes dirty*. It is difficult to paraphrase the idiomatic expression *hati pun serta-merta kotor* as literally as possible into an idiomatic expression in English, as one cannot talk about **a dirty heart*, or **a heart becoming dirty* in English. From an allocentric perspective a heart can contain only good things but not bad things (Wierzbicka 1992: 47–48). However, one can talk about *a dark mind, dark thoughts* or *a dirty mind* in English as in the following examples:

(11) a. *I came up with the theory that Saddam Hussein's dark mind was convinced that his power lay in his special weapons ...*

b. *... and then the dark thought that one day the hard and stable US dollar will weaken.*

c. *... if you believe there's anything sleazy about big party donations, you've got a dirty mind, and risk corrupting politics with your smutty suspicions.*

From an allocentric point of view the Indonesian LIVER can constitute good and bad qualities – unlike the English HEART. This comparison demonstrates that in terms of MORAL VALUES the Indonesian LIVER is more equivalent to the English MIND than to the English HEART. That is, the Indonesian LIVER is the seat of mental activities. In example (10c), *hati* occurs as the counterpart of *pikiran* 'thought'. The dichotomy HATI-PIKIRAN is also combined with the metaphors for morality: BEING MORAL IS BEING CLEAN, BEING IMMORAL IS BEING DIRTY. In this dichotomy, the LIVER is understood as the seat of emotion only, not as the seat of both emotion and thought. However, the LIVER is also conceptualized as something that can influence thought, zest for life and morality. Example (10c) reflects the cultural model widespread in Indonesia that the liver is the most important organ for human beings because it can influence and control other aspects of life. In example (10d), the LIVER concept is combined with the concept of NURANI (LIGHT), which is believed to be the light coming from gods. With the help of this light, human beings are believed to be able to see clearly and understand better (UNDERSTANDING IS SEEING). *Tertegun hati nuraninya* refers to the core of human reflection considering something deeply after being surprised by it.

3.6. Using light and depth of *hati* to denote divine qualities

Unlike the previous examples (10), the LIVER in the following examples (12) constitutes divine rather than human qualities. It is combined with the concept LIGHT or with the concept DEPTH. *Lighted liver* or *deep liver* is conceptualized as the inner voice or innermost feeling of human beings which is inside us and tells us what is right and wrong. It reflects the cultural belief that the liver is the seat of the true human self and innermost feelings or thoughts. In English, this notion is defined as conscience.

(12) a.

... namun	*soal*	*menyoblos,*		*harus*	*sesuai*
but	matter	punch		must	appropriate
dengan	*hati*	*nurani*	*yang*	*paling*	*dalam...*
with	liver	lighted	COMP	most	deep

'... as for voting (by punching ballot), however, it should come from the bottom of (our) heart... (lit. it should correspond with the deepest, lighted liver)'

b.

Kebenaran	*datang*	*dari*	*lubuk*	*hati ...*
truth	come	from	deep.pool	liver

'Truth comes from the bottom of (our) heart (lit. from the deep pool of the liver) ...'

c.

... direnungkan	*dalam*	*lubuk*	*hati*	*nurani*
thought	in	deep.pool	liver	lighted
yang	*paling*	*dalam*		
COMP	most	deep		

'... (it should be) considered in mind (lit. in the depths of one's lighted liver)...'

d.

Tugas	*yang*	*diembannya*	*merupakan*	*tugas*	*mulia*
duty	COMP	carried.out.3SG	constitute	duty	noble
yang	*merupakan*	*panggilan*		*hati*	*nurani*
COMP	constitute	call		liver	lighted
terhadap	*saudara-saudara*	*yang*	*terkena*	*musibah*	
to	brother.sister-RED	COMP	affected	disaster	

'The duty he carried out was a noble duty which was a call from the depths of his heart (lit. a call from the lighted liver) concerning the comrades affected by the disaster.'

e.	... *yaitu*	*"bisikan"*	*hati*	*nuraninya*	*Sebab,*	*bisikan*
	that.is	whisper	liver	lighted.3SG	because	whisper
	itulah	*yang*	*"menghubungkan"*		*dia*	*dengan*
	DET.PART	COMP	connect		3SG	with
	Sang		*Penciptanya*			
	honorific.epithet		creator.3SG			

'...that is the whisper from the depths of his heart (lit. from his lighted liver). Because, it's the whisper that "connects" him with his Creator.'

In (12a) HATI is understood as something that can tell us which political party we should choose. At first glance, *hati nurani yang paling dalam* looks similar to the English idiomatic expression *from the bottom of my heart*, defined in the SOED as "with the sincerest or deepest feeling". Yet, in terms of voting for a political party, HEART is normally not understood in English as the organ that can tell us which political party we should choose. If an Indonesian person should choose a political party according to their *hati nurani yang paling dalam* 'deepest pool of the lighted liver' it means that they should choose according to their own choice, belief and feeling, and not according to what other people tell them. It is quite common in Indonesia for people, especially those who are poor and uneducated, to be bribed by certain political parties during electoral campaigns. This association of the liver with a person's belief, feeling and choice corresponds with the traditional belief in liver divination, when the liver of a sacrificial animal is believed to reflect the character of its owner. Furthermore, the LIVER in (12a) is combined with the concept DEPTH. Hence, the Indonesian LIVER – like the English HEART – is understood as a CONTAINER, the BOTTOM of which is the main part or the core. Consequently, the *deepest lighted liver* means the core of human belief and feeling. In (12b) truth is believed to come from *lubuk hati* 'the deep pool of the liver'. Again, the liver is understood here as a CONTAINER, from which truth emanates. Example (12b) gives evidence that we can still find a certain degree of ancient cultural belief in the present-day Indonesian language in which the liver is regarded as the source of truth coming from the gods or divine spirits. In (12c) the LIVER is also combined with the concept DEPTH. Interestingly, the Indonesian speaker understands this *deep room* as a place in which we can reflect upon something deeply. In English, the HEART is not understood as the place in which we think, but rather the HEAD/BRAIN (cf. Niemeier, this volume). As the seat of mental activities, the LIVER is combined with the concept LIGHT, *nurani*, an entailment of the common metaphorical concept

UNDERSTANDING IS SEEING. With the help of this divine light, one is *enlightened*. In example (12d) the noble duty to help other people is understood as *the call of the liver*. Example (12e) is further linguistic evidence that the LIVER is conceptualized as something with divine quality. *Bisikan hati nurani* 'a whisper from the lighted liver' is understood as something that can connect a human being with God.

3.7. Conceptualization of *hati* as the seat of religious belief

In Section 2, in which the tradition of liver divination and the old religious beliefs in Indonesia were discussed, it was demonstrated that the LIVER is associated with divine quality and that it can influence human beings in terms of their moral or religious values. The following examples show how this belief still holds in the present-day Indonesian language.

(13) a.

... *umat*	*Buddha*	*selalu*	*berpegang*	*teguh*	*pada*
community	Buddha	always	hold.on.to	firmly	on

proses	*karma*	*dan*	*lurus*	*hati.*
process	karma	and	straight	liver

'... Buddhists should always hold on firmly to karma and their sincere belief (lit. straight liver).'

b.

... *orang*	*yang*	*bersih*	*hati,*	*taat,*
people	COMP	clean	liver	obedient

dan	*mulia*	*dunia*	*akhirat ...*
and	noble	world	the.hereafter

'... true believers (lit. people with clean liver), who are obedient and noble in this world and the hereafter ...'

c.

Iman	*itu*	*adanya*	*di*	*hati.*
belief	DET	located	in	liver

'Religious belief is located in your heart (lit. in the liver).'

d.

... *ia*	*harus*	*memiliki*	*hati*	*yang*	*bersih*	*dan*
3SG	must	have	liver	COMP	clean	and

menghindari	*kelakuan-kelakuan*	*buruk,*	*seperti*	*korupsi.*
avoid	behaviour-RED	bad	like	corruption

'... he must have a clean heart (lit. liver) and avoid bad behaviour such as corruption.'

e. *Aa* *berulang-ulang* *mengingatkan* *agar*
NAME repeat-RED remind in.order.to
tetap *menjaga* *hati* *karena* *baik* *buruknya*
always look.after liver because good bad
perbuatan, *tergantung* *dari* *hati.*
action depend.on from liver
'Aa kept reminding us to look after ourselves (lit. to look after the liver), because the goodness or badness of our action depends on our moral (lit. liver).'

Being conceptualized as the seat of moral or religious value, the Indonesian LIVER in the above examples is combined with the concepts STRAIGHTNESS and CLEANNESS. The fixed idiomatic expressions *straight liver* and *clean liver* can both refer to religious and moral values. Thus, in terms of religious values, *straight liver* (13a) and *clean liver* (13b) mean piety and sincerity to God. In terms of moral value, *clean liver* (13d) means honesty or sincerity to other people. In (13c) the LIVER is understood as the religious core of human beings. As I have discussed in the previous section of this paper, the association of the liver with human qualities, which are believed to have come from the hands of gods, is one of the most important aspects of liver divination. The religious and moral meaning of the liver manifested in Indonesian language is a reflection of the cultural belief of Indonesian people that this organ is linked with gods and divine spirits. In example (13e), the LIVER is conceptualized as an organ which human beings should take care of and which can influence human attitudes and behaviour. This conceptualization of the LIVER reflects the ethnic belief that the liver, being the seat of the living soul, has to be taken care of by human beings in order not to come into conflict with it.

3.8. *Hati* as the seat of inner thought

As I mentioned earlier in this paper, HATI is not only understood as the seat of emotion but also as the seat of inner thought. The Indonesian LIVER is understood as a place in which people think and do mental activity. In this way, HATI is similar to the English concept of MIND. English speakers understand MIND AS A CONTAINER, in which they think and reason about the world around them. The following sentences are instances of the conceptualization of HATI as MIND or as the seat of mental activity.

(14) a.

... *dalam*	*hati*	*kecil*	*saya*	*akui...*
inside	liver	small	1SG	admit

'... in my mind (lit. in my small liver) I admit... '

b.

Sebab	*rakyat*	*tentu*	*mencatat*	*dalam*	*hati, ...*
because	people	for.sure	note	inside	liver
Parpol		*mana*	*sebenarnya*	*yang*	*kelak*
political.parties		which	really	COMP	will
konsisten.					
consistent					

'Because the people will surely note in their mind (lit. in their liver) ... which political parties will really be consistent.'

c.

Sialan,	*serunya*	*dalam*	*hati.*
damn.it	shout.3SG	inside	liver

'Damn it, he shouted silently (lit. in the liver).'

d.

Kupejamkan		*mata*	*sambil*	*mengucap*	*doa*	*dalam*
1SG.close		eye	while	say	pray	inside
hati,	*mohon*	*kekuatan*		*Tuhan.*		
liver	ask	strength		God		

'I closed my eyes and said a prayer silently (lit. in the liver), asking for God's help.'

e.

Ketika	*itulah*	*saya*	*berjanji*	*dalam*		*hati, suatu*
as	DET.PART	1SG	promise	inside		liver a
waktu	*saya*	*akan*	*memberi*	*pelajaran*	*atas*	*perbuatannya*
time	1SG	will	give	lesson	for	act.3SG

'That was the time when I promised myself (lit. in the liver), that I would teach him a lesson for his act.'

f.

Aku	*mengutuk*	*dalam*	*hati*
1SG	curse	inside	liver

'I cursed silently/to myself (lit. in the liver).'

g.

Saya	*bertanya*	*dalam*	*hati ...*
1SG	ask	inside	liver

'I asked myself (lit. inside my liver).'

One important aspect relating to HATI understood as the seat of mind or mental activity is that all the mental activities that take place in the Indonesian LIVER are understood as those that take place without the knowledge of other people. Before I discuss this aspect of HATI and compare it with the concept MIND in English, I need to discuss first the concept MIND in English. Wierzbicka (1992: 44–47) points out that English culture has experienced a change in the folk concepts of MIND and SOUL. The older English *mind* was linked with emotions and therefore has a spiritual as well as a psychological dimension (e.g. *noble mind, ignoble mind, innocent mind* or *generous mind*), whereas the modern MIND is linked almost exclusively with the intellect and focuses on thinking and knowing, not on feeling or wanting (Wierzbicka 1992: 45). The older English SOUL is understood as "a very comprehensive concept, combining religious, psychological, and moral aspects, seen as one and jointly opposed to the material body" (Wierzbicka 1992: 46). The present-day English SOUL has become "restricted to the purely religious sphere" (Wierzbicka 1992: 46). Wierzbicka claims that English culture has experienced a separation of THINKING and FEELING that results in another change of view of certain aspects of human beings. The psychological and moral aspect of humans which used to be understood as SOUL (as opposed to BODY) is now reduced exclusively to the psychological aspect which is understood as MIND (as opposed to *body*). Compared with HATI, Wierzbicka points out that the modern MIND – unlike the old SOUL – seems to lack depth (e.g. we cannot say **from the bottom of my mind*) so that the old SOUL seems to be more inscrutable than the modern MIND. In short, there is one component of the old SOUL that the modern MIND does not imply:

> other people cannot know what things happen in that part of a person

In her comparison between the old English SOUL and the modern MIND, Wierzbicka (1992: 46) points out that:

> modern psychology and the more recent cognitive science seem to be predicated on the assumption that what is called the 'human mind' *can* be studied … The *soul* used to be a very comprehensive concept, combining religious, psychological, and moral aspects (italics in original)

In the light of this knowledge, the conceptualization of the Indonesian LIVER as the seat of the mind or of mental activity is closer to the meaning of the old English SOUL. I discussed earlier in this paper that HATI is understood as representing several abstract notions of emotion and other human qualities such as the thinking and moral/religious aspects of a human being. Interestingly, HATI is not seen as "one of two parts" of the human being. Unlike HATI, the BODY in English language is seen as "one of two parts" of the human being as in the dichotomy BODY and SOUL. Indeed, HATI is a body part itself, yet manifests both the THINKING and FEELING aspects as well as the psychological and the moral/religious aspects of human beings. In English culture THINKING and FEELING are separate. Furthermore, HATI is seen as deep, so that other people cannot see or know what it is. In other words, other people cannot see or hear or know all the mental activities that take place in HATI. In the above examples, if an Indonesian speaker *admits something* (14a), *notes something* (14b), *shouts* (14c), *says a prayer* (14d), *promises something* (14e), *curses* (14f) or *asks something* (14g) in their liver, no one else knows or sees or hears these activities. These mental activities are not carried out physically, but mentally. If an Indonesian speaker *curses in their liver*, other people can neither see nor hear nor know that they are cursing. As already discussed in the previous section of this paper, in ancient ethnic religion the concept of the LIVING SOUL is understood as a kind of person inside a human being who experiences what is happening in that human's dreams. The linguistic manifestation of this belief can be found in the present-day Indonesian language: HATI is understood as a place where human beings can experience EVENTS that can neither be seen nor heard by other people. The linguistic examples of HATI as MIND or MENTAL ACTIVITY in the present-day Indonesian language show how firmly the ethnic religious belief is still entrenched in Indonesian culture despite the spread of world religions in that country.

4. Conclusion

The conceptualization of HATI in Indonesian as the seat of emotional and mental activities is not something arbitrary, but reflects a cultural model common among Indonesian people. Its origin can be found in the ancient Indonesian ritual of liver divination and in ancient Indonesian cultural belief. This study shows correspondences between this cultural model and linguistic expressions in the present-day Indonesian language.

According to the tradition of liver divination, the liver of sacrificial animals is regarded as the seat of life and soul. Shaped by gods and divine spirits, the liver is regarded as the seat of truth and the core of the human being. Consequently, it reveals the core of human character, attitudes, emotions and thought. In the language, the liver organ is conceptually mapped into the target domains of abstract notions of HUMAN CHARACTER, ATTITUDE, EMOTION, THOUGHT and MORAL VALUES.

According to ancient ethnic religions, the liver is regarded as the seat of the LIVING SOUL, which is an animistic concept of a kind of person residing within a human being. Coming from the hand of gods, the LIVING SOUL is independent of the human being in which it dwells and can influence them or be in conflict with them. The voice of the LIVING SOUL is regarded as the inner voice of the human being and the source of truth, which cannot be heard or known by other people. The LIVING SOUL refers to human and divine qualities. In the language, the liver organ is conceptually transferred into abstract notions such as EMOTION, THOUGHT, MORAL or RELIGIOUS VALUES, INNER VOICE, ATTITUDE, COURAGE or DETERMINATION. In this way, the Indonesian LIVER is understood as the seat of both EMOTION and MIND. In short, HATI is the organ that FEELS and also THINKS.

Although today the tradition of liver divination has almost vanished and the ethnic religions have been replaced by the world religions, the present-day Indonesian language still preserves this cultural model that was once used by Indonesian people to structure and understand the world. This cultural model has thus survived three changes. The first is the replacement of the old ethnic religions by the world religions, which also caused the tradition of liver divination to vanish. The second is the impact of Western influence, especially from English-speaking culture, in terms of the conceptualization of body parts where the separation of emotional and mental activities is associated with the heart and head/brain. Third is the spread of modern scientific knowledge that the brain controls both emotional and mental activities.

The study of metaphorical concepts can reveal the cultural model of a society. In the case of the conceptualization of the LIVER in Indonesian, we have seen that this cultural model is based on old traditions and ancient religion, which are no longer practised and long forgotten in modern life. Nevertheless, this cultural model is still preserved in the language and thus influences its people's way of thinking.

Acknowledgements

This paper is based on one chapter of my dissertation, which is a contrastive study on metaphorical concepts in German and in Indonesian. I wish to thank David Gil

(Max Planck Institute for Evolutionary Anthropology, Leipzig), Yeshayahu Shen (Tel Aviv University) and Uri Tadmor (Max Planck Institute for Evolutionary Anthropology, Jakarta Field Station) for their comments and suggestions on the first presentation of this paper at the "Workshop on Indonesian Linguistics" on 26th August 2004 in Leipzig. They acquainted me with the topic of liver divination. I wish to express my gratitude to Janet Hoskins (Department of Anthropology, University of Southern California) for the correspondence in which she gives me insightful suggestions on the topic of liver divination. I am indebted to Imran Ho Abdullah (School of Language Studies and Linguistics, National University of Malaysia) for reading this paper at the "International Cognitive Linguistic Conference" on 19th July 2005 in Seoul, which I was not able to attend myself. I also thank the editors of this volume, Farzad Sharifian, René Dirven, Susanne Niemeier and Ning Yu for their comments and suggestions on earlier drafts of this paper.

Notes

1. The following grammatical acronyms and abbreviations are used in the glosses: COMP complementizer; DET determiner; NEG negation; lit. literally; PART particle; PL plural; POSS possessive; RED reduplication; SG singular; 1 first person; 3 third person.
2. The concept SOUL is very culturally specific and varies greatly even among the Western cultures (Wierzbicka, 1992). The use of the concept SOUL in this paper should therefore be regarded in its broadest sense.
3. Data sources for English examples: Daily Mirror (www.mirror.co.uk), Daily Mail (www.dailymail.co.uk), The Times (www.timesonline.co.uk), The Independent (www.independent.co.uk), The Guardian (www.guardian.co.uk), Morning Star (www.morningstaronline.co.uk), Daily Telegraph (www.telegraph.co.uk), Financial Times (http://news.ft.com/home/uk), BBC (http://news.bbc.co.uk), The Observer (http://observer.guardian.co.uk).
   Data sources for Indonesian examples: Kompas (www.kompas.com), Berita Sore (www.beritasore.com), Indo Media (www.indomedia.com), Suara Pembaruan (www.suarapembaruan.com), Media Indo (www.mediaindo.co.id), Pikiran Rakyat (www.pikiran-rakyat.com), Harian Analisa (www.analisadaily.com), Riau Pos (www.riaupos.com), Waspada (www.waspada.co.id), Harian SIB (www.hariansib.com).
4. As a result of Dutch colonisation (from 1641–1941) Indonesian language uses Dutch loan words for modern medical terms.
5. The translation of the word *senang* as 'happy' here is based on the Indonesian-English Dictionary (Echols and Shadily 1992). Nevertheless, it is important to bear in mind that both *senang* and *happy* are actually very culturally specific

words, and that they have no exact equivalents in other languages (cf. Wierzbicka 1992; Goddard 1997; Stefanowitsch 2003).

References

Fischer, Hans

1965 *Studien über Seelenvorstellungen in Ozeanien.* München: Klaus Renner Verlag.

Geeraerts, Dirk/Stefan Grondelaers

1995 Looking back at anger. Cultural traditions and metaphorical patterns. In *Language and the Construal of the World,* John Taylor & Robert E. MacLaury (eds.), 153–180. Berlin: Mouton de Gruyter.

Goddard, Cliff

1997 Cultural values and 'cultural scripts' of Malay (Bahasa Melayu). In *Journal of Pragmatics.* 27: 183–201.

2001 'Hati': A key word in the Malay vocabulary of emotion. In *Emotions in cross-linguistic perspective,* Jean Harkins and Anna Wierzbicka (eds.), 167–195. Berlin/New York.:Mouton de Gruyter.

Hoskins, Janet

1993 Violence, sacrifice, and divination: Giving and taking life in eastern Indonesia. In *American Ethnologist.* 20 (1): 159–178.

Howell, Signe

1996 A life for life? Blood and other life-promoting substances in Lio moral system. In *For the Sake of our Future. Sacrificing in Eastern Indonesia,* Signe Howell (ed.), 92–110. Leiden: Research School CNWS.

Jastrow, Morris

1907 *The Liver in Antiquity and the Beginnings of Anatomy.* Transactions College of Physicians. 3. s., XXIX: 117–138 Not too sure what kind of reference this is.

Lakoff, George/Mark Johnson

1980 *Metaphors We Live By.* Chicago: The University of Chicago Press.

1999 *Philosophy in the Flesh. The Embodied Mind and its Challenge to Western Thought.* New York: Basic Books.

Niemeier, Susanne

This vol. To be in control: Kind-hearted and cool-headed. The head-heart dichotomy in English.

Osler, Sir William

1921 *The Evolution of Modern Medicine.* New Haven: Yale University Press.

Quinn, Naomi
1987 The cultural model of American marriage. In *Cultural models in language and thought,* Dorothy Holland and Naomi Quinn (eds.), 173–192. Cambridge: Cambridge University Press.

Schefold, Reimar
1980 *Spielzeug für die Seelen. Kunst und Kultur der Mentawai-Inseln (Indonesien).* Zürich: Museum Rietberg Zürich.

Sibeth, Achim
1990 *Batak – Mit den Ahnen leben. Menschen in Indonesien.* Stuttgart, London: Edition Hansjörg Mayer.

Sneddon, James Neil
1996 *Indonesian: A Comprehensive Grammar.* London: Routledge.

Stefanowitsch, Anatol
2003 Happiness in English and German: A metaphorical-pattern analysis. In *Language, culture, and mind,* Achard, Michel and Suzanne Kemmer (eds.). Stanford: CSLI should be full name of publisher.

Stöhr, Waldemar
1965 Die Religionen der Altvölker Indonesiens und der Philippinen. In *Die Religionen Indonesiens,* Stöhr, Waldemar/Zoetmulder, Piet. (eds.), 1–221. Stuttgart/Berlin/Köln: Mainz. W. Kohlhammer Verlag.

Warneck, Lic. Joh.
1909 *Die Religion der Batak. Ein Paradigma für die animistischen Religionen des Indischen Archipels.* Leipzig: Dieterich'sche Verlagsbuchhandlung Theodor Weicher.

Wierzbicka, Anna
1992 *Semantics, Culture, and Cognition. Universal Human Concepts in Culture-Specific Configurations.* New York/Oxford: Oxford University Press.

Yu, Ning
This vol. The Chinese heart as the central faculty of cognition.

Dictionaries

Duden, Deutsches Universalwörterbuch
1989 Drosdowski, Günther (ed.). Mannheim, Leipzig, Wien, Zürich. Dudenverlag.

An Indonesian-English dictionary
1992 Echols, John M./Hassan Shadily (eds.). (Third edition). Jakarta. Gramedia.

Shorter Oxford English dictionary on historical principles
2002 Trumble, William (ed.). (Fifth edition). Oxford. Oxford University Press.

Contrastive semantics and cultural psychology: English *heart* vs. Malay *hati*

Cliff Goddard

Abstract

This is a contrastive analysis of two ethnopsychological constructs (English *heart*, Malay *hati*), using the natural semantic metalanguage (NSM) approach to semantic description (Wierzbicka 1996). Rejecting the use of English-specific meta-terminology, such as *mind*, *cognition*, *affect*, etc., as both ethnocentric and inaccurate, the study seeks to articulate the conceptual content of the words under investigation in terms of simple universal concepts such as FEEL, THINK, WANT, KNOW, PEOPLE, SOMEONE, PART, BODY, HAPPEN, GOOD and BAD. For both words, the physical body-part meaning is first explicated, and then the ethnopsychological sense or senses (it is claimed that English *heart* has two distinct ethnopsychological senses). The chapter also reviews the phraseology associated with each word, and in the case of English *heart*, proposes explications for a number of prominent collocations: *a broken heart*, *listening to your heart*, *losing heart* and *having your heart in it*. The concluding discussion makes some suggestions about experiential/semantic principles whereby body-parts can come to be associated with cultural models of feeling, thinking, wanting and knowing. At a theoretical level, the study seeks to draw links between culturally-informed cognitive semantics, on the one hand, and the field of cultural psychology, as practised by Richard Shweder and associates.

Keywords: lexical semantics, NSM, emotion, body-parts, English, Malay, cognitive linguistics, ethnopsychology

1. Introduction

The Natural Semantic Metalanguage (NSM) literature includes significant studies of ethnopsychological concepts in different languages, including Anna Wierzbicka's (1992, 2005) seminal studies of Russian *duša* and comparable concepts in other European languages, Rie Hasada's (2000) study of *kokoro* and other emotional concepts in Japanese, Kyung-Joo Yoon's (2006) work on Korean *maum* and *kasum*, and my own study of *hati* in Malay (Bahasa Melayu) (Goddard 2001). The present chapter builds on this earlier study of Malay *hati*. My goal is to develop a comparably detailed semantic analysis of the English word *heart* and related expressions, and then to compare the resulting "semantic portraits" of the two concepts in fine detail. Although the methodology is that of contrastive semantics, the exercise can also be viewed as a study in cultural psychology (Shweder and LeVine eds. 1984; Stigler, Shweder and Herdt eds. 1990; Shweder 2003, 2004; Cole 1996; Wierzbicka 2003), a field which has important affiliations with cognitive linguistics, and more particularly with the "culturally aware" cognitive linguistics which has been developing in recent years (cf. Palmer 2006; Sharifian and Palmer eds. 2007).

Cultural psychology has been characterised by one of its founders Richard Shweder (2003: 27) as the study of "the distinctive mentalities and modes of psychological functioning of members of different communities". One of the assumptions behind cultural psychology is that the semantic constructs of different languages provide culture-specific cognitive frameworks (cultural models, in the sense of Holland and Quinn (1987)) through which people interpret their interpersonal experiences, encode them in memory, and negotiate them in narratives of life experience. Everyday words and expressions thus form part of "the implicit meanings (the goals, values, and pictures of the world) that give shape to psychological processes" (Shweder 2003: 28). I will return to the relationship between contrastive semantics and cultural psychology in Section 5, which also reflects on some of the conceptual and methodological issues facing the two fields.

1.1. Semantic approach

The NSM methodology of semantic description involves explication in terms of a tightly constrained metalanguage of universal semantic primes and elementary syntax (Wierzbicka 1996; Goddard and Wierzbicka eds. 2002; Goddard 1998). An explication is essentially an extended reductive paraphrase, which ought to be substitutable *salvo sensu* into the contexts of use of the original expression. The full inventory of semantic primes is given in the Appendix, in English and Malay versions. Aside from semantic primes, NSM explications can also include certain "semantic molecules", i.e., complex and sometimes language-specific lexical meanings, functioning as units in the semantic structure of yet more complex concepts (Goddard 1998: 254–255, 2006a; Wierzbicka 2006b, 2007a). Semantic molecules are marked as such in explications with the notation [M].

Of particular relevance to the present study are three of the semantic primes in the "mental predicates" grouping, i.e., THINK, WANT and FEEL.

Along with KNOW, SEE and HEAR, and various other primes, they form the basis for a culture-neutral semantic metalanguage for cognition-related meanings (Goddard 2007, In press). Unlike ordinary English words such as *mind*, *emotion*, *memory*, and the like, they have been empirically established as plausible candidates for universally lexicalised meanings, and, furthermore, they have proven themselves as practical tools for the analysis of ethnopsychological concepts in numerous languages (cf. Harkins and Wierzbicka eds. 2001; Wierzbicka 1999; Enfield and Wierzbicka eds. 2002; Goddard 2003).

NSM researchers hold that semantic explications are not mere analytical conveniences, but represent cognitive or conceptual realities for the speech communities in question. That is, a semantic explication is intended as a representation of a conceptual model which is distributed across the members of a particular language-and-culture community. The claim to conceptual reality is made *prima facie* plausible by the fact that explications are framed in terms which are transposable into the languages concerned; that is, the explications do not depend on any "alien" terms and concepts deriving from technical or academic English. Needless to say, this fact in itself does not guarantee the accuracy of any particular explication. The tests for accuracy depend on the extent to which the explication "mimics" and accounts for the semantic properties of the original expression (how well it substitutes into natural contexts of use, its relations with other words, entailments, implications, and so on), and the extent to which it satisfies the intuitions of native speakers.

1.2. English body-part terms: *heart* and *head*

Since we are undertaking to explicate ethnopsychological concepts which are in a polysemic relationship with literal body-part meanings, it will be useful to see explications for two body-parts. The explication for *head* follows Wierzbicka (2007b), with the addition of the final component: 'when someone thinks about something, something happens in this part'. Unlike the other components, which are likely to be widely shared across languages, this final component represents a language-specific aspect of the English "ethno-theory" of thinking. It accounts for common expressions such as *What's going on in his head?* (in the sense 'What is he thinking?'), for numerous fixed expressions, such as *a good head*, *wrong-headed*, etc., and it helps account for the existence of a second polysemic extension, namely, "cognitive" uses of English *head*, as explored in Niemeier (this volume).

[A1] *someone's head*
a. one part of this someone's body
b. this part is above all the other parts of this someone's body
c. this part is round [M]
d. when someone thinks about something, something happens in this part

The second explication, for body-part *heart*, is adapted from one presented in Wierzbicka (1992). Being an internal body-part, the "topographical" components for *heart* naturally refer to its location inside the body (specifically, inside the upper part of the body). The bulk of the explication consists of functional components, linking the heart of a living person with some continuous activities and movements, which can be potentially audible.

[A2] *someone's heart$_1$*
a. one part of this someone's body
b. it is inside the upper [M] part of this someone's body
c. when someone is living, something is happening in this part all the time
d. because of this, something is moving in this part all the time
e. someone can hear this if they do some things

It might be questioned whether the *heart* explication in [A2] ought not include an additional component linking it with people's feelings, as a kind of counterpart to the final component of [A1], which linked the *head* with thinking. The point is, however, that ordinary speakers of present-day English truly believe that some brain activity accompanies thinking, i.e., they believe that something really does happen inside people's physical *heads* when they think. On the other hand, they surely do not believe that anything happens inside people's physical *hearts* when they feel things. The numerous expressions linking the human *heart* with feelings are best treated as examples of a second, polysemic meaning of *heart*, which will occupy us for much of the chapter.

1.3. Explicating English *mind*

Before embarking on our investigation of the emotional-moral *heart* in English, it is also useful to consider the meaning of another English-specific ethnopsychological concept, namely, *mind*. Cross-linguistically, English *mind* is unusual in its "rationalistic" character, i.e., its focus on thinking and knowing to the exclusion of feeling (Wierzbicka 1992, 2005; cf. Yu 2003). This contrast is clear enough if we compare expressions such as *a good mind* and *a good heart*, and sundry other contrasts between *head* and *heart* (Niemeier this volume). The modern English sense of *mind* can be explicated as shown in [B]. According to this explication, the term suggests a dualistic way of thinking about a person, in which the *mind* is an invisible counterpart of a person's body, on account of which a person can think and can know things, and in which something happens when a person thinks. This explication is closely modelled on Wierzbicka (1992, 2005), except for the addition of the final component. This has been added to depict the dynamic aspect of the *mind* concept, reflected in common expressions such as *What's going on in his mind?*, and in phrases such as *the workings of the mind*. Notice that this component creates a conceptual association between the *mind* and the *head*, as one would expect.

[B] *someone's mind (a person's mind)*
a. one of two parts of this someone (one part is the body, this is the other part)
b. people cannot see this part
c. because someone has this part, this someone can think about things
d. because someone has this part, this someone can know things
e. when someone thinks about something, something happens in this part

Space does not permit us to discuss the semantics of the English *mind* concept any further at this point. It does bear comment though, that although the English-specific character of this concept is well established, it continues to be a key word in the discourses of philosophy, linguistics, psychology, and cognitive science, as one can see from the countless journals and books with titles such as *Mind*, *Language and Mind*, *Philosophy of Mind*, *How the Mind Works*, *Mind and Brain*, *Psychology: Mind, Brain and Culture*, and so on. With a few exceptions, even those scholars who are aware of the language-specific character of the "mind concept" (including some leading cognitive linguists), seem convinced that the implicit Anglocentrism is benign. I will return to this issue in section 5.

2. English *heart*$_2$ and related expressions

2.1. Explicating *heart*$_2$

Compared with *mind*, English *heart* is semantically complex, more polysemous, and more phraseologically active – the legacy of a longer and much richer history. A selection of relevant examples, mostly drawn from the Cobuild Word Bank of English, is given in (1)–(7) below. As we look through these examples, it is important to recognise that although one may choose to regard them as "metaphorical" in a loose sense, the expressions in question are certainly not "active metaphors" (Goatly 1997). They are used without metalexical awareness, as shown by the fact that it would be anomalous to append a metalexical tag such as *so to speak*, or *as it were*, to these uses of *heart* (Goddard 2004). They are therefore best regarded as representing an additional stable and conventionalised meaning of the lexeme *heart*, which I will term *heart*$_2$ (cf. Niemeier this volume, 2000).

Example (1) shows that having *a good heart* is seen as a virtue and a potentially redeeming characteristic. Native speakers understand that the sentence implies that the subject is capable of doing good things for other people and has a capacity for compassion. Conversely, example (2) shows that to have *no heart* is bad, and is associated with indifference to others and with a lack of love and care.

(1) *Sure, he's been in a bit of trouble but he's got a good heart.*

(2) *He was a ruthless man. No heart, no love for anyone.*

The examples in (3) and (4) show *heart* connected with something like compassion or "fellow feeling". Those in (5) are associated with empathy, the capacity to be touched by and to share the suffering of others. The expressions in (6a)–(6d) are pairs of converses (or near converses) designating character traits associated with kindness, generosity, and the like.

(3) a. *I didn't have the heart to tell him.*
b. *Have a heart! He's just a kid.*

(4) a. *The thieves must have been heartless to pick on sick young children.*
b. *It wasn't that Dwight was cold or heartless or lacked feeling – not at all.*

(5) a. *My heart went out to him.*
b. *It touched my heart.*

(6) a. *warm-hearted, cold-hearted*
b. *soft-hearted, hard-hearted*
c. *heart of gold, heart of stone*
d. *heart-warming, heart-rending*

The expressions in (7) show that the *heart* can be associated with "valanced" feelings (i.e., with good feelings such as joy, or bad feelings such as sadness), but not with feelings which are neutral in affective tone, such as surprise and interest. This observation is due to Wierzbicka (1992).

(7) a. *His heart was full of joy/bitterness/sadness.*
b. *His heart was full of *surprise/*interest.*

Like Stubbs (2001), I believe that systematicity in phraseology is reflective of underlying semantic patterns or schemas. The semantic explication for *heart*$_2$ given in [C] below provides such a schema. Equally, the explication can be viewed as a cultural model (Holland and Quinn eds. 1987), i.e., as a model of the semantic structure of a particular ethnopsychological construct which is taken for granted in innumerable fixed and semi-fixed expressions of present-day English and which is reinforced and recapitulated as expressions like these are uttered and interpreted on a daily basis. As Stubbs (2001: 166) remarks, referring to Foucault's notion of discursive formation: "A community's value system is built up and maintained, at least partly, by the recurrent use of particular phrasings in texts". Much the same can be said of a community's belief system, and of its characteristic ways of thinking.

Let us work through explication [C] below, component by component. Components (a)–(b) depict the *heart*$_2$ as a part of a person, a part which cannot be seen. Notice that the initial component does not say 'one of two parts' (as with *mind*), because *heart* does not invoke the same dualistic opposition to *body* as does *mind*. The next two components characterise the *heart*$_2$ as the source and locus of multiple and variable interpersonal feelings, in component (c), and especially of good feelings towards other people, in component (d).

Continuing down the explication, components (e)–(f) characterise the *heart*$_2$ as the source and locus of the capacity to want that other people not undergo bad things, and of the desire to 'do good things for other people'.

These components constitute the special affiliation between the *heart*$_2$ and interpersonal virtues such as *kindness*, *compassion*, *generosity*, etc., and their converses.

The final set of components spell out the precise nature of the link between this ethnopsychological concept and the physical body-part *heart*$_1$. They say that people can think of the *heart*$_2$ as like a part of the body, a part which is located inside a person's body in the vicinity of the *heart*$_1$. The wording of these components has been carefully crafted in order to mimic, in an appropriately nuanced way, the exact nature of the implicit relationship between these two English concepts, ethnopsychological *heart*$_2$ and physical *heart*$_1$. (Later, when we turn to a contrastive study of the comparable Malay concept *hati*$_2$, we will see that its relationship to the body-part *hati*$_1$ 'liver' is somewhat different.) Evidence for the reality of the "locatedness" of *heart*$_2$ is provided by the stylised depictions of a heart-shaped organ as a representation of the *heart*$_2$ (e.g., the so-called "love heart") situated in the chest area, and by the way that English speakers will touch the chest area when referring to "heart-felt" emotions.

[C] *someone's heart*$_2$
a. one part of this someone
b. people cannot see this part
c. because someone has this part, when this someone thinks about other people, this someone can feel many things
   these things can be good things, these things can be bad things
d. because someone has this part, this someone can feel good things towards other people
e. because someone has this part, this someone can want other people not to feel bad things
f. because someone has this part, this someone can want to do good things for other people
g. people can think about this part like this:
h.   it is like a part of someone's body
i.   this part is near the place where this someone's heart$_1$ [M] is

As mentioned, the *heart*$_2$ concept is implicit in a number of "phraseological clusters". Though it would be impossible here to review them all, let alone to explore each in full semantic detail, we can look briefly at four of them.

2.2. *A broken heart$_2$*

This collocation, traceable at least as far back in English as John Donne's celebrated 1635 poem 'The broken heart', appears in various expressions, such as to *die of a broken heart, mend a broken heart, nurse a broken heart, the pain of a broken heart,* and in related formations such as *to break someone's heart, heart-break, heart-broken* and *broken-hearted.*

From the Cobuild corpus it is apparent that its most frequent use is probably in relation to failed romantic love (especially in a vast number of love songs); but one can also speak of *a broken heart* (being *heart-broken,* etc.) in relation to grief or other loss, as shown in examples (8) and (9).

(8) *If you feel like your heart has broken, you're not alone. Just about everyone experiences the type of grief we call heartbreak at one time or another – and some people seem to have their hearts broken many times throughout their lives. Sometimes it feels like all those songs about broken hearts were written just for you!*

(9) *I looked after my father until he died of a broken heart because of the death of my younger sister.*

For the expression *a broken heart,* I would advance the explication in [D]. Essentially, it says that the subject dwelt for some time on the thought that something very bad has happened to him or her, on account of his or her strong feelings towards someone else (for example, suffering betrayal by one's sweetheart, or the loss of a beloved relative). The subject cannot accept this as inevitable, but thinks: 'it didn't have to be like this'. These aspects are covered in components (a)–(d). This particular combination of thoughts results in the experiencer feeling something very bad for some time, and this particular bad feeling is seen as somehow special – as qualitatively different to other kinds of bad feelings which people can have in response to bad things happening: see components (e) and (f). On account of this particular emotional situation, there is an internal event ('something bad happened inside this someone' – component (g)), which has the consequence that the subject's capacity to feel very good feelings towards other people is blocked: they can no longer feel such strong positive feelings towards other people as they could before – components (h) and (i).

[D] *Someone (X) has a broken heart*
a. someone (X) thought like this for some time:
b. "something very bad happened to me
c. because I felt something very good towards someone else
d. it didn't have to be like this"
e. because of this, this someone felt something very bad for some time,
f. not like people feel something bad when other bad things happen
g. because of this, something bad happened inside this someone
h. because of this, this someone can't feel very good things towards other people anymore
i. like they could feel very good things towards other people before

It must be added that I am not claiming that every usage of a "broken heart"-type expression will fall under this explication. For example, the expression *It's breaking my heart* is more "dramatic" and more focused on a particular moment in time. For example, a sentence like *To miss the game is breaking my heart* does not imply that the speaker has (or will have) a *broken heart*. The expression *It's breaking my (his, her,* etc.) *heart* requires a separate explication. The same goes for the related word *heart-breaking* (in expressions like *a heart-breaking experience*, *the heart-breaking climax of the play*).

2.3. *Listening to your heart*

A second phraseological cluster associated with *heart*$_2$ is the set of expressions: *to listen to one's heart*, *to follow one's heart*, *to know in your heart* and *heart-searching*. All these expressions convey the idea of a person wanting to find some kind of moral or spiritual direction or certainty, and, "consulting their feelings" to obtain this guidance, rather than relying on purely rational thought. One of the hallmarks of this cluster is the contrast between *heart* and *head* (in the "cognitive" sense, loosely corresponding to *mind*); cf. Niemeier (this volume). A couple of examples follow:

(10) *Don't walk away, listen to your heart, hear what it says cause your heart won't lie. When the right person comes along you will know it in your heart.*

(11) *That decision was a small, first step in following my heart ... After that I listened to my heart and took its advice more seriously.*

I propose the explication in [E] for a typical member of this cluster: *to listen to one's heart*. It depicts a person in doubt about 'what it is good to do', i.e., someone facing a moral choice, a life decision, or the like. This person deliberately attends to how he or she feels when thinking about the relevant matters, in the belief that knowing how he or she feels will lead to knowing what to do. In the event, doing this does lead the person to deciding what to do.

[E] *Someone (X) listened to his/her heart*
a. at this time someone (X) thought like this:
b. "I don't know what it is good to do
c. I want to know what I feel when I think about these things
d. because if I know what I feel, I will know what it is good to do"
e. after this someone thought about it like this for some time, this someone thought like this:
f. "I know now what I want to do"

2.4. *Losing heart, taking heart* and *(one's) heart sinks*

Without proposing a formal explication for the pair of expressions *to lose heart* (cf. *disheartened)* and *to take heart*, we can gloss them (roughly) into reductive paraphrases as follows. *To lose heart* is 'to think at one time that something bad for someone' can't be prevented from happening after all, 'to feel something bad because of it', and no longer have the will to keep trying. A couple of examples follow.

(12) a. *I didn't lose heart then and I mustn't now. I won't; I can't give up ...*
b. *The French could have fought on, but had lost heart.*

To take heart is 'to think that something bad for people doesn't have to happen after all, to feel something good because of it', and to maintain the will to keep trying.

(13) a. *Shareholders can take heart from the payment of a 2p interim dividend ...*
b. *We need to take heart from the Manchester United game, and build on it ...*

c. *On a crumpled piece of paper, it said: take heart the world knows you are alive.*

In the same abbreviated fashion one can gloss the expression *(one's) heart sinks*, roughly, as: 'to think at one time that something bad will happen which will be bad for someone, when one didn't think like that before, and to feel something bad because of it'.

(14) a. *My heart sank when I found a syringe hidden in my boyfriend's apartment.*

b. *It made my heart sink to read about the ten-year-old Russian tennis prodigy Anna Kournikova who is being hyped as a future world champion.*

3. *Heart$_3$* and related uses

A different and distinct meaning, which I designate *heart$_3$*, is found in verbal expressions such as *to have one's heart in something* and *to put one's heart into something*, and in nominal expressions like *big-hearted* and *lots of heart*. The meaning *heart$_3$* is related to *courage* (which itself comes from Old Fr. *corage*, from Latin *cor* 'heart'), and to words such as *encourage* (Middle English, from Fr. *encourager*).

(15) *They showed they have the heart of a fine team. England showed heart but ...*

(16) *We have the heart and energy to do the job.*

(17) *He was an outstanding player who always put his heart into every game.*

(18) *He was a good athlete and a good kid. ... He had lots of heart and there was absolutely no quit in him..*

It should be clear from the examples that this meaning is different from *heart$_2$*, because the explication of *heart$_2$* given in [C] would not be substitutable into these contexts. A formal indication that we are dealing here with a distinct polysemic meaning is provided by the fact that the expression *heart* *and soul* can often substitute for *heart*$_3$, but not for *heart*$_2$. Finally, a clinching argument for polysemy is that the expression *a big heart* is ambiguous between the two meanings. It can mean either generous and/or compassionate (i.e., related to *heart*$_2$) or having a lot of determination, spirit, passion, etc. (i.e., related to *heart*$_3$).

An explication for the expression *to have (one's) heart in it*, associated with *heart*$_3$, is given in [F]. It depicts someone with a goal intended to be of high benefit to someone, understanding and accepting the need to do certain things for some time. The *heart*$_3$ concept, furthermore, envisages that such an attitude can bring about a kind of internal change in a person, such that when faced with setbacks or difficulties, the person does not lose faith in his or her abilities, and does not give up hope. The entire scenario is viewed as good for someone.

[F] *Someone (X) has his/her heart*$_3$ *in it*
a. someone (X) thinks like this for some time:
b. "I want something to happen because it will be very good for someone
c. because of this, I have to do some things for some time
d. I want to do these things"
e. when this someone thinks like this, they feel something
f. because of this, something happens inside this someone
g. because of this, when bad things happen, this someone doesn't think like this:
h. "I can't do this anymore, this good thing will not happen"
i. it is good if someone can be like this

Having reviewed the semantics of the *heart* concepts in present-day English, we are in a position now to undertake comparison with its nearest counterpart in Malay (Bahasa Melayu), the national language of Malaysia.

4. Malay *hati*

Hati is a Malay cultural key word (Goddard 1995, 2001) – extremely common both as an independent item, and in numerous fixed expressions for emotions, moods, attitudes and personal traits. Like *heart*, the word *hati* can refer to a particular organ of the human body – the liver, but unlike with *heart*, there is no Malay iconography featuring stylised representations of the liver as a symbol of the corresponding ethnopsychological construct (see below).

It is not necessary or practical here to undertake a very detailed investigation of Malay *hati*. Rather we will recapitulate the main findings of Goddard (2001), with limited exemplification. Further examples can be found in Siahaan's (this volume) treatment of *hati* in neighbouring Indonesian (Bahasa Indonesia), which can be regarded as a part of the same language as Malay. The main ideas are: (i) that *hati* is broader and more holistic in its scope than *heart*, combining thinking and wanting, as well as feeling, especially in relation to other people, (ii) that there is a moral ambivalence to the *hati*, which is seen as a potential source of wrong-doing and error, as well as of morally correct actions. Its relation to the physical body-part (the liver) is also different in some ways to that displayed by the English word *heart*$_2$.

4.1. Phraseology of the *hati*

As evidenced by its phraseology, the Malay *hati* is more cognitive and more dynamic than the *heart*. Memories and thoughts, for example, can be located in the *hati*, and there are a number of fixed expressions which are best rendered into English with *mind*, rather than with *heart*. For example, the expression *senang hati* [lit. easy *hati*] refers to 'peace of mind, an untroubled attitude' (cf. Goddard 1997); *sehati* 'one *hati*' describes a general alignment of views and attitudes – people who are all of the one mind, while those who are *hati lain-lain* 'different *hati*' think and feel differently and cannot get on harmoniously. The *hati* may also be the locus of desire or intention, especially if the desire or intention concerns another person. Not only may one speak of *niat hati* 'desire in one's *hati*', *hasrat hati* '*hati*'s desire', *kehendak hati* '*hati*'s wish, desire', but also of the *hati* actively wanting or intending to do something. Likewise, the *hati* can "speak" to the experiencer, voicing an emotional reaction or urging some course of action; cf. the frequent use of speech-act verbs such as *kata* 'say', *bisik* 'whisper', and *tanya* 'ask', with the *hati* as subject.

Like *heart*, the *hati* is the realm of feelings, especially sustained feelings about other people, such as *cinta* 'romantic love', *cemburu* 'jealousy, envy', and *sedih* 'sorrow'. The expression *perasaan hati* (where *perasaan* 'feelings' is a nominalisation of *rasa* 'feel'), corresponds closely to *feelings of the heart*. That is, it designates feelings which are relatively longstanding, involve evaluation, and are directed toward another person (as opposed to transient or impersonal feelings such as *terkejut* 'being startled', *takut* 'fear'), cf. Heider (1991), Goddard (1995). Phraseologically, the *hati* can stand in for the full person when deep emotional reactions are concerned. That is, deep emotional reactions can be attributed directly to a person's *hati*: one says 'my *hati* is sad' or 'this *hati* is sad', 'it hurt my *hati*' or 'it hurt this *hati*'.

Looking now in more detail at some common collocations involving *hati*, it can be noted that the most common pattern is [adjective + *hati*], cf. Abdullah (1990). In general, adjectives tend to follow nouns in Malay, but it seems that the [adjective + *hati*] pattern is more like a "locus" relationship than a simple attributive one. That is, the preposed adjective does not "describe" the *hati*, but rather indicates a feeling or attitude which is being located in the *hati* (somewhat similar in construction to English phrases such as *sick at heart*). The locus construction with *hati* also seems to "personalise" the state being described. Examples showing feelings and moods in this construction are given in (19) below. For explications of a sample of these expressions, see Goddard (2001).

(19) *besar hati* 'proud, thrilled with' (*besar* 'big'), *kecil hati* 'feeling hurt' (*kecil* 'small'), *geli hati* 'amused, tickled' (*geli* 'ticklish'), *iri hati* 'envious, spiteful' (*iri* 'envious'), *panas hati* 'angry, worked up' (*panas* 'hot'), *sejuk hati* 'calmed, reassured' (*sejuk* 'cool, cold'), *susah hati* 'troubled' (*susah* 'difficult'), *senang hati* 'contented, relaxed' (*senang* 'easy'), *hancur hati* 'devastated' (*hancur* 'crushed'), *puas hati* 'satisfied' (*puas* 'sated'), *sakit hati* 'annoyed, resentful' (*sakit* 'sick, pain'), *putus hati* 'despairing' (*putus* 'broken asunder'), *lapang hati* 'content' (*lapang* 'vacant, open')

The examples in (20) show aspects of a person's moral character described via the [adjective + *hati*] pattern.

(20) *baik hati* 'kind, good-hearted' (*baik* 'good'), *busuk hati* 'vicious, mean' (*busuk* 'putrid'), *tulus hati* 'sincere, honest' (*tulus* 'pure, unadulterated'), *murah hati* 'generous hearted', (*murah* 'generous'), *rendah hati* 'humble, modest' (*rendah* 'low'), *tinggi hati* 'proud' (*tinggi* 'high'), *putih hati* 'sincere, honest' (*putih* 'white'), *suci hati* 'pure of motive' (*suci* 'morally pure'), *tabah hati* 'determined' (*tabah* 'resolute, persevere'), *lembut hati* 'tender-hearted' (*lembut*

'tender'), *tawar hati* 'lost interest' (*tawar* 'bland'), *keras hati* 'stubborn' (*keras* 'hard').

There are also a large number of verbal expressions with *hati* as the complement noun, as illustrated in (21).

(21) *jaga hati* 'to be protective of (someone's) feelings' ('watch over *hati*'), *ambil hati* 'to charm (someone), win over' ('get *hati*'), *suka hati* 'do as you please, do whatever you feel like' ('please *hati*'), *tidak sampai hati* 'can't stand to (do something)' ('not reach *hati*'), *cuba hati* 'test (someone's) honesty or commitment' ('try *hati*'), *ubah hati* 'change attitude, switch allegiances' ('change *hati*').

Taken together, examples like those in (19)–(21) are evidence of the semantic richness of the *hati* concept, which embraces feelings, attitudes, and intentions. The same breadth of concept is indicated by the numerous traditional sayings (*peribahasa*) which mention the *hati*. A clear theme among these is the ambivalence of the *hati*, which can just as easily lead one into error as into virtue. The saying in (23) may seem striking from a European point of view (especially if one is unconsciously identifying *hati* with *heart*), but it gains its meaning in the Malay cultural context from the idea that impulsive and unreflective actions (undertaken at the "urging" of the *hati*) can easily lead one into sin and disaster (cf. Goddard 2001).

(22) *Biar buta mata, jangan buta hati.*
'Let the eyes be blind, (but) not the *hati* be blind.'

(23) *Ikut rasa binasa, ikut hati mati.*
'Follow feelings suffer, follow *hati* die.'

4.2. Explicating *hati*$_2$

I propose the semantic explication in [G]. Leaving aside for the moment the final set of components, which spell out the relationship with the physical body-part *hati* 'liver', the content of this explication can be explained as follows. The initial pair of components are the same as for English *heart*$_2$, i.e., they designate an invisible part of a person. Component (c) states that sustained thinking about another person can be associated with 'something happening' in this part. This already gives *hati* a more dynamic quality than English *heart*$_2$ (while at the same time linking *hati*, in this respect, with English *mind*). The components in (d) say that such internal events can give rise to multiple feelings – feelings which can be either good or bad. These feelings in turn can have motivational consequences, i.e., they can lead to the person wanting to do something, which can again be either good or bad, as set out in component (e). Here we see the link between the *hati* and the "moral will". Component (f) recognises the active phenomenology of the *hati*: that it can be experienced as 'saying something' to a person. Component (g) captures an aspect not mentioned up to this point, but evidenced by numerous Malay sayings and collocations, namely, that the workings of the *hati* are inaccessible to other people.

[G] someone's *hati*$_2$ [Malay]
a. one part of this someone
b. people cannot see this part
c. when someone thinks about someone else, something can happen in this part
d. when this happens, this someone can feel many things
these things can be good things, these things can be bad things
e. because this someone can feel these things, this someone can want to do many things
these things can be good things, these things can be bad things
f. when something happens in this part of someone, this someone can think about it like this: "something inside me is saying something to me"
g. other people can't know what happens in this part of someone if this someone doesn't want them to know
h. people think about this part of someone like this:
i. "it is like a part of someone's body
j. it is somewhere inside the middle [M] of this someone's body"

In explication [G], the final set of components (h)–(j) address the relationship between the ethnopsychological *hati*$_2$ and the body. They state that people think about the *hati*$_2$ in terms of an analogy with the body: 'it is like a part of someone's body'; and more specifically, like a part located 'somewhere inside the middle [M] of this someone's body'. Notably, there is no direct reference here to the physical *hati*$_1$ 'liver', either by way of direct comparison with the liver or by way of comparison with the location of the liver.

In framing the components in this fashion, I am adopting a conservative position. There is some cultural-internal evidence for an association between the ethnopsychological *hati*$_2$ and the body, and for the notion that the hypothetical internal body-part is thought of as being somewhere in the middle of the body. Malay consultants will say as much. But consultants do not speak about any direct connection with the physical liver, nor do they describe the internal location of the *hati* by reference to the physical liver (presumably, in part because the location of one's liver is a lot less salient, from an experiential point of view, than that of the heart). Furthermore, as mentioned above, one never sees stylised depictions of the liver as a symbol of the ethnopsychological *hati*$_2$. I have therefore framed the relevant components in a deliberately non-specific fashion.

In traditional Malay folk belief, the bodily *hati* was regarded as playing an important role in both physical and psychological functioning. In relation to traditional healing practices, Peletz (1996: 206) remarks that: "Some Malays refer to the liver as the 'ruler' (*raja*) of the human body and note that it 'governs' or 'regulates' (*merintah*) the rest of the body, much like a ruler or commander governs his army." Siahaan (this volume) provides an account of the roots of the archaic cultural model "almost forgotten and hardly practised" (as she puts it), which is still reflected in present-day Indonesian and Malay. She documents an ancient tradition of liver divination, and an ancient cultural model according to which a person harbours within their body a "living soul" somewhat independent of the person and linked in a mystical fashion with "the hand of gods".

Siahaan's study, along with others in this volume, and earlier work by Geeraerts and Grondelaers (1995), among others, show that much of the ethnopsychological phraseology of any particular period of time (its collocations, catch phrases, cliches and idioms) can be thought of as "echoing" the beliefs, values and cultural models of the past. In their forms, they are reflective of cultural tradition and of older cultural models, rather than present-day ones (cf. Stubbs 1996, 2001). This does not mean, however, that the meanings of such expressions have stood still. On the contrary, there is every reason to believe that the meanings of fixed expressions and phraseological patterns shift more quickly than their forms, adapting and evolving in tandem with changes in cultural beliefs and values. In any case, what is clear from the semantic analyses undertaken in the present study is that the semantics of present-day English *heart* and present-day Malay *hati* no longer mirror the archaic cultural models which have given rise to the forms of these words and to their numerous collocations.[1] The present-day *heart* and *hati* expressions can be regarded, therefore, as "new wine in old bottles".

5. Reflections on conceptual and methodological issues

In this concluding section, I would like to reflect on some conceptual and methodological issues facing us when we undertake to dissect (so to speak) the Malay *hati*, Russian *duša*, Korean *maum*, Chinese *xin*, Persian *del*, and other comparable ethnopsychological concepts from languages other than present-day English (cf. Yu this volume; Sharifian this volume). How serious are the dangers of taking the "familiar" English categories of *mind* and/or *heart* as common reference points? How serious are the dangers of employing English-specific terminology in our descriptions and explanations of non-English concepts? And what are the alternatives?

This is where NSM research has something unique to offer. For among contemporary semantic theories, NSM alone has problematicised ethnocentrism as a methodological problem and developed a well-theorised solution (cf. Goddard 2006b). NSM research, furthermore, has produced a solid body of empirical data on lexical universals (cf. the studies in Goddard and Wierzbicka eds. 1994, 2002; Goddard 2001, 2003, Goddard ed. 2008). From this body of work it has become clear that there is a small set of fundamental concepts which are shared across all cultures and lexicalised in all languages. By fashioning semantic descriptions in terms of this shared "intersection of all languages", one can free the formal representation of meaning from terminological ethnocentrism and at the same time achieve a degree of precision and analytical detail which cannot be reached by other methods (Goddard In press). As the distinguished anthropologist Roy D'Andrade (2001: 246) remarks, the natural semantic metalanguage "offers a potential means to ground all complex concepts in ordinary language and translate concepts from one language to another without loss or distortion in meaning". Semantic primes THINK, KNOW, WANT and FEEL are among the shared fundamental concepts in the world's languages, and can therefore provide the basis for a culture-neutral metalanguage for contrastive semantics and cultural psychology (Goddard 2007; Shweder 2004; Wierzbicka 2003), whereas *reason*, *emotion*, *intellect*, *heart* – and *mind* – are not.

In this connection, the tradition in cognitive linguistics of referring to a presumed universal, or at least widespread, conceptual metaphor THE MIND IS THE BODY (Lakoff and Johnson 1980, 1999) can be seen to be seriously flawed. Extensive cross-linguistic research indicates that one term of this metaphor, namely BODY, stands for a simple concept which is universal and present as a lexical meaning in all languages (Goddard and Wierzbicka eds. 2002; Wierzbicka 2007b; cf. Majid, Enfield and van Staden eds. 2006). But, as mentioned earlier, the opposite is the case with the term *mind*, which represents a concept that is semantically complex and exquisitely culture-bound.

It seems to me that the basic claim behind the proposed THE MIND IS THE BODY metaphor is a claim about links between ways of thinking about "cognitive" processes and states (thinking, knowing, wanting, feeling) and the human body: i.e., as a claim that all or most cultures have cultural models which represent these cognitive processes and states as functions or attributes of certain parts or "quasi-parts" of the body. On this interpretation, the word MIND is being used in this context as a kind of shorthand or cover-term for any selection of semantic components involving semantic primes THINK, KNOW, WANT and FEEL. Be that as it may, the formula THE MIND IS THE BODY cannot possibly represent a "conceptual metaphor" in any psychologically realistic sense for speakers of most languages of the world.

To say this in no way denies the critical insight that the concept of the BODY, and the shared human experience of living an embodied life, provide a common conceptual and experiential base for ways of thinking and speaking about cognitive functions, i.e., for ways of thinking and speaking about how people think, know, want and feel about the world around them (and, perhaps above all, about the other people in the world around them). Rather, it means that we must better articulate the multiple relationships which are involved, and I would like to make some suggestions in this respect. To begin with, I would suggest that the presence in all languages and in all human conceptual systems of the concept BODY is partly responsible for the prevalence in languages of nominal expressions designating non-physical parts of a person, i.e., parts akin to the *heart*, *mind*, *soul*, and so on. Why? Because in a sense the semantic primes BODY and SOMEONE together pose a kind of conceptual problem: given that a person has a body, and given that there is evidently more to a person than a body alone – what else is there? Ethnopsychological constructs are answers to this question, and, as amply evidenced by the studies in this volume, a multiplicity of different answers (different cultural solutions) are provided in the languages and cultures of the world.

As for the nature or structuring of these ethnopsychological constructs, again the body evidently provides a possible model. It is clear that people everywhere in the world think of their bodies as having certain parts: visible external parts such as eyes, ears, heads, hands, legs, which have obvious functions, and also certain internal parts which are not visible and whose functions are less clear (Wierzbicka 2007b). This set-up provides a model for ethnopsychological constructs in at least four ways, which can be used separately or in combination. First, an ethnopsychological entity can be "defined" in an exclusionary or complementary relationship with the body, vis-à-vis the person as a whole. Examples are concepts such as present-day English *mind* and the *soul* concept of late Middle English (roughly, Shakespeare's English), discussed by Wierzbicka (1992). The ethnopsychological entity is conceived as 'one of two parts' of a person, with the other part being the body.

Second, the ethnopsychological entity can be either identified with or likened to a part of the body, normally an internal body-part. Either way there are implications for the location within the body, in addition to the partonymic or quasi-partonymic relationship as such.

Third, cognitive functions (thinking, knowing, wanting, feeling) can be attributed to the ethnopsychological entity by analogy with the more obvious functionality of external body-parts, i.e., the fact that we see with our eyes, hear with our ears, manipulate things with our hands, and so on.

Fourth, and more specifically, there is an inherent semantic affiliation between feeling and the body, because, according to NSM research, languages universally provide a combinatorial frame whereby people can speak of feeling something in particular parts of the body, i.e., languages universally provide ways of speaking about and characterising feelings "localised" in the body. (This provides the semantic basis for sensation concepts like *hunger* and *thirst*, and for pain and "pain-like" concepts, for sensory properties such as *sweet* and *sour*, *hot* and *cold*, among others, cf. Goddard and Wierzbicka 2007)

To return to my main theme, however: to explore how the languages and cultures of the world exploit these multiple connections between the body and embodiment on the one hand, and cognitive functions (thinking, knowing, wanting, feeling) on the other, requires analytical tools which are both accurate and non-ethnocentric. The NSM methodology of reductive paraphrase in simple universal meanings provides such a methodology.

Appropriately analysed, the ethnopsychological constructs of individual languages are revelatory of widely differing cultural models of cognitive experience, moral reasoning, and social values. Because they are embodied in the words and phrases of everyday talk and because they represent "experience-near" concepts, such models find their way into people's self-understandings and into their life narratives (Geertz 1983; Bruner 1990). Contrastive semantic analysis of ethnopsychological constructs can therefore make a crucial contribution both to cognitive linguistics in its quest to understand the relation of language and cognition (Langacker 1988, 1999), and to cultural psychology in its quest to understand "the distinctive mentalities and modes of psychological functioning of members of different communities" (Shweder 2003).

Acknowledgements

An earlier version of this chapter was presented at the workshop on conceptualisations of the "heart" held at the International Cognitive Linguistics Conference, Seoul, South Korea, July 2005. I would like to thank Anna Wierzbicka for detailed comments and suggestions on the explications, and the editors of this volume for their useful comments.

Appendix

Semantic primes – English exponents, with Malay equivalents (Goddard 2002)

Substantives:	I *aku*, YOU *kau*, SOMEONE *seseorang*, SOMETHING/THING *sesuatu/benda*, PEOPLE *orang*, BODY *badan*
Relational substantives:	KIND *jenis*, PART *bahagian*
Determiners:	THIS *ini*, THE SAME *sama*, OTHER/ELSE *lain*
Quantifiers:	ONE *satu*, TWO *dua*, MUCH/MANY *banyak*, SOME *beberapa*, ALL *semua*
Evaluators:	GOOD *baik*, BAD *buruk*
Descriptors:	BIG *besar*, SMALL *kecil*
Mental predicates:	THINK *fikir*, KNOW *tahu*, WANT *mahu*, FEEL *rasa*, SEE *nampak*, HEAR *dengar*
Speech:	SAY *kata*, WORDS *kata-kata*, TRUE *benar*
Actions, events, movement, contact:	DO *buat*, HAPPEN *terjadi*, MOVE *bergerak*, TOUCH *sentuh*
Location, existence, possession, specification:	BE (SOMEWHERE) *berada*, THERE IS *ada*, HAVE *ada*, BE (SOMEONE/SOMETHING) *ia(lah)*
Life and death:	LIVE *hidup*, DIE *mati*
Time:	WHEN/TIME *bila/masa*, NOW *sekarang*, BEFORE *sebelum*, AFTER *selepas*, A LONG TIME *lama*, A SHORT TIME *seke-jap*, FOR SOME TIME *beberapa lama*, MOMENT/IN ONE MOMENT *saat*
Space:	WHERE/PLACE *mana/tempat*, HERE *sini*, ABOVE *atas*, BELOW *bawah*, NEAR *dekat*, FAR *jauh*, SIDE *sebelah*, INSIDE *dalam*
Logical concepts:	NOT *tidak*, MAYBE *mungkin*, CAN *boleh*, BECAUSE *kalau*, IF *sebab*
Intensifier, augmentor:	VERY *sangat*, MORE *lagi*
Similarity:	LIKE/AS *macam*

Notes:

- Primes exist as the meanings of lexical units (not at the level of lexemes)
- Exponents of primes may be words, bound morphemes, or phrasemes
- They can be formally complex
- They can have combinatorial variants (allolexes)
- Each prime has well-specified syntactic (combinatorial) properties.

Notes

1. To say this is not to deny continuity with the past, but continuity is not the same as identity: it implies change. Reconstructing the pathways of such changes is one of the most formidable tasks facing historical semantics. Arguably, NSM semantics has an important contribution to make to this task, given that the need for a principled and disciplined method of semantic analysis is, if anything, even more pressing in historical semantics than in synchronic studies, cf. Wierzbicka (2006b; Bromhead In press). I would also not want to deny the possibility that well-established phraseological patterns may exert some kind of conservative influence on ongoing semantic change.

References

Abdullah Hussain
1990 *Kamus Simpulan Bahasa*. Kuala Lumpur: Dewan Bahasa dan Pustaka.
Bromhead, Helen
In press *The Reign of Truth and Faith: Epistemic Expression in 16th and 17th Century English*. Berlin: Mouton de Gruyter.
Bruner, Jerome A.
1990 *Acts of Meaning*. Cambridge, MA: Harvard University Press.
Cole, M.
1996 *Cultural Psychology: A Once and Future Discipline*. Cambridge, MA: Harvard University Press.
D'Andrade, Roy
2001 A cognitivist's view of the units debate in cultural anthropology. *Cross-Cultural Research* 35 (2): 242–257.
Enfield, N. J., and Anna Wierzbicka (eds.)
2002 *Pragmatics and Cognition*. Special issue on "The Body in the description of Emotion", 10 (1/2).
Geeraerts, Dirk, and Stephan Grondelaers
1995 Looking back at anger. Cultural traditions and metaphorical patterns. In *Language and the Construal of the World*, John R. Taylor, and Robert E. MacLaury (eds.), 153–180. Berlin: Mouton de Gruyter.
Geertz, Clifford
1983 *Local Knowledge. Further Essays in Interpretive Anthropology*. New York: Basic Books.
Goatly, Andrew
1997 *The Language of Metaphors*. London/New York: Routledge.

Goddard, Cliff
1995 'Cognitive mapping' or 'verbal explication'?: Understanding love on the Malay archipelago. *Semiotica* 106 (3/4): 323–354.
1997 Cultural values and 'cultural scripts' of Malay (Bahasa Melayu). *Journal of Pragmatics* 27 (2): 183–201.
1998 *Semantic Analysis: A Practical Introduction.* Oxford: Oxford University Press.
2001 *Hati*: A key word in the Malay vocabulary of emotion. In *Emotions in Crosslinguistic Perspective*, Jean Harkins, and Anna Wierzbicka (eds.), 167–195. Berlin: Mouton de Gruyter.
2002 Semantic primes and universal grammar in Malay (Bahasa Melayu). In *Meaning and Universal Grammar – Theory and Empirical Findings*, Vol. I, Cliff Goddard, and Anna Wierzbicka (eds.), 87–172. Amsterdam/Philadelphia: Benjamins.
2003 'Thinking' across languages and cultures: Six dimensions of variation. *Cognitive Linguistics* 14 (2/3): 109–140.
2004 The ethnopragmatics and semantics of 'active metaphors'. *Journal of Pragmatics* 36: 1211–1230.
2006a Natural Semantic Metalanguage. In *Encyclopedia of Language and Linguistics,* 2d ed., Keith Brown (ed.), 544–551. Oxford: Elsevier.
2006b Verbal explication and the place of NSM semantics in cognitive linguistics. In *Cognitive Linguistics Investigations: Across Languages, Fields, and Philosophical Boundaries*, June Luchjenbroers (ed.), 189–218. Amsterdam/Philadelphia: Benjamins.
2007 A culture-neutral metalanguage for mental state concepts. In *Mental States. Volume II: Language and Cognitive Structure*, Andrea C. Schalley, and Drew Khlentzos (eds.), 11–35. Amsterdam/Philadelphia: John Benjamins.
In press Universals and variation in the lexicon of mental states. In *Words and the World: How Words Capture Human Experience*, Barbara Malt, and Phil Wolff (eds.). New York: Oxford University Press.

Goddard, Cliff (ed.)
2008 *Cross-Linguistic Semantics*. Amsterdam/Philadelphia: John Benjamins.

Goddard, Cliff, and Anna Wierzbicka
2007 NSM analyses of the semantics of physical qualities: *Sweet, hot, hard, heavy, rough, sharp* in cross-linguistic perspective. *Studies in Language* 31 (4): 765–800.

Goddard, Cliff, and Anna Wierzbicka (eds.)
1994 *Semantic and Lexical Universals – Theory and Empirical Findings.* Amsterdam/Philadelphia: John Benjamins.

2002 *Meaning and Universal Grammar – Theory and Empirical Findings.* Vol. I and II. Amsterdam/Philadelphia: John Benjamins.

Harkins, Jean, and Anna Wierzbicka (eds.)
2001 *Emotions in Cross Linguistic Perspective*. Berlin: Mouton de Gruyter.

Hasada, Rie
2000 An Exploratory Study of Expression of Emotions in Japanese: Towards a Semantic Interpretation. PhD thesis, Australian National University.

Heider, Karl G.
1991 *Landscapes of Emotion: Mapping Three Cultures of Emotion in Indonesia*. Cambridge: Cambridge University Press.

Holland, Dorothy, and Naomi Quinn (eds.)
1987 *Cultural Models in Language and Thought*. Cambridge, MA: Cambridge University Press.

Lakoff, George, and Mark Johnson
1980 *Metaphors We Live By*. Chicago: University of Chicago Press.
1999 *Philosophy in the Flesh: The Embodied Mind and Its Challenge to Western Thought*. New York: Basic Books.

Langacker, Ronald A.
1988 An overview of cognitive grammar. In *Topics in Cognitive Linguistics*, Brygida Rudzka-Ostyn (ed.), 3–48. Amsterdam: Benjamins.
1999 Assessing the cognitive linguistic enterprise. In *Cognitive Linguistics: Foundations, Scope and Methodology*, Theo Janssen, and Gisela Redecker (eds.), 13–59. Berlin/New York: Mouton de Gruyter.

Majid, Asifa, N. J. Enfield, and Miriam van Staden (eds.)
2006 Parts of the body: Cross-linguistic categorisation. Special issue of *Language Sciences* 28 (2/3).

Niemeier, Susanne
2000 Straight from the heart – metonymic and metaphorical explorations. In *Metaphor and Metonymy at the Crossroads: A Cognitive Perspective*, Antonio Barcelona (ed.), 185–213. Berlin/New York: Mouton de Gruyter.
This vol. To be in control: *Kind-hearted* and *cool-headed*. The head-heart dichotomy in English.

Palmer, Gary
2006 When does cognitive linguistics become cultural? Case studies in Tagalog voice and Shona noun classifiers. In *Cognitive Linguistics Investigations: Across Languages, Fields, and Philosophical Boundaries*, June Luchjenbroers (ed.), 13–45. Amsterdam: John Benjamins.

Peletz, Michael G.
1996 *Reason and Passion: Representations of Gender in a Malay Society.* Berkeley: University of California Press.

Sharifian, Farzad, and Gary B. Palmer (eds.)
2007 *Applied Cultural Linguistics: Implications for Second Language Learning and Intercultural Communication.* Amsterdam/Philadelphia: Benjamins.
This vol. Conceptualizations of *del* 'heart-stomach' in Persian.

Shweder, Richard A.
2003 *Why Do Men Barbeque? Recipes for Cultural Psychology.* Cambridge, MA: Harvard University Press.
2004 Deconstructing the emotions for the sake of comparative research. In *Feelings and Emotions: The Amsterdam Symposium*, Antony S. R. Manstead, Nico Fridja, and Agneta Fischer (eds.), 81–97. Cambridge: Cambridge University Press.

Shweder, Richard A., and Richard LeVine (eds.)
1984 *Culture Theory: Essays on Mind, Self and Emotion.* Cambridge: Cambridge University Press.

Siahaan, Poppy
This vol. Did he break your *heart* or your *liver*? A contrastive study on metaphorical concepts from the source domain ORGAN in English and in Indonesian.

Stigler, J., R. A. Shweder, and G. Herdt (eds.)
1990 *Cultural Psychology: Essays on Comparative Human Development.* New York: Cambridge University Press.

Stubbs, Michael
1996 *Text and Corpus Analysis: Computer-assisted Studies of Language and Culture.* Oxford: Basil Blackwell.
2001 *Words and Phrases: Corpus Studies of Lexical Semantics.* Massachusetts: Blackwell Publishers.

Wierzbicka, Anna
1992 *Semantics, Culture and Cognition.* Oxford: Oxford University Press.
1996 *Semantics: Primes and Universals.* Oxford: Oxford University Press.
1999 *Emotions Across Languages and Cultures.* Cambridge: Cambridge University Press.
2003 A conceptual basis for cultural psychology. *Ethos* 21: 205–231.
2005 Empirical universals of language as a basis for the study of other human universals and as a tool for exploring cross-cultural differences. *Ethos* 33 (2): 256–291.
2006a The semantics of colour: A new paradigm. In *Progress in Colour Studies. Volume I: Language and Culture*, C. P. Biggam and C. J. Kay (eds.), 1–24. Amsterdam/Philadelphia: John Benjamins.

2006b *English: Meaning and Culture*. New York: Oxford University Press.
2007a Shape and colour in language and thought. In *Mental States. Volume II: Language and Cognitive Structure*, Andrea C. Schalley, and Drew Khlentzos (eds.), 37–60. Amsterdam/Philadelphia: John Benjamins.
2007b Bodies and their parts: An NSM approach to semantic typology. *Language Sciences* 29 (1): 14–65.

Yoon, Kyung-Joo
2006 *Constructing a Korean Natural Semantic Metalanguage*. Seoul: Hankook Publishing Co.

Yu, Ning
2003 Chinese metaphors of thinking. *Cognitive Linguistics* 14 (2/3): 141–165.
This vol. The Chinese heart as the central faculty of cognition.

Guts, heart and liver: The conceptualization of internal organs in Basque

Iraide Ibarretxe-Antuñano

Abstract

This paper studies the conceptualizations of four Basque internal organs: bihotz 'heart', gibel 'liver', urdail 'stomach' and heste 'intestine', and their use in compound words and idiomatic expressions. In the first part, it examines the main conceptualizations associated with these internal body organs. It is argued that there are at least three different groups of meanings in these words: (i) internal reference: the word refers to a specific internal organ and to the processes related to it; (ii) external reference: the internal organ acts as a focalized centre and is used to metonymically describe a specific physical – internal and external – area of the body, and (iii) emotions. The second part of this paper focuses on the cultural models and embodiment of some of the conceptualizations of these internal body parts, with special attention to those related to the group of emotions. It also discusses the main differences and similarities found in these internal organs and their conceptualizations in comparison with other Basque body parts, especially those related to the mind and intellect. Data come from monolingual and bilingual dictionaries as well as from idiomatic dictionaries and compilations of idioms and proverbs.

Keywords: body parts, Basque, embodiment, metaphor, metonymy, compositional polysemy.

1. Basque internal body organs and their conceptualizations

This section examines the conceptualizations of four internal body parts in Basque: the heart, the liver, the stomach and the intestines. What is interesting about these body parts is that, despite their different roles and locations in the body, they are used in similar ways as far as their conceptualization is concerned. That is to say, the different uses of these internal organs follow similar paths of semantic extension. All are used to refer to illnesses related to them (Section 1.1), to other internal and external body parts (Section 1.2.1), to locations in space (Section 1.2.2), and very interestingly, to emotions (Section 1.3).

Later on in the paper, I will explore why these internal organs are conceptualized in such specific ways. I will argue that the answer lies in their physical embodiment.[1] These conceptualizations are based on both individuals' human bodily experiences and society's cultural/spiritual/ ethnomedical traditions and beliefs. But firstly I will list the different conceptualizations that are found in these body parts, and briefly focus on the cognitive mechanisms that can explain how these extended senses are derived from the physical meaning of the corresponding internal body parts.

I will use three cognitive tools: metaphor, metonymy, and compositional polysemy. I understand these terms as follows: metaphor, a mapping between two different conceptual domains (Lakoff and Johnson 1980, 1999), or across mental spaces (Fauconnier and Turner 2002); metonymy, a mapping within the same conceptual domain (Barcelona 2000; Kövecses and Radden 1998); and compositional polysemy (Ibarretxe-Antuñano 1999), a mechanism that explains how different semantic extensions of a lexical item are obtained through the interaction of the semantic content of both the lexical item itself and its different co-occurring elements. The weight of the semantics of these elements in the creation of polysemes is not always the same; it varies according to the degree of semantic influence of these elements on the overall meaning.

All the different senses, as well as the examples, are extracted from monolingual and bilingual dictionaries (Agud and Tovar 1992; Azkue 1905; Bostak Bat 1996; La Gran Enciclopedia Vasca 1976; Mitxelena 1987–2005; Morris 1998; Mugica 1981; Sarasola 1984–1995) and idiomatic compilations (Garate 1998; Mokoroa 1990; Stürze et al. 2002) (see reference section).

Let us start with the first group of meanings.

1.1. Internal reference

In this section, I examine those meanings of the internal body parts that refer to the internal organ itself and to the processes related to it. All the internal body parts analyzed in this paper show this type of conceptualization. They refer both to (i) areas or parts related to the internal organ, and (ii) illnesses related to the internal organ. In most cases, these two semantic extensions are obtained by means of compounding, e.g. *Ipurteste* (arse.intestine) 'rectum' or affixation, e.g. *Gibeleko* (liver.adn)[2] 'liver complaint'. There is, however, a case where the internal organ itself refers to the illness: if we use *urdail* 'stomach' together with the verb *ukan* 'have', it means 'stomach ache'. This is an example of the metonymy ORGAN FOR ILLNESS.

Table 1 summarizes the *internal reference* meanings in these internal organs.

Table 1. Internal reference in internal body parts

Body part	Areas or parts	Illness
Bihotz 'heart'	*Bihotz-sare* 'pericardium'; *Bihotz-ginhar* 'myocardium'; *Bihotz-mamia* 'myocardium'	*Bihotz-antsia* 'angina'; *Bihotzeko-min* 'angina'; *Bihotz-sueri* 'carditis'
Gibel 'liver'	*Gibelmin* (liver.pain) 'gall, bile'	*Gibeletik egon* (liver.abl be) 'have a liver complaint'; *Gibelmin* (liver.pain) 'liver complaint'; *Gibel-arri* (liver-stone) 'gallstone'; *Gibeleko* 'liver complaint'; *Gibelekoak aurpegia horitu dio* 'His liver complaint has yellowed his face'
Urdail 'stomach'	*Urdabilaren ahoa* 'pit of the stomach'	*Estamagu min* 'stomach ache'; *Urdaileko min* 'stomach ache'; *Urdaileri* 'gastritis'; *Urdaila ukan* (stomach have) 'stomach ache'
Heste ' intestine'	*Heste itsu* (intestine blind) 'caecum'; *Heste lodi* (intestine fat) 'large intestine'; *Heste luze* (intestine long) 'small intestine'; *Heste mehar* (intestine thin) 'small intestine'; *Hestegorri* (intestine.red) 'oesophagus'; *Ipurteste* (ass.intestine) 'rectum'; *Ondeste* (back.intestine) 'rectum'	*Hestereango* 'diarrhoea'; *Hesteriko* 'diarrhoea'; *Hestemin* (intestine.pain) 'intestinal complaint'

Another possible interpretation for this type of example is to propose that they are cases of a metaphorical projection from shapes onto attributes onto functions (Garai 2001; Lakoff and Turner 1989). That is, illness is attributed to the organ due to a conceptualization of the organ as cause. This metaphorical projection chain is necessary in order to explain 'misleading' localizations of some illnesses such as heartburn (Basque *Bihotz-erre*) for oesophageal ache, and *okerreko samatik joan* (wrong.adn neck.abl go) 'choke'. These types of metaphorical projections are not trivial; quite the contrary, they are basic mechanisms for cultural models parallel to that of the *great chain of being* metaphor (Lakoff and Turner 1989), where a hierarchical organization is also involved. These are important for interpreting other secondarily metaphorized contexts, such as those provided in proverbs, where two body organs, i.e. their cultural conceptualizations, are opposed.

1.2. External reference

In this section, I deal with those meanings included under the generic name of *external reference*. In this group, we find that internal organs are used to describe (i) other body parts, either internal or external; (ii) a physical area or location (the internal organ acts as a focalized centre and as such, it is used to metonymically describe a specific physical – internal and external – area of the body); and (iii) metaphorical and metonymic extensions related to the internal organ itself, or derived from the metonymical interpretation described in (ii).

1.2.1. Reference to other internal and external organs

All four of the internal organs studied are used to refer to other body parts, both internal, e.g. *urdail* 'womb, uterus,[3] stomach', *gibelarai* 'lung', and external, e.g. *hesteontzi* 'belly'. These are examples of PART FOR PART metonymy, because one part of the body (internal) is used to refer to another part of the body (internal or external). What is even more interesting is the fact that these body parts can also refer to whole internal and external areas of the body, not just to another specific organ. For example, *gibel* also means 'back', *hesteak* 'guts', and *gibelarrai* 'entrails (heart, liver, spleen, lungs)'. These extensions can be explained by means of the PART FOR WHOLE metonymy, where one part of the body (internal) is used to refer to a whole area in the body (both external and internal). Table 2 summarizes these metonymic extensions.

As can be seen in Table 2, *Gibel*, apart from meaning 'back',[4] can also be used to refer to 'bottom'. This semantic extension can be explained in two different ways. On the one hand, one could argue that it is another case of PART FOR PART metonymy, because the liver is used to describe an external body part, the bottom. On the other hand, the meaning of 'bottom' could be derived from the metonymic meaning of 'back', and consequently, 'bottom' would be the result of WHOLE FOR PART metonymy. The latter is perhaps more plausible, since in many languages, the word 'back' is used in order to avoid the taboo term 'arse'.

Table 2. Reference to other internal organs

Body part	Other internal / external organs
Bihotz 'heart'	*Bihotz-koilara* 'pit of the stomach'; *Bihotzerre* 'heartburn'; *Bihotztar* 'heartburn'; *bihotz-estaldura* 'bloated feeling in the stomach'
Gibel 'liver'	*Gibelerrai* 'entrails, guts (heart, liver, spleen, lungs)', 'lungs'; *Gibel-zilo* 'anus'; *Gibelalde* (liver.side) 'back', 'bottom'; *Gibel-haize* (liver-air) 'fart'; *Gibehats* 'fart'
Urdail 'stomach'	*Urdail* 'womb, uterus'; *Urdail* 'breastbone'
Heste 'intestine'	*Hesteak* (intestines) 'guts'; *Hesteontzi* (intestine.container) 'belly'

1.2.2. Reference to a physical area or location

Two of the body parts studied in this paper, *bihotz* 'heart' and *gibel* 'liver', develop this semantic extension of 'physical area' or 'location' as shown in Table 3 below. The word *bihotz* can be used in the sense of 'core, or inside of something'. For example, the bone marrow is *hezurraren bihotza*, literally 'the heart of the bone', the inside part of the bread is *ogiaren bihotza* 'the heart of the bread', *neguaren bihotza* 'the heart of the winter'. The word *gibel* can convey the meaning 'back side, rear side', as in *etxearen gibelean* (house.gen *gibel*.loc) 'at the rear part of the house', and in other compound words such as *gibeleratu* (*gibel*.all.verbal.suffix) 'go to the back', *gibel egin* (*gibel* make) 'move backwards', and *gibel-so* (*gibel*-look) 'look back'.

These semantic extensions can be taken as examples of the metonymy ORGAN FOR LOCATION, since the organ comes to designate the area where it is located. It is interesting to note that, whereas in the case of *bihotz*, the heart refers to the real position of this organ in the body, that is, inside the body, in *gibel*, the semantic development is a little more complex. The body part *gibel* 'liver' is metonymically used to refer to the 'back', and once the word acquires this meaning, it develops the sense of 'rear part, backside'. As discussed in Section 1.3 below, this path of semantic extension is extremely crucial for the explanation of the different conceptualizations of *gibel* as the seat of emotions, since each extension of *gibel*, as 'liver' or 'back', develops its respective set of meanings.

Table 3. Reference to a physical area

Body part	Physical area or location
Bihotz 'heart'	CORE, INSIDE OF SOMETHING *Ogiaren bihotza* 'bread crumb'; *Sagarraren bihotza* 'apple's core'; *Hezurren bihotza* 'bone marrow'
Gibel 'liver'	BACK SIDE *Gibel egin* (liver make) 'move backwards, return'; *Gibeleratu* 'return'; *Gibelaz zuhaitz bati* (liver.inst tree one.dat) 'with the back to the tree'; *Gibela eman* (liver give) 'have sb's back to sb'; *Gibelalde* (liver.side) 'back side, rear part'; *Gibelalde bikoitza* (liver.side double)'big bottom'; *Gibelera* (liver.all) 'backwards'; *Gibelaka* 'from behind'; *Gibelalderatu* (liver.side.all) 'go backwards'; *Gibeldegi* 'rear part of the house'; *Gibel-so* 'look back'; *Gibelean pot egin* (liver.loc kiss) 'kiss sb's arse'; toponymy 'back side of a mountain' *Jaizkibel, Gibele*
Urdail 'stomach'	—
Heste 'intestine'	—

1.2.3. Reference to metaphorical and metonymic interpretations

In the case of *bihotz*, its spatial conceptualization as 'core' or 'inside of something' gives rise to examples where the metaphor IMPORTANT IS CENTRAL is at play. For instance, in the expression *hiriaren bihotza* (city.gen heart), the meaning of *bihotza* goes beyond the spatial interpretation of 'centre'; it also implies that this is the most important part of the city, where everything happens.

The word *gibel*, in its metonymical sense of 'back', also conveys a wide variety of metaphorical meanings: 'delay, be late' as in *gibeldu* 'be late', 'push back, withdraw' as in *gibelarat hartu* (liver/back.all take), 'withdraw', 'dissuade, keep away from' as in *zerbaitetik gibelarazi* (something.abl liver/back.causative) 'keep away from something', 'last position' as in *gibelsolas* (liver/back-talk) 'epilogue', 'negative result, consequence' as in *gibelondo* 'bad result', and 'neglect' as in *gibel eman* (liver/back give). The body part *gibel* is considered the opposite to another body part *begi* 'eye' in their spatial meaning (Mitxelena 1987-2005); the former means 'back side' and the latter can be used to mean 'front', 'in the direction of'. This opposition is also reflected in their corresponding metaphorical meanings. *Gibel* conveys the meaning 'neglect', which could be paraphrased as 'not face something', and *begi* 'face', *begitarte* (eye.between) 'face' (Ibarretxe-Antuñano and Garai 2004). Those interested in the relation between face and attention, and the conceptualization of volition as change of direction should consult Garai (2001).

It is important to point out that in some of these metaphorical extensions, the word *gibel* implies a default negative value or evaluation. For example, in the case of *gibelondo* 'result, consequence', the default interpretation is that the consequence is bad, as in example (1).

(1)	*Ekintz*	*horrek*	*bere*	*gibelondoak*	*ekarriko*	*ditu*
	action	that.erg	its	consequence.abs.pl	bring.fut	aux

'That action will bring bad consequences'

In example (2), on the other hand, the meaning of *gibelondo* is understood as 'good consequence'.

(2)	*Badakigu*	*zer*	*gibelondo*	*ona*	*duten*
	already.know.1pl	what	consequence	good.abs	aux.comp
	Zestoako	*urek*			
	Zestoa.adn	water.erg.pl			

'We know what good the waters of Zestoa will do us'

However, this positive connotation is not triggered by the word *gibelondo* itself, but by the insertion of the adjective *on* 'good'. Example (2) is clearly a case of compositional polysemy, because this interpretation is only possible thanks to the semantics of the adjective *on* in conjunction with the word *gibelondo*.

Table 4. Reference to metaphorical and metonymic interpretations

Body part	Metaphorical and metonymic interpretation
Bihotz 'heart'	*Hiriaren bihotzean* 'at the city centre'; *Ogiaren bihotza* 'bread crumb'; *Sagarraren bihotza* 'apple's core'; *Neguaren bihotzean* 'in the depths of winter'; *Lurraren bihotzeraino joan ziren* 'they went to the core of the Earth'; *San Bizenti hotza, neguaren bihotza* 'Saint Vincent cold, heart of winter'; *Hezurren bihotza* 'bone marrow'
Gibel 'liver'	NEGLECT → *Gibela eman* (liver give) 'turn one's back to sb', 'be neglectful about sth, neglect sth'; *Lehen arazoak agertu bezain pronto, gibel eman zion* 'The minute the first problems started to appear, he neglected them'; *Gibel egiok ekaitzari* 'turn your back to bad weather'
	DELAY, HOLD OFF → *Gibelarazi* (liver.causative) 'delay, hold off'; *Gibelatu* 'delay, hold off'; *Gibeldu* 'be late'; *Gibeldura* 'delay'; *Gibelapen* 'delay'; *Gibelamendu* 'delay'; *Gibelatzaile* 'delayer'; *Gibelatzapen* 'delay'; *Gibeldari* 'late'; *Gibelsko* 'late'
	PUSH BACK → *Gibelarazi* (liver.causative) 'push back'; *Gibelatu* 'push back'; *Han amerikar soldaduak gibelarazi zituzten* 'there the American soldiers were pushed back'; *Etsaiak giberalat hartu zuen* 'the enemy withdrew'
	DISSUADE → *Gibelarazi* (liver.causative) 'dissuade'; *Gibelatu* 'dissuade'; *Gure eginbideetatik gibelarazten gaituztenak* 'those who keep us from our duties'
	LAST ONE → *Gibelsolas* (liver.talk) 'epilogue'; *Gibelengo* 'last one'
	NEGATIVE RESULT → *Gibelondo* 'bad result'; *Ekintza horrek bere gibelondoak ekarriko ditu* 'that action will bring bad consequences'; *Badakigu zer gibelondo ona duten Zestoako urek* 'we know what good the waters of Zestoa will do us'
Urdail 'stomach'	*Urdailbete* 'full stomach'
Heste 'intestine'	*Hestebete* (intestine.full) 'saussage'; *Hesteki* 'saussage'; *Odoleste* (blood.instestine) 'black pudding'
	Hestebete (intestine.full) 'bellyful, be stuffed'; *Egun hartan txokolatez eta pastelez egindako hestebetea ahaztu ezinik* 'unable to forget how he had stuffed himself silly with chocolate and cakes'; *Hestekada* 'bellyful', 'feast'; *Hesteak beste* (intestines other) 'stuff one's guts'

Urdail and *heste* share one semantic extension, 'be stuffed (full of food)'. In both cases, this meaning can be taken as an example of the metonymy ACTIVITY FOR RESULT. Both terms, *urdail* 'stomach' and *heste* in the metonymic sense of 'belly', refer to the place where food goes when we eat. When food fills up the space, the sensation that we have is precisely that of being 'filled up' (see also Grady's 1997 experientially based correlative metaphors).

In the case of *heste*, we find a further semantic development, the meaning 'sausage', which can be explained via the metonymy CONTAINER FOR PRODUCT, since the intestines of some animals, e.g. pigs, are traditionally used in the manufacture of these products.

Table 4 summarizes the metaphorical and metonymic semantic extensions of these four internal body parts.

1.3. Emotions

Another common path of semantic extension in internal body parts is that of emotions. These internal organs are used to express all sorts of different negative (anger, disdain, cowardice) and positive (love, mercy) emotions. Sometimes the internal organ is conceptualized as the seat or place where these emotions are found as in *poza handia bihotzeratzen digu* (happiness big.abs heart.all.hab aux) 'it makes our hearts rejoice',[5] but at other times the internal organ is understood as an object that can be manipulated and characterized in different ways as in *bihotz eman* (heart give) 'encourage', *gibelhandi* (liver.big) 'easygoing, mellow, happy-go-lucky' and so on.

1.3.1. The heart

The conceptualization of the heart in Basque shows a wide and rich array of different types of emotions, both positive and negative. Among the former, the heart is associated with feelings in general as in *Bihotzak-egin* (heart.erg-make) 'be moved', love as in *Bihotz-bihotzez* (heart-heart.inst) 'with all my love, from my very heart', encouragement as in *Bihotzari egin* (heart.dat make) 'encourage', generosity as in *Bihotz-zabal* (heart-wide) 'generous', and mercy as in *Bihotz-biguin* (heart-soft) 'tender, compassionate'. Among the latter meanings, the heart is related to cruelty as in

Table 5. *Bihotz* as the seat of emotions

Positive emotions	
Emotion, feelings in general	*Bihotzetik diotsut hau* 'I'm telling this from the heart'; *Bihotzeratu* (heart.all.tu) 'carry to the heart'; *Tristura handia bihotzeratzen digu* 'it makes our hearts ache with great sadness'; *Poza handia bihotzeratzen digu* 'it makes our hearts rejoice'; *Bihotzondo* (heart.side) 'bottom of one's heart, place of feelings, spirit'; *Bihotzaldi* (heart.time) 'feeling, emotion'; *Bihotzak-egin* (hearts-make) 'be moved'; *Bihotza-ttipitu* (heart-make.small) 'be moved'; *Bihotz-ukigarri* (heart-touchable) 'exciting'; *Bihotz-ikara* (heart-tremble) 'emotion'
Love	*Emakume hark bihotz ebatsi zion* 'that woman stole his heart'; *Bihotza hautsi* (heart break) 'break sb's heart'; *Bihotzeko laguna* (heart.adn friend) 'loved friend'; *Bihotz-zale* (heart-fond) 'passionate'; *Bihotz-bihotzez* (heart-heart.inst) 'with all my love'; *Ene bihotzeko kutuna!* 'My precious darling!'; *bihozño* 'darling'; *bihoztxo* 'darling'
Encouragement, courage	*Bihotz izan* 'hearten, encourage'; *Bihotz eman* 'encourage, give sb heart'; *Bihotz hartu* 'take heart'; *Bihotza har ezazu, ez dezazula gal itxaropena* 'take heart and don't lose hope'; *Bihotz galdu* 'lose heart'; *Izan bihotz, ez beldurtu* 'have courage, don't be frightened'; *Bihotzak garaia darama, ez ezpatak* (heart.erg victory brings neg sword.erg) 'courage obtains victory, not the sword'; *Bihotz-altxagarri* (heart-arouser) 'encouraging'; *Bihotzari egin* (heart.dat make) 'encourage'
Generosity, magnanimity	*Bihotz-zabal* (heart-wide) 'generous'; *Bihotzhandi* (heart.big) 'generous, giving, magnanimous'; *Bihotzbero* (heart.warm) 'generous'; *Bihotz-hun* (heart-good) 'generous'; *Bihotz-zindo* (heart-firm) 'generous'
Mercy, compassion, benevolence	*Etzaika batere bihotzik hausten* 'he doesn't have any compassion'; *Bihozbera* (heart.soft) 'merciful, kind'; *Bihotz-biguin* (heart-soft) 'tender, compassive'; *Etzaika batere bihotzik hausten* 'he doesn't have any compassion'; *Bihotz onez* (heart good.inst) 'benevolently'; *Bihozbigun* (heart-soft) 'tender-hearted'; *Bihotz-guri* 'compassive'; *Bihoztiar* 'compassive'
Braveness	*Bihotztoi* 'braveness'; *Bihoztun* (heart.possessor) 'brave, valiant, bolt, gallant'

Negative emotions	
Cruelty, heartlessness, lack of pity	*Bihotzbeltz* (heart.black) 'cruel, wicked'; *Bihozgogor* (heart.hard) 'hard-hearted, cruel'; *Bihozgabe* (heart.without) 'heartless, cruel'; *Bihotzgaixto* (heart.bad) 'cruel'
Cowardice	*Bihotz-apal* (heart.weak) 'weak, coward'; *Bihozgabe* (heart.without) 'coward'; *Bihotz-mehe* (heart-thin) 'coward'; *Bihotz-urri* (heart-scarce) 'coward'; *Bihotza erori* (heart fall) 'lose one's nerve'; *Bihozka* 'coward'
Sadness	*Bihotza erori zitzaion hori ikustean* 'his heart sank when he saw that'; *Bihozkatu* 'sadden'; *Biotzil* (heart.die) 'sadness'
Anguish, grief	*Bihozmin* (heart.pain) 'heartbreak, anguish, grief'; *Bihozmin eman* (heart.pain give) 'grieve sb'; *Bihotz-erdigarri* (heart-divider) 'heartbreaking'; *Bihotz-herstura* (heart-tightening) 'anguish'; *Bihotz-miko* (heart-bit) 'anguish'; *Bihotzterri* 'grief'; *Bihotza hil* (heart die) 'grieve'
Remorse, regret	*Bihotz zimiko* (heart pinch) 'remorse, regret'
Discourage	*Bihozgabetu* 'discourage'; *Galerarengatik bihozgabetu zen* 'He became disheartened over the loss'; *Bihozkatu* 'lose courage'; *Bihotz-aphaltü* 'discourage'
Anger	*Bihotzerra* (heart.anger)'anger'

Bihozgabe (heart.without) 'heartless, cruel', cowardice as in *Bihotz-mehe* (heart-thin) 'coward', sadness as in *Biotzil* (heart.die) 'sadness', anguish as in *Bihozmin* (heart.pain) 'grief', remorse as in *Bihotz zimiko* (heart pinch) 'remorse, regret', discouragement as in *Bihozgabetu* (heart.without.verbal suffix) 'discourage', and anger as in *Bihotzerre* (heart.anger) 'anger'. Table 5 summarizes all the emotions associated with *bihotz* 'heart'.

A major difficulty in analyzing the meanings of *bihotz* 'heart' is that it is a concept that has been widely used by Christian culture, and consequently, most written texts are biased by these religious beliefs. A direct consequence of this religious bias is the idea of 'softening' somebody's heart through words, which is related to persuasion (and gender as in *Mintzoemeak bihotz gogorra bera diro* 'the soft [=woman's] speech can soften a hard heart') and to forgiveness (see Garai 2001, 2002, for a thorough explanation of these concepts). 'Softening' is opposed to rigidity, and to the way divine law is enforced.

There are two issues that are difficult to explain: on the one hand, the relationship between severity and rigidity at the experiential level, or the association of mercy with forgiveness (*misericordiam = bihotz bera*); and on the other, the reasons why the heart is conceptualized as an object with these particular attributes-states and manipulability for religious indoctrination. Depending on the Christian creed one embraces, the way divine law is applied is different. In the Catholic frame, the sinner has to look for an intermediary advocate in order for the law not to be enforced to the full extent of its severity, the background scene being the Final Judgement. In contrast, in the Protestant frame forgiveness is granted via the family relationship we all have with God, but the volition has to be 'malleable', the background scene being the medieval contract of homage vassalage.

We cannot forget either that the heart in the Roman world was considered not only the seat of feelings but also that of thought (see Malina 1993). This conceptualization has left some traces in the English vocabulary in words such as *accord*, *concord*, both derived from Latin *cor-cordis* 'heart'.[6] Although traces of the pre-roman Basque cultural model are hard to find, it might be possible to also identify the heart with the self (see note 5). In any case, this issue is one that deserves significantly more research before we can be sure of the correctness of this assertion.

It might seem surprising that the same body organ conveys opposite emotions such as courage and cowardice, love and anger, mercy and cruelty. However, I would like to argue that this is due to compositional polysemy, and that the heart itself is only associated with positive emotions. In other words, *bihotz* 'heart' only conveys negative feelings when this organ is used in conjunction with certain words whose semantics are responsible for these negative interpretations. Let us examine a couple of examples.

One of those 'double' conceptualization pairs involving *bihotz* is that of courage and cowardice. If we look at the expressions that lexicalize these two emotions, we find that only one word, *bihotz*, is necessary to trigger the concept of courage. Heart (=courage) is understood as an object that can be manipulated in different ways, and as such it can be given to other people as in *bihotz eman* (heart give) 'encourage' or it can be lost as in *bihotz galdu* (heart lose) 'lose heart'. On the contrary, in the case of cowardice, it is the semantics of words like *apal* 'weak', *mehe* 'thin', *gabe* 'without', and *urri* 'scarce' that make it possible for the heart to be related to this negative emotion.

Cruelty is another case of the negative emotion assigned to *bihotz* as illustrated by the following four examples: *Bihotzbeltz* 'cruel, wicked'; *Bihozgogor* 'hard-hearted, cruel'; *Bihozgabe* 'heartless, cruel'; *Bihotzgaixto* 'cruel'; *Bihotzgaldu* 'cruel'. In all these expressions, *bihotz* is accompanied by a word whose meaning is negative: *beltz* 'black',[7] *gogor* 'hard', *gabe* 'without', *gaixto* 'bad', and *-galdu* 'lose'. When these words are used together with *bihotz*, the positive character of the latter gives way to a negative interpretation.

In these examples of compositional polysemy, it is important to look for the semantic relational network that may be established for the second members of these compounds. For instance, the interchangeability of *bihotzgaldu, bihozgogor, bihozgabe* and so on may be due to the following two reasons: (i) the possible collocational meaning acquired in *galdu* in other compounds, as in *odolgaldu* (blood.lose) 'corrupted'; and (ii) the interchangeability of the second element in many of these compound words, as in *zorigaizto, zorigogor, zorigaldu, zorigabe* 'unlucky', and *lotsagaldu, lotsagabe* 'insolent, rude'. A certain degree of synonymy is acquired by these elements in context.

1.3.2. The liver

In the case of *gibel* 'liver', the emotions associated with this organ are all negative: distrust as in *Gibel-beldur* (liver-fear) 'mistrust, distrust', disdain as in *Gibelondo* (liver.side) 'disdain, scorn', aversion as in *Gibelgogo* 'aversion, antagonism, hostility', listlessness as in *Gibelkoi* 'lazy, indolent', withdrawn character as in the variant meaning of *Gibelkoi* 'shy, bashful', bitterness, spite as in *Gibelez gibel* (liver.inst liver) 'get angry with sb, back to back'. Table 6 summarizes these meanings.

Unlike in the case of heart, where many of the negative emotions associated with that organ are the result of compositional polysemy, and not of the conceptualization of the heart itself, *gibel* 'liver' seems to bring about the negative connotations by itself. Let us draw a contrastive example to illustrate this point. In both organs, it is possible to create a compound with the organ and the word *ondo* 'side'. Although the structure of these words is the same, the connotations associated with the resulting compound words are opposite. In the case of *bihotzondo* (heart.side), the meaning is positive since this word means 'place of feelings, emotion, spirit', but in the case of *gibelondo* (liver.side), the meaning is negative 'disdain, scorn'. Based on these examples, it seems possible to hypothesize that some internal organs are associated with positive emotions whereas others are related to negative ones.

Table 6. *Gibel* as the seat of feelings and attitudes

Negative feelings and attitudes	
Listlessness, lethargy, laziness	*Gibelhandi* (liver.big) 'slow, sluggish, carefree'; *Halako gibel-handirik ez dut behin ere ezagutu* 'I've never met such a carefree person'; *Horrek zeuzkak gibelak!* (that.erg have liver.pl) 'Is he ever slow!'; *Gibelkoi* 'lazy, indolent'
Mistrust, distrust	*Gibel-beldur* (liver-fear) 'mistrust, distrust'; *Gibel-asmo* (liver-intention) 'ulterior motives'; *Gibel-korda* 'ulterior motives'; *Argi eta garbi mintzatzen zitzaion, inolako gibelasmorik gabe* 'He talked to him upfront, without ulterior motives'; *Gibeljale* (liver.eater) 'detractor, slanderer, backbiter'; *Bekaizgo gibel-jalearen atzaparretan eroria* 'caught in the hands of an envy's detractor'; *Gibeljoko* (liver.game) 'dirty trick'; *Gibel-athe* (liver-door) 'surreptitiously'; *Gibel-atherik gabeko gizona* (liver-door.part without.adn man) 'upfront man'; *Gibelaz* (liver.inst) 'against'; *Gibeletik ibili* (liver.abl walk) 'act surreptitiously'
Disdain, scorn	*Gibelondo* (liver.side) 'disdain, scorn'; *Badirudi ere gibelondo edo mesprezio gehiago egiten zaiola* 'It also seems that he is disdained or scorned'
Aversion, antagonism, hostility	*Gibelgogo* 'aversion, antagonism, hostility'; *Erlijioarekiko gibelgogoa antzeman zitekeen Frantziako Iraultza garaian* 'One could smell the aversion against religion during the French Revolution'
Introvert, withdrawn character	*Gibelkari* 'introvert'; *Gibelkoi* 'shy, bashful' *Gibelgaldu* (liver.lose) 'nervous, gutless'; *Ni horretarako gibel-galdu hutsa nauk* 'I chickened out of that'; *Gibeldu* 'hide'
Bitterness, rancour, spite	*Gibelmin* 'bitterness, rancour, spite'; *Ez da atzo goizeko gauza indu eta mahometarren alkar ezin-ikusia, gorrotoa eta asarrea India aldean... Amaren bularrekin batera artzen dute bata be-stearekiko gibelmin ori* 'It is not a recent thing that the Hindus and the Muslims cannot stand each other, their hate and bitter-ness in India...they breath this rancour from when they are in their mother's belly'; *Gibelmindu* 'make angry'; *Gibelez gibel* (liver.inst liver) 'get angry with sb'

1.3.3. The stomach and the intestines

The conceptualizations of the last two internal organs are not as rich and diverse as *bihotz* and *gibel* as far as the number and variety of emotions they can convey. As shown in Table 7, there is only one emotion per organ. *Urdail* 'stomach' is related to hysteria and hypochondria as in *urdaileko* (stomach.adn) 'hysterical',[8] and *heste* 'intestine' is associated with annoyance as in *Hesteak jan* (intestines eat) 'annoy'.

Table 7. Urdail and *heste* as the seat of feelings and attitudes

Emotions	
Hypochondria, hysteria	*Urdaileko* 'hysterical'; *Urdabileko* 'hysterical', 'hysteria', 'women's nervous breakdown'; *Urdailen* 'hysteria'; *Urdaildu* 'lose one's patience, get exasperated'
Annoyance	*Hesteak jan* (intestines eat) 'annoy'

2. Cultural models and embodiment in the conceptualization of emotions in Basque internal body organs

In the previous section, I analyzed some of the main conceptualizations associated with the four internal body organs under investigation in this paper. These body parts are related to a wide variety of emotions, both positive and negative. Here, I would like to focus on the question of why these internal organs are linked to this specific range of emotions. In other words, I will explore the cultural models that underlie the conceptualizations of these internal organs in the domain of emotion.

In order to understand what the basis for these conceptualizations is, we must make reference to two different, but complementary, types of explanations. On the one hand, we need to take into account two cultural models of the body: Descartes' dualism, and the theory of the four humours. On the other hand, we cannot forget how we experience and understand our own body and all the functions and processes related to it. In other words, the notion of embodiment (Johnson 1987), that is, the role of the human body in our cognitive experience of the world, is also fundamental to this analysis.

In the case of *bihotz*, the conceptualization of the heart as the seat of emotions, in opposition to the head/mind/brain as the seat of thought and reason, may find its explanation in the philosophical tradition of dualism. The seventeenth-century philosopher René Descartes proposed the concept of mind-body dualism, which implied that human behaviour could be understood as resulting from both a free, rational soul and from automatic, nonrational processes of the body. The dualism of the heart (emotions) and the head (intellect) is found in other languages entrenched in the same culture. Spanish and English, for example, have expressions like *Romperle el corazón a alguien* 'to break sb's heart', *Metérsele una cosa a alguien en la cabeza* 'to get sth into sb's head', and *Tener cabeza para algo* 'to have a good head for sth'. However, the opposition between heart and intellect predates Cartesian dualism. In Basque, for instance, the following example is attested in written texts in 1596, *bihotza eta zentzuna dira guduan beharrago iskiloa baino* (heart.abs and sense.abs are war.loc necessary.more strength more) 'wisdom (heart and intellect) is better than strength' (Lakarra 1997). Therefore, we would need to look farther back in time (perhaps in Classical Greece) to find the ultimate origin of this dichotomy.

In Basque, we find examples with similar meanings in the case of the two body parts *bihotz* 'heart' and *buru* 'head', as in *buruargi* (head.light) 'intelligent', *buruan izan* 'to have in mind', and so on (see Ibarretxe-Antuñano 2002). In this sense, we can argue that Basque associates "emotion" and "intellect, mind" with the same organs as do other European languages. In other words, Basque also conceptualizes the rational and irrational sides of the body as two separate entities. However, this is not the whole story.

Basque is a language that has always been in contact with other languages and cultures. However, until the twentieth century, its transmission was almost exclusively oral. Its written tradition is relatively young compared to that of other European languages. Moreover, the earlier pre-twentieth-century texts were almost always composed explicitly for religious purposes (Oyharçabal 2001). For this reason it is sometimes very difficult to distinguish between conceptualizations that are native to the Basque language and its culture (prior to contact, say, with Christianity), and those conceptualizations that have been acquired through contact with other languages and cultures.

There are, nevertheless, some expressions and usages that now and then reveal and corroborate that Basque used to have its own conceptualizations and unique worldview. These words are usually ubiquitous, very much entrenched in culture and thus, very difficult to explain and translate into other languages. These can be called *cultural words*. One of these is the word *gogo*.[9] This word covers a wide variety of different meanings that range from the emotional as in *gogoa berotu* (*gogo*.det heat) 'encourage', *gogo galdu* (*gogo* lose) 'to become discouraged', *gogobizi* (*gogo*.life) 'ardent desire, passion' to the intellect as in *gogoargi* (*gogo*.light) 'intelligent', *gogo euki* (*gogo* have) 'to have in mind, remember'. Notice that the same expressions with *gogo* can be constructed with *bihotz* and *buru* with similar meanings. For example, there are the following pairs: *gogoa galdu* 'become discouraged' and *bihotz galdu* 'lose heart'; *gogoargi* 'intelligent' and *buru-argi* 'intelligent'. What changes is the conceptualization of the processes described in those expressions.

Contrary to *bihotz* 'heart' and *buru* 'head', *gogo* does not seem to separate the rational from the irrational; quite the opposite, this word harmoniously unites these two "apparent" contrary concepts in one.[10] The result is a kind of *primitive thought* (Ibarretxe-Antuñano 2008), where there is an intellectual reasoning process, but one based on intuition and emotion; or to put it in another way, an intellectual reasoning process prior to any distinctions between feelings and thought – which, in fact, implies that reason and feelings are not differentiated at all.[11]

In the case of *gibel*, it is necessary to distinguish between those expressions that arise from the conceptualization of *gibel* as 'back', and those that arise from *gibel* as 'liver (bile)'. As discussed in Section 1, *gibel* is metonymically used to refer to the back. It is this secondary conceptualization that gives rise to the expressions for negative feelings presented in Section 1.3.2: listlessness, mistrust, disdain, aversion, and introverted character. This means that these expressions are to be explained on the basis of embodiment. Human beings rely on their eyes much more than on any other sense (cf. Classen 1993; Howes 1991; Sekuler and Blake 1994). The eyes provide the most reliable information and give us the highest degree of security. This is why we do not trust things that we do not see, and why we have expressions such as *seeing is believing* (cf. Danesi 1990; Ibarretxe-Antuñano 1999; Sweetser 1990). Consequently, anything that falls outside the scope of our eyes is generally considered negative, or at least, untrustworthy. All of the emotions mentioned above reflect the unreliable and treacherous character of the 'back', understood as the area hidden from the eyes.

The conceptualization of *gibel* as liver, and more concretely as bile, gives rise to the last type of emotion mentioned in Section 1.3.2: bitterness, rancour and spite. The conceptual basis of this emotion can be found in the ancient physiological-psychological theory of the four humours, still current in the European Middle Ages. The foundations of this theory were laid by Hippocrates, who established that the body and its vital processes are regulated by four fluids: blood, black bile, yellow bile, and phlegm. In turn, each of them is associated with variations in temperament. A person's character is then determined by the preponderance of one of these fluids. It was thought that emotional stability and general health depend on a balance among the four bodily humours; an excess of one may produce a particular illness or an exaggerated personality trait. Therefore, a person with too much blood is supposed to have a sanguine temperament. An excess of black bile produces a melancholic temperament. Abundance of yellow bile results in a choleric temperament. An oversupply of phlegm is believed to produce a phlegmatic temperament. Each of these four humours is related to a specific body organ and also to one of Empedocles' cosmic elements (for a discussion, see Geeraerts and Grondelaers 1995; Turner 1990, among others). Table 8 summarizes the relations among the four humours.

Table 8. The theory of the four humours and its correspondences (see Hoskin 1962; Jones 1979; Loudon 1997 for more information)

Humour	Tempera-ment	Characteristics	Organ	Cosmic element
BLOOD	Sanguine	ENTHUSIASM: warm and pleasant, optimistic, excit-able, happy, generous	Heart / Blood	AIR (warm and moist)
BLACK BILE	Melan-cholic	MELANCHOLY: de-pressed, sad, lazy, glutton-ous, sentimental	Spleen	EARTH (cold and dry)
YEL-LOW BILE	Choleric	ANGER: quick to react, hot tempered, irritability, jaundiced view of life, violent	Liver (gall bladder)/ stomach	FIRE (warm and dry)
PHLEG M	Phlegmatic	APATHY: stolid, dull, undemonstrative, slow moving, apathetic, pale	Brain / bladder	WATER (cold and moist)

As shown in Table 8, the liver (gall bladder that produces bile) is associated with the yellow bile humour, which, in turn, is related to a choleric temperament. This is in accordance with the last type of emotion, bitterness, rancour, and spite, which is associated with this internal body organ in Basque.

The relation of *urdail* 'stomach' to hysteria and restlessness is not easy to discern. A possible explanation would be to propose a direct physical link between this type of behaviour and the ulcers that are usually provoked by a constant state of restlessness. Remember that *urdaileko* means 'dyspepsia'. In other words, the explanation for this conceptualization of *urdail* would be based on a physical embodiment. However, this hypothesis needs to be confirmed. It has also been suggested by one of the reviewers that the connection between *urdail* and hysteria is the result of erudite/literate medical traditions and translations. In early medical writings, it was believed that the womb was a dangerous, aggressive, insatiable organ that could move freely about internally, and was even capable of leaving a woman's body and wandering about on its own. These characteristics led to witchcraft-like connections, and the womb came to be linked to a "malady" of women (see Allison and Roberts 1994; Norman 2000). Although this explanation is possible, the womb is usually called *sabel* or *umetoki* (baby.place) or *umontzi* (baby.container) in Basque (see note 3), and therefore this explanation is too weak to be used as the grounding of such a semantic extension.

3. Conclusions

In this paper, I have analyzed linguistic expressions involving four internal organs in Basque: *bihotz* 'heart', *gibel* 'liver', *urdail* 'stomach' and *heste* 'intestine'. These body parts have given rise to different conceptualizations, both physical and metaphorical. All of them can be used to refer to other body parts (*urdail* 'stomach' for 'womb') or to smaller parts of themselves (*heste lodi* 'large intestine'), both internal (*hesteak* 'instestine' and 'guts') and external (*gibel* 'liver' and 'back'). In the case of heart and liver, we also find that these organs can metonymically refer to spatial areas and locations, 'core or inside' and 'back side, rear', respectively. These spatial readings, in turn, are the basis of several metaphorical semantic extensions in these two organs. Finally, I have shown that these internal organs are related to a wide range of emotions. The heart is associated with both positive (love; encouragement, courage; generosity, magnanimity; mercy, compassion, benevolence; bravery) and negative emotions (cruelty, heartlessness, lack of pity; cowardice; sadness; anguish, grief; remorse, regret; discouragement; anger). The liver is connected to negative emotions (listlessness, lethargy, laziness; mistrust, distrust; disdain, scorn; aversion, antagonism, hostility; introversion, withdrawn character; bitterness, rancour, spite). The stomach is linked to hysteria and hypochondria and the intestines to annoyance.

In the last part of this paper, I proposed several factors that may contribute to explaining why these organs are linked to these specific emotions. I distinguished between two types: on the one hand, those related to culture such as Cartesian dualism and the theory of the four humours; and on the other, those related to embodiment. These factors interact with each other and therefore the explanation of the emotions related to a single organ may require reference to cultural and embodiment factors. In the case of the heart, it is necessary to bring forward traditional Cartesian dualism, where the heart is viewed as the seat of emotions and the mind/head as the seat of the intellect. This dichotomy is reflected in Basque *bihotz* 'heart' and *buru* 'head'. Nevertheless, there is a cultural word in Basque, *gogo*, which does not fit into this dualism, because it is conceptualized as a kind of "primitive (irrational) thought", that is, an intellectual reasoning process based on intuition and emotion. The emotions relating to the liver can be divided into two groups: one related to the conceptualization of liver as back, and the other related to the conceptualization of the liver as bile (produced by the gall bladder). The former can be explained by means of embodiment. Things that we do not see, that are beyond the scope of our eyes, are felt to be untrustworthy and treacherous. The latter is based on the theory of the four humours, specifically on the yellow bile humour, which produces a choleric temperament and is located in the gall bladder. The stomach might be another case of embodiment, but explanations on this basis are only possible hypotheses about the conceptualizations of this internal organ, and as such, further research into this area is needed.

Acknowledgements

I would like to express my deepest gratitude to Farzad Sharifian for his editorial work and patience, and to Roslyn Frank for her thorough and helpful comments. The research presented in this paper is part of a collaborative research project with Koldo J. Garai on Basque conceptual structure.

Notes

1. The term "embodiment" has several different interpretations in both psychological and anthropological studies (see Chrisley and Ziemke 2002; Wilson 2002; Ziemke 2003). In this paper, embodiment is always used in the cognitive linguistic tradition of Johnson's (1987) work. Johnson's definition of embodiment is quite problematic, however, especially with regard to the role of culture as a determining factor in the interpretation of what embodiment is. Although Johnson includes social and cultural factors in his definition of embodiment, these two elements have been largely ignored or perhaps, pushed into the background, in many cognitive linguistic studies. This situation is changing in current cognitive linguistic research. A proper discussion of the interaction between culture and embodiment lies outside the scope of this paper, but the interested reader can consult the following references: Dirven, Frank, and Pütz (2003), Ziemke, Zlatev, and Frank (2007), and Frank, Dirven, and Ziemke (in press).
2. Abbreviations used are as follows: abs: absolutive; abl: ablative; adn: adnominal marker; all: allative; aux: auxiliary; comp: complementiser; dat: dative; det: determiner; erg: ergative; fut: future; gen: genitive; hab: habitual; inst: instrumental; loc: locative; neg: negation; part: partitive; pl: plural.
3. *Urdail* can be used to refer to 'womb, uterus' but this semantic extension is not frequent. In fact, 'womb' and 'uterus' are usually called *sabel*, *umetoki* (baby.place) and *umontzi* (baby.container) in Basque rather than *urdail*.
4. A reviewer asks why *gibel* means 'back' in the first place, if it is located under the right rib cage in humans. I can only say that *gibel* comes to represent all that area that is not the front part of the body in Basque. When asked where the liver is located, Basque speakers will place their hands on the back. The liver is conceptualized as located at the back and as such, it has come to represent that whole area.
5. This is an interesting example because if we look at the literal translation of this sentence 'it brings joy to our hearts', one could argue that the heart is metonymically understood as the self, which, in turn, is interpreted as a container.
6. I have found only a couple of Basque examples in Azkue (1905) where *bihotz* is translated as 'mind' as in *bihotzean zituenak atera diotza* (heart.loc has.what take.out.perf aux) 'He took out all of what he had in his mind', and *Idukak bihotzean hartua* (aux heart.loc taken) 'Keep it in your mind'. Since I have not found any other examples or dictionary entries for this sense, I prefer to exclude it from the table of meanings for now, that is, until I find more information about this possible extended meaning.
7. See Perurena (1992), for more information on colour in Basque.
8. *Urdaileko* also means 'dyspepsia'.

9. Many of the examples I have gathered for *bihotz* might end up being an elaboration of *gogo* as far as 'self' and 'seat of understanding' are concerned, both metonymically related in space if the etymology proposed by Gorostiaga is correct (Agud and Tovar 1992: 675 [91]–676 [92]).
10. It is interesting to notice the definition of this word that Azkue (1905) gives in his dictionary: "con esta palabra se designan todas las potencias y operaciones del alma [this word designates all the capacities and operations of the soul]". Although the word "soul" is employed in this definition, we have to take into account that the "soul" described in *gogo* is closer to the sense of "spirit" than to that of "soul" in the religious sense, which is *arima* in Basque religious texts.
11. Those interested in a detailed description and application of *gogo* should consult Aldapeko (1951), Agud and Tovar (1992), and Garai (1995: 5–6).

References

General

Aldapeko
1951 'Gogo' dala-ta. *Euzko Gogoa* 10: 36–37.

Allison, D. B., and M. S. Roberts
1994 On constructing the disorder of hysteria. *Journal of Medical Philosophy* 19 (3): 239–259.

Barcelona, Antonio (ed.)
2000 *Metaphor and Metonymy at the Crossroads. A Cognitive Perspective.* Berlin/New York: Mouton de Gruyter.

Chrisley, Ronald, and Tom Ziemke
2002 Embodiment. In *Encyclopedia of Cognitive Science,* Lynn Nagel (ed.), 1102–1108. London: Macmillan Publishers.

Classen, Constance
1993 *Worlds of Sense: Exploring the Senses in History and Across Cultures*. London: Routledge.

Danesi, Marcel
1990 Thinking is seeing: Visual metaphors and the nature of abstract thought. *Semiotica* 80 (3/4): 221–237.

Dirven, René, Roslyn M. Frank, and Martin Pütz (eds.)
2003 *Cognitive Models in Language and Thought: Ideologies, Metaphors, and Meanings*. Berlin: Mouton de Gruyter.

Fauconnier, Gilles, and Mark Turner
2002 *The Way We Think. Conceptual Blending and the Mind's Hidden Complexities*. New York: Basic Books.

Frank, Roslyn M., René Dirven, and Tom Ziemke (eds.)
in press *Body, Language and Mind Vol. 2: Interrelations between Biology, Linguistics and Culture.* Amsterdam/Philadelphia: John Benjamins.
Garai, Koldo J.
1995 Euzko Gogoa. Diaspora garaiko aldizkari literarioa (1950–1959). Unplublished manuscript. University of Deusto.
2001 Epaiketa eta endogamia Nafarroako Margaritaren *Le Miroir de l'amé pécheresse*-n [Anima Bekatariaren Mirailan]. *Lapurdum* VI: 119–193.
2002 Limurketa eta esanguraren arteko harremanak Euskal atsotitzetan. *Gogoa* II (2): 227–287.
Geeraerts, Dirk, and Stefan Grondelaers
1995 Looking back at anger: Cultural traditions and metaphorical patterns. In *Language and the Construal of the World,* John Taylor, and Robert E. McLaury (eds.), 153–180. Berlin/New York: Mouton de Gruyter.
Grady, Joseph
1997 Foundations of meaning: Primary metaphors and primary scenes. Ph.D. Diss., Department of Linguistics, University of California, Berkeley.
Hoskin, M. A.
1962 *History of Science.* New York: Science History Publications.
Howes, David (ed.)
1991 *Varieties of Sensory Experience: A Sourcebook in the Anthropology of the Senses.* Toronto: University of Toronto Press.
Ibarretxe-Antuñano, Iraide
1999 Polysemy and metaphor in perception verbs: A cross-linguistic study. Ph.D. Diss., Department of Theoretical and Applied Linguistics, University of Edinburgh.
2002 The conceptual structure of Basque *buru* 'head'. *Fontes Linguae Vasconum* 91: 463–491.
2008 Cultural words in metaphor and translation: The case of Basque *gogo.* Paper presented at the 7th International Conference on Researching and Applying Metaphor (RaAM7). Metaphor in Cross-Cultural Communication. May 29–31, 2008. University of Extramadura Cáceres (Spain).
Ibarretxe-Antuñano, Iraide, and Koldo J. Garai
2004 Basque body parts and their conceptual structure: the case of *oin* 'foot' and *begi* 'eye'. In *Linguagem, Cultura e Cognição: Estudos de Lingüística Cognitiva,* Augusto Soares da Silva, Amadeus Torres, and Miguel Gonçalves (eds.), 269–291. Coimbra: Almedina.

Johnson, Mark
1987 *The Body in the Mind. The Bodily Basis of Meaning, Reason and Imagination*. Chicago: Chicago University Press.
Jones, W. H. S.
1979 *Philosophy and Medicine in Ancient Greek*. New York: Arno Press.
Kövecses, Zoltan, and Günter Radden
1998 Metonymy: Developing a cognitive linguistic approach. *Cognitive Linguistics* 9 (1): 37–77.
Lakarra, Joseba
1997 Euskararen historia: arazo zahar, bide berri. *Anuario del Seminario de Filología Vasca 'Julio de Urquijo'* [International journal of Basque language and linguistics]. Donostia: Gipuzkoako Foru Aldundia.
Lakoff, George, and Mark Johnson
1980 *Metaphors We Live By*. Chicago: Chicago University Press.
1999 *Philosophy in the Flesh. The Embodied Mind and its Challenge to Western Culture*. New York: Basic Books.
Lakoff, George, and Mark Turner
1989 *More Than Cool Reason. A Field Guide to Poetic Metaphor*. Chicago: Chicago Univesity Press.
Loudon, Irvin
1997 *Western Medicine. An Illustrated History*. Oxford: Oxford University Press.
Malina, B. J.
1993 *The New Testament World: Insights from Cultural Anthropology*. Louisville, KT, Westminster: John Knox Press.
Norman, George
2000 'The devil hath laughed at the physicians': Witchcraft and medical practice in seventeenth-century New England. *Journal of the History of Medicine and Allied Sciences* 55 (1): 5–36.
Oyharçabal, Beñat
2001 Statut et evolution des letters basques durant les XVIIème et XVIIIème siécles. *Lapurdum* VI : 219–287.
Perurena, Patziku
1992 *Koloreak Euskal Usarioan* [Basque colours in use]. Donostia: Erein.
Sekuler, Robert, and Randolph Blake
1994 *Perception*. New York: McGraw-Hill.
Sweetser, Eve
1990 *From Etymology to Pragmatics. Metaphorical and Cultural Aspects of Semantic Structure*. Cambridge: Cambridge University Press.
Turner, Mark
1990 *The Literary Mind*. Oxford: Oxford University Press.

Wilson, Margaret
2002 Six views of embodied cognition. *Psychological Bulletin and Review* 9 (4): 625–636.
Ziemke, Tom
2003 What's that thing called embodiment? In *Proceedings of the 25th Annual Meeting of the Cognitive Science Society*. Mahwah, NJ.: Lawrence Erlbaum.
Ziemke, Tom, Jordan Zlatev, and Roslyn M. Frank (eds.)
2007 *Body, Language and Mind Vol. 1: Embodiment*. Amsterdam/Philadelphia: John Benjamins.

Dictionaries

Agud, Manuel, and Antonio Tovar
1992 *Diccionario Etimológico Vasco*, ASJU-XXX, t. V. Donostia: Gipuzkoako Foru Aldundia.
Azkue, Resurrección María de
1905 *Diccionario Vasco-Español-Francés I-II*. Bilbao: La Gran Enciclopedia Vasca.
Bostak Bat
1996 *Diccionario Hiru Mila Hiztegia. Adorez 7*. Bilbao: Bostak Bat.
Garate, Gotzon
1998 *Atsotitzak*. Bilbao: BBK
Gran Enciclopedia Vasca
1976 *Diccionario Retana de Autoridades del Euskera*. Bilbao: Biblioteca de la Gran Enciclopedia Vasca.
Mitxelena, Koldo
1987–2005 *Diccionario General Vasco. Orotariko Euskal Hiztegia*. Bilbao: Euskaltzaindia.
Mokoroa Mugika, Justo M.
1990 *'Ortik eta Emendik' Repertorio de Locuciones del Habla Popular Vasca, Oral y Escrita, en sus diversas variedades*. Bilbao: Labayru-Etor-Eusko Jaurlaritza.
Morris, Mikel
1998 *Morris Student Plus. Euskara-Ingelesa. English-Basque*. Donostia: Klaudio Harluxet Fundazioa.
Mugica Barrondo, Placido
1981 *Diccionario Vasco -Castellano I,II*. Bilbao: Mensajero.
Sarasola, Ibon
1984–1995 *Hauta-lanerako Euskal Hiztegia*. Zarauz: Itxaropena.

Stürtze, Alizia, Maria Colera, and Imanol Barrenetxea
2002 *Eguneroko Esamoldeak (Ingelesa-Euskara-Gaztelania).* Donostia: Gaiak Argitaletxea.

C. Holistic heart-centering conceptualizations

The Chinese heart as the central faculty of cognition

Ning Yu

Abstract

This paper studies the Chinese *xin* 'heart', traditionally conceptualized as the centre for cognitive activities. In Chinese, the word *xin*, which primarily denotes the heart organ, means both "heart" and "mind" as understood in English, and by metonymic association it can also mean "thoughts; ideas; emotions; feelings". In ancient Chinese philosophy, the heart is regarded as the thinking and reasoning organ. It unifies human will, desire, emotion, intuition, reason and thought. The attribution of such supreme mental power to the heart results in a popular conceptual metaphor THE HEART IS THE RULER OF THE BODY. In traditional Chinese medicine, the heart is also said to be the "monarch" of the body, governing various types of intellectual and emotional activities. A linguistic survey shows that the cultural conceptualization of the heart as the centre for both emotional and intellectual activities is reflected in the Chinese language. A historical look at the evolution of the concept of heart in English indicates that the Chinese conception of heart is in fact quite similar to that found in the Old and Middle English periods, when the English *heart* was also conceptualized as the seat of both feeling and thought. A cross-cultural comparison of the four-humour and five-element theories reveals both similarities and differences between these two cultural systems.

Keywords: cultural conceptualization, metaphor, metonymy, heart, Chinese

1. Introduction

It is impossible to speak about Chinese *xin* 'heart' without including a number of other body parts in the discussion. In this respect, this paper is in line with Wierzbicka (1992), who presents a cross-linguistic study comparing some folk concepts encoded in, for example, *soul, mind,* and *heart* in English. The central message of this study is that the lexical and semantic differences of such concepts, if interpreted properly, can serve as "clues to the different cultural universes associated with different languages", or as "significant evidence for cultural history and for prevailing modern ethnophilosophy" (Wierzbicka 1992: 63). In the light of such a claim, this paper presents a historical study of the Chinese conceptualization of the heart. It is hoped that a comprehensive diachronic study will shed light on Chinese cultural history, and gain insights into the current conception of the heart as a unique cultural construct.

The Chinese concept of heart, encoded in the Chinese word 心 *xin* 'heart', lies at the core of Chinese thought. This is because the heart is traditionally believed to be the central faculty of cognition. Its importance in cognition can be seen in the following famous quotations from two representative figures of the "Learning of the Mind-and-Heart" (心学 *xinxue*), or literally the "Learning of Heart", a school of Neo-Confucianism lasting from the Song (960–1279) to the Ming (1368–1644) dynasty:[1]

(1) a. 宇宙便是吾心，吾心即是宇宙。《陆九渊集·年谱》 (Ji 1995, vol. 50: 2407)
The universe is my heart, and my heart is the universe. *Collected Writings of Lu Jiuyuan: Chronological Life*

b. 人者，天地万物之心也；心者，天地万物之主也。《王阳明集·答季明德》 (Ji 1995, vol. 54: 83)
Man is the heart of Heaven and Earth and the ten thousand things; the heart is the master of Heaven and Earth and the ten thousand things. *Collected Writings of Wang Yangming: Reply to Ji Mingde*.

Apparently, the heart is the most important thing in the whole universe, to these two philosophers. In (1a), from Lu Jiuyuan (1139–1193), the heart is metaphorically equated to the universe, thus highlighting the unity and correspondence between the macrocosm and microcosm. That is, the heart as the central faculty of cognition is the "mirror" of the universe. In (1b), from Wang Shouren (1472–1529), also known as Wang Yangming, "Heaven and Earth and the ten thousand things", that is, "Heaven and Earth and all things in between", is the literal translation of the Chinese phrase that can be interpreted as meaning, again, the "universe". The saying defines man metaphorically as the "heart" of the universe, and the heart as the cognitive and physiological centre of man. It is because the heart is conceived of as the cognitive centre of man that it can metonymically stand for man as the "master" of the universe. Regardless of what philosophical outlook these sayings manifest, the fact that the heart is regarded as the central faculty of cognition is obvious. The universe will make no sense to us without our heart functioning as the organ for thinking, knowing, and understanding.

In what follows, I will attempt to outline the Chinese conceptualization of the heart in its various aspects, based on a discussion of its roots in ancient

Chinese philosophy and traditional Chinese medicine, which constitute the foundation of the Chinese cultural model for the concept of heart. I will then provide some linguistic evidence for the impact of this "cultural conceptualization" (Sharifian 2003) on the Chinese language. Finally, I will take up a comparative perspective, outlining some similarities and differences between Chinese and English.

2. Philosophical conception of the heart

Western philosophical tradition upholds the reason-emotion dichotomy, contrasting "the reasoning faculty (the mind) with the irrational, base, physical desires and passions (the heart)". To Western thinkers beliefs and ideas come from the mind, whereas the heart is the seat of desires and emotions (Hansen 1992: 22). In ancient Chinese philosophy, however, the heart is "the site of both affective and cognitive activities" (Shun 1997: 48). The "distinction between 'heart' and 'mind' in the Western sense does not exist in Chinese philosophy" (Lin 2001: 202). In other words, the "heart" and the "mind", as known in English, are conceptualized as being one, the *xin* 'heart', which houses thoughts and feelings, ideas and emotions. That is why the Chinese term *xin* 'heart' is often glossed as "heart-mind" or "heart/mind", as well as either "heart" or "mind", in English literature on Chinese philosophy (e.g., Chan 2002a; Hansen 1992; Shun 1997; Slingerland 2003; Yearley 1990), "taking into consideration both the affective and cognitive concerns that the concept of *xin* encompasses" (Chan 2002b: 42). As Hansen (1992: 20) points out, "The common translation of *xin* as *heart-mind* reflects the blending of belief and desire (thought and feeling, ideas and emotions) into a single complex dispositional potential".[2]

Importantly, the Chinese heart is traditionally regarded as a thinking and reasoning organ. It is given at birth by the heaven, as Guanzi (c. 725–645 B.C.), a Daoist statesman and philosopher, asserted. Thus, a baby in the mother's womb will:

(2) 五月而成，十月而生。生而目视，耳听，心虑。《管子·水地》 (Zhang and Zhang 1995, vol. 49: 787).
Take shape in five months, and be born in ten months. After the birth, the eyes will see, the ears will hear, and the heart will think. *Guanzi: Water and Earth.*

It is taken for granted that humans are born with the heart to think, as much as with the eyes to see and the ears to hear.

In Chinese thought, the heart is the organ of moral intelligence and distinguishes between right and wrong, between good and bad, and thus guides action, conduct, and behaviour (Hansen 1992). For instance, the famous ancient Confucian philosopher, Mencius (c. 372–289 B.C.), distinguished the heart, the thinking organ, from the organs of ears and eyes, which do not think and which tend to be deceived by external things when in contact with them. He argued that the function of the heart organ is to think, and a good sense of understanding and reasoning can only be obtained through thinking.

(3) 耳目之官不思，而蔽于物。物交物，则引之而已矣。心之官则思，思则得之，不思则不得也。《孟子•告子上》
(Zheng et al. 1993: 418)
Such organs as ears and eyes do not think, so they tend to be deceived by external things. When in contact with external things, they are often led astray. The organ of heart is for thinking. If you think, you will get it; if you do not, then you will not get it. *Mencius: Gaozi,* Part I.

It is common sense that people should examine what they hear and scrutinize what they see. That is accomplished through their heart, the faculty of epistemic intelligence, as further illustrated by a conversation recorded in the *Kongcongzi*, between Confucius (c. 551–479 B.C.) and his grandson, Zisi:

(4) 子思问于夫子曰："物有形类，事有真伪，必审之，奚由？"子曰："由乎心，心之精神是谓圣，推数究理，不以物疑。"《孔丛子·记问》 (Ji 1995, vol. 49: 390)
Zisi asked Confucius, "Objects vary in shape and kind, and things may be true or false. Therefore, one has to examine them. But through what?" Confucius said, "Through the heart. The spirits of the heart are regarded as holy. Through inferring and reasoning, the heart studies the whys and wherefores of things, and will not be confused by them". *Kongcongzi: Recorded Questions.*

In Confucius' words, *jingshen* 'spirits' refers to the mental aspects of a person, in contrast to the physical form. The locus of that "mental half" is the heart. Through thinking and reasoning the heart can gain insights into things that are otherwise confusing. Thinking results in knowing and understanding. That is why Confucius referred to the mental aspects of a person as "holy".

As "the most valuable part of the self", the heart is conceptualized metaphorically as "the ruler of the other parts" (Slingerland 2003: 229). This is because the heart is seen as the central faculty of cognition with supreme mental power and the whole body is under its command and control. For instance, another famous Confucian philosopher, Xunzi (c. 313–238 B.C.), employed this HEART AS RULER metaphor (see, also, Slingerland 2003: 230–231) when he argued that the heart is "the ruler of the body" and "the master of the spiritual light":

(5) 心者，形之君也，而神明之主也，出令而无所受令。
《荀子·解蔽篇》 (Ji 1995, vol. 49: 362)
The heart is the ruler of the body and master of the spiritual light, who issues commands but does not receive commands. *Xunzi: Dispelling Blindness.*

In this sentence, the "spiritual light" refers to the "spirits" (as in 4) or the totality of mental aspects of a person. It is this "spiritual light" that enables people to "see". It is interesting to note that this MENTAL POWER AS LIGHT metaphor is but an entailment of the conceptual metaphor THINKING, KNOWING, or UNDERSTANDING IS SEEING.[3]

The following quote from the *Wenzi* also characterizes the relationship between the heart, the body, and the "spirits", i.e. the "spiritual light", by means of metaphors.

(6) 心者形之主也，神者心之宝也。《文子·九守》
(Ji 1995, vol. 49: 82)
The heart is the master of the body, and the spirits are the treasures of the heart. *Wenzi: Nine Things to Be Preserved*

A human being consists of two distinct halves: the physical form (形 *xing*), i.e. the body, and the mental aspect (神 *shen*), i.e. the "spirits". The heart, although it is a part of the body, is unique in that it is also the "master" of the body. It is in this unique position because it "owns" the "spirit-treasures", namely the supreme mental power with which it controls and commands the rest of the body. That is to say, the mental half, or the "mind", of a person is located in the heart.

As the ruler, the heart has authority over other parts of the body. This kind of sovereign-subject relationship is brought out by the following passage from the *Xunzi*:

(7) 耳目鼻口形能，各有接而不相能也，夫是之谓天官。心居中虚以治五官，夫是之谓天君。《荀子·天论篇》
(Ji 1995, vol. 49: 352)
Ears, eyes, nose, mouth, and body each have the capacity to provide sense contact, but their capacities are not interchangeable. They are called the heavenly officials/organs. The heart dwells in the central cavity and governs the five officials/organs, and hence it is called the heavenly ruler. *Xunzi: On Heaven.*

The heart takes the central position of the body both in the physical and metaphorical sense. It regulates and coordinates the functions of other parts of the body. It is worth mentioning that the Chinese word 官 *guan* means both "organ", such as sense organs of eyes and ears, and "official", "officer", or "office". Therefore, "heavenly officials" and "five officials" in the above passage may be better alternative English translations because they fit the HEART AS RULER metaphor.

3. Medical characterization of the heart

Traditional Chinese medicine uses a GOVERNMENT metaphor system to define the internal organs of the human body (see also Wang et al. 1997, Yu 2003a). In the earliest existent medical text in China, *The Elementary Questions* (《素问》) of *Yellow Emperor's Internal Classic* (《黄帝内经》), the *zang* and *fu* organs are all assigned specific offices in the empire government. Given in (8) are some examples (Ji 1995, vol. 65: 10; translations adopted from Wiseman and Feng 1998):

(8) a. 心者，君主之官也，神明出焉。
The heart holds the office of monarch, whence the spiritual light emanates.

b. 肝者，将军之官，谋虑出焉。
The liver holds the office of general, whence strategies emanate.

c. 胆者，中正之官，决断出焉。
The gallbladder holds the office of justice, from which decision emanates.

d. 脾胃者，仓廪之官，五味出焉。
The spleen and stomach hold the office of the granaries, whence the five flavours emanate.

e. 肺者，相傅之官，治节出焉。
The lung holds the office of assistant, whence management and regulation emanate.

f. 肾者，作强之官，伎巧出焉。
The kidney holds the office of labour, whence agility emanates.

As we can see, the heart is the master of the body, as much as a king or emperor is the master of a nation. Functionally speaking, "the heart governs the spiritual light" (心主神明), carrying out thinking, storing memory, producing emotions, commanding will, governing perception, and evolving dreams (Wang et al. 1997: 181).

Uniquely in traditional Chinese medicine, the five *zang* organs are all related to mental functions, falling into the pattern of the five-element categorization. This is because they produce and store essential elements of the body such as blood, *qi*, fluid, and essence,[4] which are taken as the physical basis for mental vitality. It is believed that the operations of the five *zang* organs result in mental activities. The spirits, which include all mental activities, are classified as five different types, known as the "five spirits" (五神 *wushen*): *shen* (神) 'spirit', *hun* (魂) 'ethereal soul', *po* (魄) 'corporeal or animal soul', *yi* (意) 'intention or purpose', and *zhi* (志) 'will'.[5] It is claimed that the five different types of spirits are stored separately in the five *zang* organs, but are all commanded by the heart as the "grand master" of the five *zang* and six *fu*:

(9) 人之身，居要者心也。而心之神明散寄于五脏，待感于五官。(from Cai 2000: 331)
What takes the critical position in the human body is the heart. The spiritual light of the heart, however, is stored separately in the five *zang* organs, and depends on the perception of the five sense organs.

That is, the heart governs all the spirits but stores them separately in the other four *zang* organs as well as in itself. Specifically, the heart stores the spirit, the liver stores the ethereal soul, the lungs store the corporeal or animal soul, the spleen stores the intention or purpose, and the kidneys store the will. Thus, in traditional Chinese medicine, the five *zang* organs are all related to mental functions. This is because they all produce and store the essential elements of the body, which constitute the physiological foundation of mental functions. In fact, an important quality that distinguishes between the *zang* and the *fu* organs is that the former are related to mental functions but the latter are not. However, the gallbladder, one of the six *fu* organs, is an exception. As shown in (8c), the gallbladder, holding the "office of justice", is believed to be responsible for making judgments and decisions, and to be the seat of courage.[6]

Also according to traditional Chinese medicine, the five *zang* organs produce five *qi*, which result in five emotions: *nu* (怒) 'anger', *xi* (喜) 'joy', *si* (思) 'overthinking', *bei* (悲) 'sorrow', and *kong* (恐) 'fear'.[7] They are each associated, as in the five-element categorization, with a *zang* organ: anger with the liver, joy with the heart, overthinking with the spleen, sorrow with the lung, and fear with the kidney. It is said in the medical text:

(10) 肝在志为怒，心在志为喜，脾在志为思，肺在志为忧，肾在志为恐。 (Wang et al. 1997: 45)
The emotion related to the liver is anger, to the heart is joy, to the spleen is overthinking, to the lung is sorrow, and to the kidney is fear.

While these emotions arise from the functions of their corresponding *zang* organs, their excessiveness will hurt each of them accordingly. For instance, sudden excessive joy can hurt the heart by making the spirit inside unsteady or dispersed so that the person can hardly concentrate on anything. However, because the heart is the "governor" of all mental states and activities, it reacts to other emotions as well. Of all the internal organs, in fact, the heart is the most sensitive to emotional changes. This is because emotional changes, as dynamic "forces" that "move" around, originate in the heart but affect other *zang* organs depending on their kinds. Thus, for example, when anger stirs in the heart, it will affect the liver; when fear shakes in the heart, it will affect the kidney. As Zhang Jingyue in the Ming Dynasty (1368–1644) argued:

(11) 心为脏腑之主，而总统魂魄，并赅意志，故忧动于心则肺应，思动于心则脾应，怒动于心则肝应，恐动于心则肾应，此所以五志唯心所使也。…… 情志之伤，虽五脏各有所属，然求其所由则无不从心而发。—明代张景岳《类经》
(Wang et al. 1997: 87)
The heart is the master of the *zang-fu,* presiding over the ethereal and corporeal souls, and directing purpose and will. Therefore, when sorrow moves in the heart, the lung will respond, when overthinking moves in the heart, the spleen will respond, when anger moves in the heart, the liver will respond, and when fear moves in the heart, the kidney will respond. That is why the five emotions are operated only by the heart. … Wounds inflicted by emotions, although they each belong to one of the five *zang*, if their causes are traced, all come from the heart. *Classified Canon* by Zhang Jingyue of the Ming Dynasty.

Apparently, traditional Chinese medicine also conceptualizes the heart as the "master" or "ruler" of the body, and as the locus of intellectual activity, as well as emotional activity. This view was adopted from ancient Chinese philosophy (see Wang et al. 1997: 181). In the Ming (1368–1644) and Qing (1644–1911) dynasties, however, the functions of the brain as the organ for mental activities came to be recognized. Today, the relationship between the heart and the brain is still under debate in Chinese medicine (see Wang et al. 1997: 180–182, 929–931). One view sees the brain as the house for mental activities, but the heart as ultimately controlling it. A different view argues that the notion of "heart" in traditional Chinese medicine cannot be equated to the heart organ. Instead, it is a combination of heart and brain. These views can be summarized by the following statement:

(12) 神明之体藏于脑，神明之用发于心也。(Wang et al. 1997: 182)
The substance of the spiritual light is stored in the brain, whereas the application of the spiritual light arises from the heart.

That is, the brain stores the substance of the "spiritual light", but the heart deploys and utilizes it. In short, Chinese medicine today still holds that the heart takes the dominant and central position in the body: it is still the "emperor" or "monarch" of the body.[8]

4. Linguistic expression of the heart

In the Chinese language, the word 心 *xin*, which primarily denotes the heart organ, refers to it as the seat of thought and feeling. This fact is clearly demonstrated in the ideographic writing system of Chinese. Thus, many Chinese characters for words related to thought and feeling contain the "heart" radical as their semantic component. Here are a few examples: 思 *si* 'think; consider; deliberate; think of; long for; thought; thinking', 想 *xiang* 'think; ponder; think back; try to remember; recall; recollect; consider; miss', 虑 *lü* 'consider; ponder; think over; concern; worry', 念 *nian* 'think of; miss; thought; idea', 感 *gan* 'feel; sense; feeling', and 情 *qing* 'feeling; affection; sentiment; passion'.[9] In what follows, I analyze linguistic expressions that demonstrate that the heart is conceptualized as the seat of both thought and feeling in Chinese.

4.1. Heart as a container

In Chinese, the heart as the seat of thought and feeling is often understood as a container, and the word for "heart" can refer to thoughts and feelings inside this container by the conceptual metonymy CONTAINER FOR CONTAINED. The linguistic expressions in modern Chinese analyzed in this section constitute the modern manifestation of the historical claim that "the heart governs the spiritual light (心主神明)". Look at the examples in (13).

(13) a. 这是发自深心的喜悦。

Zhe	*shi*	*fa*	*zi*	***shen-xin***	*de*	*xiyue.*
this	is	arise	from	deep-heart	MOD	joy

'This is the joy that has arisen from the depth of the heart (lit. the deep heart).'

b. 一把火猛然袭上心头，他也火大起来。

Yi-ba	*huo mengran*	*xi*	*shang*	***xin-tou,***	*ta*
one-CL	fire suddenly	charged	up.onto	heart-top	he
ye	*huo da-qilai.*				
also	fire grow-start				

'A flame of fire suddenly rose in his heart (lit. charged up onto the heart-top), and he was also flying into a rage (lit. his fire was also starting to flame up).'

c. 亡国之恨，丧家之悲，败身之辱尽埋心底。
Wang-guo zhi hen, sang-jia zhi bei,
conquered-nation MOD hatred lost-home MOD sorrow
bai-shen zhi ru jin mai ***xin-di****.*
ruined-body MOD humiliation all bury heart-bottom
'The hatred of a conquered nation, the sorrow of a lost home, and the shame of a ruined reputation are all buried beneath the bottom of the heart.'

As can be seen, (13a) is about the emotion of joy. This is true emotional stirring because it has arisen from the "depth of the heart". One's "deep heart" is the locus of the true self and the seat of inmost thoughts and feelings. Example (13b) is about the emotion of anger. Apparently, it is an instantiation of the metaphor ANGER IS FIRE.[10] As these two examples show, inactive emotions or feelings are located "deep in the heart". When they are activated, they go up, toward the "top of the heart". Contrastively, (13c) shows that when people want to suppress their feelings, they "bury" them "beneath the bottom of the heart". This example involves mixed feelings of hatred, sorrow, and humiliation, which are all contained in the heart.

In Yu (2002), which is a study of Chinese body-part terms in the expression of emotions, I note that the internal organs other than the heart are linked with one or two emotions only. For example, the spleen is linked with anger, the liver with anger and sadness, intestines with anxiety and sadness, the stomach with anxiety, the lung with sadness, and the gallbladder with fear. In contrast, the heart is linked with various emotions and feelings, including, in addition to those mentioned with other internal organs, happiness, relief, admiration, vexation, disappointment, disgust, hatred, and guilt. This linguistic phenomenon is certainly consistent with the claim of Chinese medicine that emotions and feelings come originally from the heart, which "governs the spiritual light".

4.2. Heart container mapped onto other containers

What follows are examples showing that the heart as a container is mapped onto some other containers.

(14) a. 他第一次把掏心窝的话都和她说了。
Ta diyi-ci ba tao ***xin-wo*** *de hua*
he first-time PRT scoop heart-nest MOD words
dou he ta shuo-le.
all to her say-PER
'For the first time he told her all his inmost thoughts and feelings (lit. he said to her all the words scooped out of his heart-nest).'

b. 进城几年了，乡亲们的嘱托他一直记在心间。
Jin cheng ji nian le, xiangqin-men de
enter city several years PER fellow-villagers MOD
zhutuo ta yizhi ji zai ***xin-jian****.*
advice he always remember in heart-room/inside
'Having lived in the city for several years, he always bears in mind (lit. in the heart room or inside his heart) the fellow villagers' advice.'

c. 我把环保种子播撒在他的心田里。
Wo ba huan-bao zhongzi bosa zai tade
I PRT environment-protection seeds sowed at his
xin-tian *li.*
heart-field in
'I sowed the environment-protection seeds in his heart (lit. heart-field).'

d. 我以为只要自己心地清白，就可以避免肇害于人。
Wo yiwei zhiyao ziji ***xin-di*** *qing-bai, jiu*
I thought as.long.as self heart-land clear-white then
keyi bimian zhao hai yu ren.
can avoid do harm to other.people
'I thought as long as I maintained a clear and clean heart (lit. my heart-land remains clear and pure), I could avoid doing harm to others.'

In (14a), the heart is conceptualized as a particular kind of container, a "nest". In this case, the words spoken are all "scooped out of the heart-nest": they express, again, the "inmost thoughts and feelings" from the locus of the true self. In (14b), the heart is conceptualized as a "room" container. This example has to do with memory. When the advice is remembered, it is actually stored inside the "heart-room". In (14c), the heart is conceptualized as "a plot of field" with its own boundaries. In this example, ideas or thoughts are "seeds", and to communicate ideas or thoughts to other people, or to educate them, is to "sow the seeds" into their "heart-field". The seeds sowed into the field will grow, blossom, and bear fruits, and those fruits are the "fruits" of earlier education. Example (14d), where the heart is conceptualized as "a stretch of land", reflects the traditional view in ancient Chinese philosophy that the heart is the locus of moral sense and moral character. Here the CONTAINER metaphor is linked to the metaphors of LIGHT and PURITY. That is, being moral and ethical is "having a pure and clean heart". This example also involves or implies a pair of common conceptual metaphors for morality: BEING MORAL IS BEING CLEAN, and BEING IMMORAL IS BEING DIRTY.

4.3. Heart as the centre of thoughts and ideas

The examples below demonstrate that in Chinese the heart is the centre of thought and the locus of the "mind". It is where the intellectual activities take place.

(15) a. 可是这件事他也只能放在心里想，不敢说出来。

Keshi *zhe-jian* *shi* *ta* *ye-zhi-neng* *fang* *zai*
however this-CL matter he can-only put at
xin *li* *xiang,* *bu* *gan* *shuo* *chulai.*
heart in think not dare speak out
'However, he can only put this matter in his heart to think about, but dare not speak it out.'

b. 这个问题一直压在她的心头。

Zhe-ge *wenti* *yizhi* *ya* *zai* *tade* ***xin-tou***.
this-CL question constantly weigh on her heart-top
'This question has always been weighing on her mind (lit. heart-top).'

c. 她根本没把他的宣示放在心头。

Ta	*genben*	*mei*	*ba*	*tade*	*xuanshi*	*fang*
she	absolutely	didn't	PRT	his	declaration	put
zai	***xin-tou***.					
at	heart-top					

'She absolutely didn't pay any attention to his declaration at all (lit. She didn't put his declaration on the top of her heart in the first place).'

d. 她在心中打了个大问号。

Ta	*zai*	***xin***	*zhong*	*da-le*	*ge*	*da*	*wen-hao.*
she	at	heart	in	draw-PER	CL	big	question-mark

'She drew a big question mark in her heart.'

The examples in (15) demonstrate how the heart is conceptualized as an entity or container in relation to intellectual activities. In (15a), the person can only "put the matter in his heart to think about" but "dare not speak it out". Obviously, the heart is the container in which thoughts and ideas are processed. In (15b), when people keep thinking about something and cannot get it off their mind, that thing is then "a load weighing on the top of their heart". This is what is meant by "overthinking", which is accompanied by worry and anxiety. Example (15c) is about attention or ignorance. The lady paid absolutely no attention to the "declaration (of love)" from the man. This mental inactivity is expressed in terms of lacking physical action: she never "placed" the man's declaration "on top of her heart" for any serious consideration. It is common to conceptualize mental activity in terms of object manipulation (see, e.g., Jäkel 1995; Lakoff 1999; and Lakoff and Johnson 1999). In Chinese, however, as this and many other examples show, mental processing is achieved in terms of the heart. Example (15d) is about doubt. Doubt arises when one "draws a question mark in one's heart".

As shown by the above examples, in Chinese the heart is the locus for various intellectual activities. This is further evidenced by the following sentences in (16) and (17) that contain a compound word consisting of the body-part term for "heart" and another morpheme (also a word when used alone) indicating a mental state or activity.

(16) a. 我猜不透他的心思。

*Wo cai-bu-tou tade **xin-si**.*

I guess-not-penetrate his heart-thought

'I can't read his mind (lit. his heart-thought) / I can't figure out what's on his mind (lit. what's his heart-thought).'

b. 我没有心思去看戏。

*Wo mei you **xin-si** qu kan xi.*

I don't have heart-thought go see play

'I'm not in the mood (lit. don't have the heart-thought) to see a play.'

c. 这就了却了我的一桩心愿。

*Zhe jiu liaoque-le wode yi-zhuang **xin-yuan**.*

this then fulfil-PER my one-CL heart-wish

'This serves to fulfil a cherished desire (lit. a heart-wish/desire) of mine.'

d. 人若有疑虑，则心志不坚，会时常犹豫。

*Ren ruo you yilü, ze **xin-zhi** bu jian,*

people if have doubts then heart-will not firm

hui shichang youyu.

should often be-irresolute

'If people have doubts, then their will (lit. heart-will) is not firm, and they should often be hesitant.'

In (16a), "reading people's mind" is trying to figure out the thoughts in their heart. In (16b), the same compound "heart-thought" means "mood". When people are not in the mood to do something, their heart does not even think of it. As in (16c), people all have desires and aspirations that they want to fulfil or satisfy, and these mental states and activities again originate in the heart: they are what the heart wishes for. As shown in (16d), people's will is their "heart's will". In order to be resolute, they should have "firm will" in their heart.

(17) a. 你们不了解他的心意。

*Nimen bu liaojie tade **xin-yi**.*

you not understand his heart-intention

'You don't understand his intention (lit. heart-intention).'

b. 您的美意，我心领了。
Ninde *mei-yi,* *wo* ***xin-ling**-le.*
your good-intention I heart-get/understand-PER
'I appreciate (lit. heart get/understand) your kindness (lit. good intention).'

c. 他的话前后不一，令人心疑。
Tade *hua* *qian-hou-bu-yi,* *ling* *ren* ***xin-yi**.*
his words front-back-not-one making people heart-suspect
'His words were self-contradictory, making people suspicious (lit. making people's heart suspect).'

e. 我想谈谈学习这篇课文的心得。
Wo *xiang* *tantan* *xuexi* *zhe-pian* *kewen* *de* ***xin-de**.*
I want talk-about study this-CL text MOD heart-obtain
'I want to talk about what I have gained (lit. what my heart has obtained) from studying this text.'

f. 你应该用心做习题。
Ni *yinggai* ***yong-xin*** *zuo* *xiti.*
you should use-heart do exercises
'You should work hard at exercises (lit. use your heart to do exercises).'

As shown in (17a), people's intention is located in their heart. To appreciate others' good intention requires the heart to get or understand it, as in (17b). That is, appreciation is also a mental activity of the heart. Doubt is a state or activity in the heart (see 15d), so is suspicion, as in (17c). When people feel suspicious, their heart is suspecting. Learning is a process of receiving knowledge by the heart-container, as in (17e). What one has learned from study is the knowledge obtained by and stored in the heart. Example (17f) further illustrates the heart as the seat of intellect. When people are working hard at mental exercises, say, solving difficult math problems, they have to "use their heart". Remember that "the application of the spiritual light arises from the heart" even though "the substance of the spiritual light is stored in the brain".

4.4. Heart as the locus of one's true self

In Chinese, as the following examples show, one's disposition, temperament, and character are ascribed to the physical nature of one's heart. Since the kind of the heart determines the "nature" of a person, it is then the locus of true self or personality of that person. Look at the following examples.

(18) a. 粗心的驾驶员对于公众是一种危险。
Cu-xin *de* *jiashiyuan* *duiyu* *gongzhong* *shi* *yi*
thick-heart MOD driver to public is a
zhong *weixian.*
kind danger
'A careless (lit. thick-hearted) driver is a danger to the public.'

b. 她做什么事都很细心。
Ta *zuo* *shenme* *shi* *dou* *hen* ***xi-xin***.
she does any thing all very thin-heart
'She is meticulous (lit. thin-hearted) in whatever she does.'

c. 我们诚心向你求教。
Women ***cheng-xin*** *xiang* *ni* *qiujiao.*
we sincere-heart from you ask.for.advice
'We sincerely (lit. with a sincere heart) ask you for advice.'

d. 他昧着良心做事。
Ta *mei-zhe* ***liang-xin*** *zuo* *shi.*
he hide-DUR fine-heart do thing
'He's doing this (evil) against his conscience (lit. with his fine-heart concealed).'

The examples in (18) suggest that the heart is the locus of one's true self. Both (18a) and (18b) refer to people's disposition, temperament or character in terms of physical traits of their heart. Thus, a careless person has a "thick heart" whereas a careful person has a "thin heart". As in (18c), people who are sincere have a "sincere heart", and to do something sincerely is to do it "with a sincere heart". In (18d), *liang-xin* (fine-heart) is the Chinese counterpart for the English word *conscience*, as the Chinese heart is conceptualized as the locus of moral sense and moral character. When people do bad things against their conscience, they do so "with their fine-heart hidden or covered". It is simply a way of deceiving oneself.

4.5. Mind, brain and heart

All the preceding examples demonstrate that in Chinese the heart is generally conceptualized as the seat of feeling and thought, where mental activities and states are located. As I have observed elsewhere (Yu 2003b), however, in modern Chinese, *tou-nao* 头脑 'head and brain' can also mean "mind" as understood in English. It is also where one's thoughts and ideas are stored and one's thinking takes place, as illustrated by the examples in (19) (from Yu 2003b). Nevertheless, this usage is limited to modern Chinese in a relatively small scope. The bulk of conventionalized expressions, including compounds, idioms, and idiomatic sayings, demonstrate that the heart, rather than the brain, is the locus of the "mind" as known in English.[11]

(19) a. 我们要有冷静的头脑。

Women	*yao*	*you*	*lengjing*	*de*	***tou-nao***.
we	should	have	cool-quiet	MOD	head-brain

'We should have a cool head (or, be sober-minded; lit. have cool-quiet head-brain).'

b. 我们应该把头脑里的错误思想清除出去。

Women	*yinggai*	*ba*	***tou-nao***	*li*	*de*	*cuowu*	*sixiang*
we	should	PRT	head-brain	in	MOD	wrong	ideas
qingchu	*chuqu*.						
clear	out						

'We should rid our minds (lit. head-brain) of erroneous ideas.'

c. 僵化思想束缚着一些人的头脑。

Jianghua	*sixiang*	*shufu*	*zhe*	*yixie*	*ren-de*
stiffened	ideas	bind	PRT	some	people's
tou-nao.					
head-brain					

'An ossified way of thinking shackles some people's minds. (lit. Stiffened ideas are binding some people's head-brain.)'

In this section, it has become clear, through analysis of linguistic evidence, that the Chinese heart is the combination of "heart" and "mind" in the Western sense. It is where emotions and thoughts are stored, processed, and manipulated. In light of the surveys in ancient Chinese philosophy and traditional Chinese medicine, it is clear that the linguistic evidence presented and analyzed in this section is actually embedded in its cultural context. The language is but a part of the cultural system (Palmer 1996). Manifested linguistically, the conceptualization of the Chinese heart is indeed a phenomenon at the cultural level of cognition (Sharifian 2003).

5. A comparative perspective from English

In this section, I take up a comparative perspective in the hope that a cross-cultural and cross-linguistic study will shed new light on the Chinese conceptualization of the heart as the centre for both emotional and intellectual activities. I start with the definitions provided in the *Longman Dictionary of Contemporary English* (LDCE, 1978). Here are the first two meanings listed under the word *heart* in this dictionary:

1. the organ inside the chest which controls the flow of blood by pushing it through the blood vessels;
2. the same organ when thought of as the centre of the feelings, esp. of kind feelings (as the brain is the centre of thought).[12]

As can be seen in sense 2, there is a perceived division of labour in contemporary English for the heart and the brain. The former is "the centre of the feelings, especially of kind feelings" whereas the latter is "the centre of thought". This division of labour is, however, still a cultural conceptualization, which is not necessarily consistent with scientific definition, as becomes obvious in light of the first meaning listed under the entry *brain*: "*med* the organ of the body in the upper part of the head, which controls thought and feeling".[13] It is obvious that the medical definition of *brain* contradicts sense 2 of *heart*, in that the brain is supposed to control both thought and feeling, that is, the division of labour suggested by sense 2 of *heart* does not exist in scientific terms. In sense 2, *brain* is defined as the "mind". The brain and the mind are the "two sides of one coin", functioning respectively as "hardware" and "software" of a computer.[14]

Now let us turn to Chinese. The following are the first two senses listed under *xin* 'heart' in *Xiandai Hanyu Cidian* (Modern Chinese Dictionary, XDHYCD, 1996), which is one of the most popular dictionaries in China:

1. the organ inside the body of human beings and other higher animals that gives impetus to the circulation of blood. The heart of a human being is in the centre, a little to the left side, of the thoracic cavity, with the shape of a circular cone and the size of one's own fist. Inside the heart there are four cavities, of which the upper two are called atriums and the lower two called ventricles. The diastoles and systoles of the atriums and ventricles circulate blood to all parts of the body. The heart is also called "the heart organ";
2. usually also refers to the organ for thinking, and to thoughts, feelings, etc.

Sense 1 is a scientific definition of the "physical heart" while sense 2 represents more of a cultural conception of the "mental heart". Significantly, in sense 2 the heart is conceptualized as "the organ for thinking" and, by the conceptual metonymy CONTAINER FOR CONTAINED, as both "thoughts" and "feelings". According to a popular Chinese-English dictionary (*Han-Ying Cidian*, H-YCD, 1995), the word *xin* is given these two English senses supposed to be equivalent to senses 1 and 2 in XDHYCD: (a) "the heart"; and (b) "heart; mind". That is, the Chinese word *xin* 'heart' comprises what are known as "heart" and "mind" in English.

That sense 2 of *xin* 'heart' in XDHYCD reflects the Chinese cultural model for the concept of heart, which is not consistent with scientific knowledge, becomes obvious in light of the definitions for *nao* 'brain' in the same dictionary:

1. the organ in the human body that is in charge of consciousness and movement of the whole body, and of such mental activities as thought and memory; the main part of the nerve system, composed of forebrain, midbrain, and hindbrain;
2. the capability of thinking, memorizing, etc.

Apparently, the brain is seen in Chinese as the organ of thinking, and the seat of intellect, in scientific terms.[15] In *Hanyu Da Cidian* (HYDCD, 2000), notably, it is pointed out that, because the ancients took the heart as the thinking organ, the term *xin* 'heart' has been traditionally used to refer to the thinking function even though modern science ascribes this function to the brain.

A comparison between English and Chinese, in terms of meanings assigned to the body-part terms for the heart at the modern stage, reveals the following similarities and differences. Firstly, the Chinese conception of the heart, as reflected in sense 2 of XDHYCD, is culturally constructed, as is the English one reflected in sense 2 of LDCE. They both are inconsistent with modern scientific knowledge, as reflected respectively in sense 1 of *nao* 'brain' in XDHYCD, and of *brain* in LDCE. Secondly, and very significantly, the Chinese heart is conceptualized as the organ of thinking, but the same thinking function has been attributed to the mind, associated with the brain, in English. The divorce between body and mind in English does not exist in Chinese thought. Traditionally, the "part" that enables a person to think and know, namely the "mind" as known in English, is located in the "heart" in Chinese, as discussed in the previous sections. Thirdly, and very interestingly, the heart in Chinese is conceptualized as the organ of all feelings, as well as thoughts and moral values, whereas the English heart is conceptualized as the organ of feelings only.

In short, it seems that English has undergone a change leading to the separation of "thinking" from "feeling", represented by the distinction between the mind and the heart, or more generally between the mind and the body. It requires further study to find out if this is the case or not, and here I make an initial attempt to that end. Let us trace back in history and get some diachronic glimpses of the conceptualization of the English heart by referring to *Shorter Oxford English Dictionary on Historical Principles* (SOED, 5th ed., 2002) and *Oxford English Dictionary* (OED, 2nd ed., 1989). There, it appears that the English heart was once conceptualized as a comprehensive structure in which all human faculties are unified and integrated, just as the Chinese heart has been conceptualized throughout history. In Old and Middle English, the heart, apart from being the bodily organ regarded as the centre of vital bodily functions or seat of life, is also the "seat of feeling and thought" (SOED) or the "seat of feeling, understanding, and thought" (OED). More specifically, in the older strata of English, the *heart* is defined as, among others things (adapted from OED):

1. = MIND, in the widest sense, including the functions of feeling, volition, and intellect
2. the seat of one's inmost thoughts and secret feelings; one's inmost being; the depths of the soul; the soul, the spirit
3. intent, will, purpose, inclination, desire
4. disposition, temperament, character
5. the seat of the emotions generally; the emotional nature, as distinguished from the intellectual nature placed in the *head* (in earlier use often referring to the physical organ; in later mostly figurative)
6. more particularly, the seat of love or affection; kindly feeling; cordiality, heartiness; susceptibility to the higher emotions; sensibility or tenderness for others; feeling
7. the seat of courage; hence, courage, spirit; the source of ardour, enthusiasm, or energy
8. the seat of the mental or intellectual faculties; often = understanding, intellect, mind, and (less commonly) memory
9. the moral sense, conscience

Of course, most of these senses are regarded as archaic or obsolete today. In modern English, they either do not exist at all, or exist to a very limited extent. For instance, *heart* can mean "intent", "will", etc., only in such idiomatic phrases as *after one's own heart*, and "memory" only in *learn by heart*.

Table 1. Historical similarity and current difference between English and Chinese in the cognitive functions ascribed to heart and brain

Cognitive functions	Old/Middle English	Ancient Chinese	Modern English	Modern Chinese
thinking	heart (brain)	heart	brain	heart (brain)
feeling	heart	heart	heart (brain)	heart

The above comparison between Chinese and English has led to the finding that the Chinese conceptualization of the heart as the seat of feeling and thought appears to be quite similar to that found in the Old and Middle English periods, when the English *heart* was also conceptualized as the centre of thought, as well as feeling. That is, the separation of the mind from the heart, or of thinking from feeling, is a later development in English.[16] In contrast, it seems the conceptualization of the heart as the seat of both feeling and thought has been more consistent and persistent in Chinese. The historical similarity and current difference between the two languages is presented in Table 1, where the parentheses indicate that the organ inside is assigned a relatively small part in the cultural conceptualization of the cognitive function, but this part has been growing owing to the increasing influence of scientific knowledge. This table, I have to stress, is very simplistic in nature, merely sketching out a possible trajectory of historical development.

The question to ask is: Why is there a difference between Chinese and English? The answer lies, it seems, in the general trends of philosophical thought. In English, the divorce between the mind and the heart took place in response to the rise of the mind-body dualism in Western thought, as promoted by Descartes (see, e.g., Lakoff and Johnson 1999: 391–414). As a result, the mind is "disembodied". In contrast, Chinese thought, dominated by Confucianism and Daoism throughout history, has always held the belief that the "heart governs the spiritual light". That is, the "mind", as known in English, has always been "embodied" in the heart in Chinese thought. That is, the change in direction of the philosophical trend in the Western world seems to have caused the change in the concept of *heart* in English, as reflected in the change in meaning of the word *heart*.[17] On the other hand, the conceptualization of the heart as the centre for all emotional and intellectual activities, and as the central faculty of human cognition, has been held relatively stable in Chinese thought. This relative stable conception of the heart is reflected in the Chinese language accordingly.

6. Four humours and five elements

In this section, I take a comparative look at two folk theories: the theory of the four humours in Western culture and the theory of the five elements (or five phases) in Chinese culture. The four-humour theory stemmed from ancient Greek medicine and philosophy and dominated Western thought until the 17th century. The five-element theory figured prominently in ancient Chinese philosophy, and it has been a guiding doctrine in traditional Chinese medicine. What I present below is an initial look and is necessarily superficial, merely pointing out some differences and similarities on the surface between these two cultural systems. The heart organ has a place in both systems. Looking at the systems provides us with a better understanding of the cultural contexts of the conceptualizations of the heart under discussion.

Both the four-humour and the five-element theories are analogical in nature. They are cultural systems of categorization, establishing interrelationships with perceived correspondences among various things or phenomena in the human body, in nature, and between human body and nature. What is of special interest is that as cultural systems they have historically exerted their influence on the English and Chinese languages respectively (see, e.g., Geeraerts and Grondelaers 1995; Fernando 1996; Yu 1995, 1998). The two systems are shown respectively in Table 2 (from Geeraerts and Grondelaers 1995: 158) and Table 3 (adapted from Wang et al. 1987: 14).

Table 2 A system of humoural correspondences

	Phlegm	Black bile	Yellow bile	Blood
Characteristic	cold and moist	cold and dry	warm and dry	Warm and moist
Element	water	earth	fire	air
Temperament	phlegmatic	melancholic	choleric	sanguine
Organ	brain/bladder	spleen	liver/stomach	heart
Colour	white	black	yellow	red
Taste	salty	sour	bitter	sweet
Season	winter	autumn	summer	spring
Wind	North	West	South	East
Planet	moon	Saturn	Mars	Jupiter
Animal	turtle	sparrow	lion	goat

Table 3. A system of five-element categorization

Elements	Wood	Fire	Earth	Metal	Water
Nature					
Location	east	south	centre	west	north
Season	spring	summer	long summer	autumn	winter
Climate	windy	hot	humid	dry	cold
Planet	Wood Planet (Jupiter)	Fire Planet (Mars)	Earth Planet (Saturn)	Metal Planet (Venus)	Water Planet (Mercury)
Animal	chicken	sheep	ox	horse	pig
Crop	wheat	millet	foxtail millet	rice	bean
Life Cycle	birth	growth	ripeness	harvest	storage
Odour	rancid	scorched	fragrant	fishy	putrid
Flavour	sour	bitter	sweet	spicy	salty
Colour	blue/green	red	yellow	white	black
Human					
Zang Organ	liver	heart	spleen	lung	kidney
Fu Organ	gallbladder	small intestine	stomach	large intestine	bladder
Sense organ	eyes	tongue	mouth	nose	ears
Tissue	tendons	vessels	flesh	skin/body hair	bones
Emotion	anger	joy	overthinking	grief	fright
Indicator	nails	complexion	lips	body hair	hair
Expression	shouting	laughing	singing	wailing	moaning
Spirit	ethereal soul	spirit	purpose	corporeal soul	will

A fundamental difference between these two systems is that the four-humour theory is primarily a medical theory whereas the five-element theory is originally a philosophical theory. Thus, in the former, the four "head" categories, under which other categories of correspondences are organized, are the four humours – phlegm, black bile, yellow bile, and blood – in the body. According to the humoural doctrine, physiologically, the four humoural fluids "regulate the vital processes within the human body"; psychologically, they "define four prototypical temperaments, i.e., a person's character is thought to be determined by the preponderance of one of the four vital fluids in his body" (Geeraerts and Grondelaers 1995: 156). For the humoural theory, "the basic therapeutic rule" is "to restore the balance of the humours, given that a disturbance of their well-balanced proportion is the basic cause of the pathological situation".

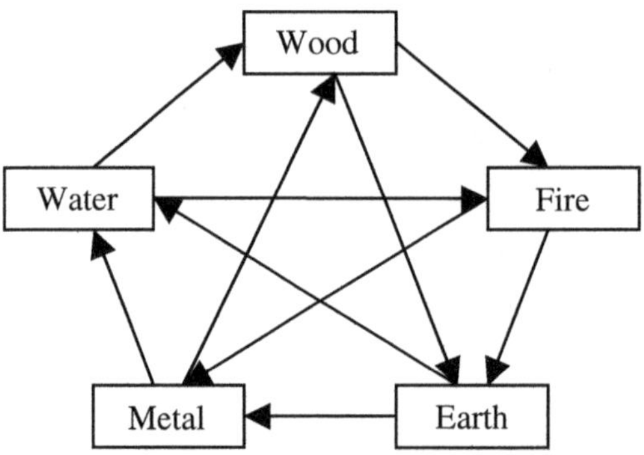

Figure 1. Relation of mutual promotion and restraint between the five elements

The five-element theory, together with the yin-yang theory, is a cosmological doctrine meant to account for the formation and operation of the universe in ancient Chinese philosophy. For that reason, its "head" categories, wood, fire, earth, metal, and water, are thought to be the five basic elements that compose the universe. These five elements are in a relation of mutual promotion and restraint, as shown in Figure 1 (adapted from Chen 1989b: 1000). In this figure, the lines forming the outer pentagon indicate the relation of promotion, and the lines forming the inner five-pointed star indicate the relation of restraint, with arrowheads indicating directions of promotion or restraint (see, also, Yu 1995, 1998). It is with this relation of mutual promotion and restraint among the five elements that the balance of the universe is achieved. The balance of the universe is destroyed when the balanced proportion of the five elements is lost, and that results in ecological problems or natural disasters. The theory of five elements has its own conceptual schematization in categorization. For instance, "wood" is associated with "spring", the beginning season of the year when it is "windy", and with the "east", the location from which the sun greets the beginning of a new day. It is associated with the "green" colour of new leaves and the "sour" taste of unripe fruit (see Wiseman and Feng 1997: 205–207 for a more detailed discussion of the correspondences in the five-element categorization).

It is interesting to note that both theories under discussion seek systematic correlations between the microcosm of the body and the macrocosm of the universe, although they go in opposite directions, as specified above. In the case of the five-element theory, Chinese medicine has borrowed it from Chinese philosophy, as a reasoning tool of analogy, to study the human body and its relationship with the natural environment. It starts with the five basic elements of the universe, and all categories of things under them fall into the pattern of five. For instance, instead of four seasons in a year, there are five, including a "long summer", in this system. Also, the six *fu* organs are reduced by one, the triple heater, to fit into this five-element schema. In contrast, the humoural theory starts with the four basic fluids in the body, and the things that come under them all fall into the pattern of four. For instance, its basic elements of the universe are fire, earth, water, and air, in contrast with the Chinese five-element categorization, i.e. wood, fire, earth, metal, and water.[18] As can be seen, both systems include adjectival modifiers such as "cold", "hot", and "dry". However, because of the difference in the starting point, the humoural theory uses them to characterize the humoural fluids whereas the five-element theory uses them to characterize the climatic patterns.

Not all categories of correspondences in each theory are included in the two tables given above. For instance, a different table in Fernando (1996: 123) lists two categories, "physiognomy" and "attributes", that are absent from Table 1. Table 2 has left out a few categories from the original in Wang et al. (1997: 14): "sound", "number", "symptomatic location", and "stress response".

In addition to those differences noted above, there are still the following differences between the two systems:

1. There are some categories that are in the one system but not in the other. Notably, the humoural theory emphasizes the correspondences between

the "temperaments" and the humours. The five-element theory, however, establishes the correspondences between the internal and external organs, and between these organs and the "emotions" and "spirits". This theory also places more emphasis on agriculture, including the categories of "crop" and "life cycle" modelled on farming seasons.

2. Although the two systems share some categories, they may have different items in those categories. Interestingly, for the category of animal, the humoural theory has wild animals whereas those corresponding to the five elements are domestic animals, typically seen in the agricultural setting in China. For the planets, the humoural theory includes the moon whereas the theory of five elements includes five major planets, named in modern Chinese after the five elements.[19] For the internal organs, saliently, the humoural theory already includes the brain whereas the five-element theory does not. Also, the former pairs the brain and bladder, and the liver and stomach, placing them in the same cell, but the latter pairs the *zang* with the *fu* organs, emphasizing their close functional relationship as pairs.
3. There are major mismatches in correspondences between these two systems. For instance, in the humoural theory, the heart is associated with air, sweetness, spring, east, and Jupiter. In the five-element theory, in contrast, the heart is associated with fire, bitterness, summer, south, and Mars, known as the Fire Planet in Chinese. Interestingly, the liver in the humoural doctrine corresponds to fire, bitterness, summer, south, and Mars, which are all aligned with the heart in the theory of five elements.

There are other similarities and differences between the two systems that cannot be discussed one by one here. The question of interest to ask is: Were these two systems in any way related or connected to each other? In other words, did the one in any way influence the other, or did they influence each other at certain points in history? The answers to these questions are not known.[20] My speculation is negative given the kinds of differences outlined above. If that is the case, however, the extent to which these two systems are similar to each other is really remarkable. It may represent some common heritage of human cognition in certain developmental stages.

7. Conclusion

The Chinese conceptualization of the heart as the central faculty of cognition was established in ancient Chinese philosophy and traditional Chinese medicine. As part of the body, the heart is understood via a socially-based metaphor: THE HEART IS THE RULER OF THE BODY. That is, the heart to the body is like the emperor or monarch to the country, with the supreme political power of the ruler mapped onto the perceived mental power of the heart, known as the "spiritual light". Being the thinking and reasoning organ, the source of "light", the heart is special and superior, "enlightening" the rest of the body. Fitting into this socially-based metaphor system, the other bodily organs are "governmental officials" performing specific functions, centrally commanded by the heart. Thus, the sense organs, the interface with the external world, separately receive information, but the heart centrally processes and stores it. The *zang* organs separately produce and store mental vitality and emotional force, but the heart centrally regulates and coordinates them. This socially-based metaphor system epitomizes the overarching philosophical belief in the unity and correspondence between the microcosm of the body and the macrocosm of the universe (天人合一, 天人相应) (see also Yu 2007a, 2007b).

A synchronic linguistic study shows that the Chinese conceptualization of the heart as the central faculty of cognition, as found in ancient Chinese philosophy and traditional Chinese medicine, is still widely manifested in the Chinese language today, despite the fact that the cognitive functions of the brain are fully recognized in scientific knowledge. The question is: Is the linguistic evidence merely a matter of language, or does it to some degree reflect the cultural conceptualization of the Chinese-speaking people? If the former is the case, the linguistic expressions analyzed in Section 4 are simply linguistic relics from the past that do not have their conceptual counterpart in the current "minds" of Chinese people. That is, they are on the same par with the oft-cited examples of English words *sunrise* and *sunset*. Although English-speaking people still use them, most people do not think or believe that the sun actually "rises" or "sets". If the latter case is true, it means that language can influence thought, given that what it suggests is not consistent with our scientific knowledge. After all, cultural conceptualization is not the same as scientific characterization.

For a comparative perspective, the English word *heart* is studied for the evolution of its meaning. In Old and Middle English, the heart appeared to be the centre of both feeling and thought. Present-day English, however, regards the heart as the centre of feeling and the mind/brain as the centre of thought, whereas in scientific terms the brain controls both feelings and thought, as well as movement. That is, the current English conceptualization of the heart as the centre of feeling is still cultural, not scientific, even according to the definition of Western medicine. In contrast, Chinese medicine today still considers the heart as commanding the rest of the body, including the brain, even though it has gained a better understanding of the cognitive functions of the brain.

The study finally compares two folk theories in Western and Chinese cultures, namely the theories of the four humours and the five elements. It is shown that they are both cultural systems of schematization and categorization, establishing correspondences within the body and between body and nature. There are important differences between them, however. Most importantly, the four-humour theory started as a medical doctrine, but no longer holds its ground in Western medicine; the five-element theory started as a philosophical doctrine, and still serves as a guiding and reasoning theory in Chinese medicine, although its limitations are noted (see Wiseman and Ellis 1996: 13–14). If the change of philosophical trend in Western thought has caused the separation of the mind from the heart in the English-speaking culture, the lack of such change in Chinese thought probably accounts for the relatively stable conceptualization of the heart as the central faculty of cognition in Chinese culture. Science will certainly cause conceptual change, but it is only one of the interactive factors in the process, others including philosophy, religion, and language.

Obviously, this study has raised more questions than it can answer. It takes cross-linguistic and cross-cultural efforts to draw a more complete picture. That is what this volume is all about.

Acknowledgments

I am profoundly grateful to René Dirven and Farzad Sharifian for their insightful comments and suggestions that have led to much improvement of this paper, and for sharing their invaluable ideas, knowledge, and experience with me. I also thank the anonymous reviewer for the useful comments on an earlier version of this paper.

Notes

1. The English translations, here and hereafter, are my own unless otherwise noted. For my translations I consulted various versions of existing translations, if such sources were available, and made modifications to varying degrees.
2. It is worth mentioning in passing that the Chinese conceptualization of the heart is quite similar to that of the Newar culture of Nepal, a neighbouring country of China. According to Parish (1991: 316), "for Newars the mind is in the heart, not the head". Although some informants mentioned the head and brain as involved in mental processes, especially thinking, Parish (1991: 347) was not sure to what extent this idea "reflects culture change and a 'modern' viewpoint influenced by outside conceptions or, simply, the notion that speech is analogous to thought".
3. THINKING, KNOWING, or UNDERSTANDING IS SEEING is but an aspect of the more general metaphor, THE MIND IS A BODY. See, e.g., Lakoff and Johnson (1999), and Yu (2003b, 2004) for further discussions. For cognitive studies of metaphor and metonymy in general, see, e.g., Barcelona (2000), Dirven and Pörings (2002), Gibbs (1994), Gibbs and Steen (1999), Lakoff and Johnson (1980, 1999), Kövecses (2002, 2005), Panther and Radden (1999).
4. In Chinese medicine, "essence" (精 *jing*) refers to "the essential substances constituting the human body and maintaining life's activities", "the kidney essence which includes the congenital essence and the acquired essence", or "reproductive essence or semen" (Zhang 2001: 450).
5. The English translations or glosses here are adopted from Wiseman and Feng (1998). It should be noted that such folk concepts are highly culture-specific, and their English translations should be taken as rough "gap fillers". It should also be noted that there exist different versions of translation. Another version, for instance, translates the "five spirits" respectively as spirit, mood, soul, idea, and will (Zhang 2001).
6. Readers are referred to Yu (2003a) for a detailed study of the Chinese understanding of gallbladder and courage. It is argued there that the culturally constructed medical characterization of the gallbladder forms the base of the cultural model for the concept of courage. In the core of this cultural model is a pair of conceptual metaphors: (a) GALLBLADDER IS CONTAINER OF COURAGE, and (b) COURAGE IS QI (GASEOUS VITAL ENERGY) IN GALLBLADDER, which is partly constitutive of the understanding of the gallbladder and courage in Chinese culture. The Chinese understanding of the gallbladder as the seat of courage contrasts with the English one that ascribes courage to the heart (see Section 5).
7. Again, these emotion concepts can be culture-specific. Notably, *si,* translated here as "overthinking" (following Larre and de la Vallée 1996), is also rendered as "anxiety" (Zhang 2001: 1006). It refers to the feeling when one is over-

whelmed by some negative thought). This sense of the word is obsolete in contemporary Chinese.

8. Even today, the role assigned to the brain is very limited in standard versions of Chinese medicine. It is said to store the marrow and be responsible for the sensorimotor system, i.e., sensual perception and physical movement (see, e.g., Wang et al. 1997: 918–919). See Farquhar (1998) for a Western anthropological-medical perspective on the lack of attention to, and interest in, the brain on the part of Chinese medicine.
9. Two things about Chinese are worth mentioning in passing. First, the "heart" radical has two variants. The canonical one, which looks the same as the word 心 *xin* 'heart', often occurs at the bottom of a character, as in 思 *si*, 想 *xiang*, 虑 *lü*, 念 *nian*, and 感 *gan*. The other one, usually called the "vertical heart radical", occurs on the left side of a character, as in 情 *qing*. Second, the single-character words often combine with other characters to form so-called compound words in modern Chinese. For instance, the preceding single-character words can form the following compounds: 思想 *sixiang* 'thought; thinking; idea; ideology', 思虑 *silü* 'consider; contemplate; deliberate', 思念 *sinian* 'think of; long for; miss', 感情 *ganqing* 'emotion; feeling; sentiment; affection', and 情感 *qinggan* 'emotion; feeling'.
10. Readers are referred to Yu (1995, 1998) for more detailed studies of Chinese expressions of anger and happiness. There I point out that Chinese seems to place more emphasis on the heart as the container of emotions than does English, for which the container is often the body. See, also, Kövecses (2000, 2002, 2005) for detailed discussions of the ANGER IS FIRE metaphor.
11. These are just linguistic observations. The questions remain as to what extent the linguistic phenomena reflect a cultural conceptualization, as opposed to scientific knowledge, and to what extent linguistic phenomena can influence cultural conceptualizations despite scientific knowledge.
12. The LDCE (2003) has made some noticeable changes in wording or definition. The second sense of *heart* now refers to "EMOTIONS/LOVE" with the definition "the part of you that feels strong emotions and feelings". That is, the heart is where emotions and feelings are actually felt.
13. The LDCE (2003) defines *brain* as "the organ inside your head that controls how you think, feel, and move". That is, the function of controlling movement is added to the brain. With regard to its difference from the heart, the brain controls feelings whereas the heart feels them. Defined as an "ORGAN", the brain is the central faculty of cognition in general; defined as "INTELLIGENCE", it represents "the ability to think clearly and learn quickly", i.e., it specializes in thinking and learning as the seat of intellect.
14. See, also, Dirven (2002), Lakoff (1999), Lakoff and Johnson (1999). Readers are referred to Dirven (2002) for a discussion of the relationship between *head,*

brain, and *mind*, and the metonymic chain *head* > *brain* > *thinking* > *mind* > *thoughts* > *intelligence.*

15. Interestingly, emotions or feelings are not mentioned here in relation to the functions of the brain in either sense 1 or sense 2 of the word *nao* 'brain'.
16. In English, as Dirven (personal communication) points out, this view of the heart as centre of higher faculties may have co-existed to some extent with an equally important folk view of the head as being in command of everything else. See Niemeier (this volume).
17. As Dirven (personal communication) points out, dualism had already been an undercurrent in Western thought, but got the upper hand after Descartes (17th century) and the rise of rationalist philosophy (18th century).
18. It is worth mentioning that the element of air, called *qi* in Chinese, is not completely out of the picture of cosmology of ancient China. On the contrary, it was regarded as a more fundamental substance than the five elements: "since ancient times the Chinese have conceived of *qi* as something indispensable to all existence, the fundamental substance from which all things rise and upon which everything depends for survival" (Zhang and Rose 2001: 26). In fact, the universe started with the undifferentiated substance of *yuan qi* 'original qi', which then divided into two opposite forces of yang and yin, separately forming Heaven and Earth. The eternal interplay between yin and yang gave rise to *qi* 'air', and gave life to the ten thousand things. This is what the founder of Daoism Laozi (c. 571–480 B.C.) meant in his *Daode Jing* when he said: "Dao gives birth to One. One gives birth to Two. Two gives birth to Three. Three gives birth to Ten Thousand Things" (see Zhang and Rose 2001: 25–27).
19. In ancient times, these five planets had different names, which are no longer widely known nowadays. The change of names for these planets apparently resulted from the influence of the five-element theory.
20. Nor is it known the exact times when these two systems came into being. It is suggested that Hippocrates (c. 460–377 B.C.) laid the foundations of the humoural doctrine but Galen (129–199 A.D.) gave it its classical form that dominated the Middle Ages (Geeraerts and Grondelaers 1995: 156–157). The five-element theory must have existed before Confucius (c. 551–479 B.C.), who made references to it in his works.

References

Barcelona, Antonio (ed.)
2000 *Metaphor and Metonymy at the Crossroads: A Cognitive Perspective.* Berlin/New York: Mouton de Gruyter.

Cai, Degui
2000 *Zhongguo Zhexue Liuxingqu* [The Pop Music of Chinese Philosophy]. Jinan, Shandong, China: Shandong People's Press.

Chan, Alan K. L. (ed.)
2002a *Mencius: Contexts and Interpretations.* Honolulu, HI: University of Hawaii Press.
2002b A matter of taste *qi* (vital energy) and the tending of the heart (*xin*) in Mencius 2A2. In *Mencius: Contexts and Interpretations*, Alan K. L. Chan (ed.), 42–71. Honolulu, HI: University of Hawaii Press.

Chen, Zelin
1989 Yin-yang wuxing [Theories of yin-yang and five elements]. In *Jiating Yixue Quanshu* [Family Medicine], Jin Wentao (ed.), 997–1002. Shanghai: Shanghai Science and Technology Press.

Dirven, René
2002 Metonymy and metaphor: Different mental strategies of conceptualization. In *Metaphor and Metonymy in Comparison and Constrast*, René Dirven and Ralf Pörings (eds.), 75–111. Berlin/New York: Mouton de Gruyter.

Dirven, René and Ralf Pörings (eds.)
2002 *Metaphor and Metonymy in Comparison and Contrast.* Berlin/New York: Mouton de Gruyter.

Farquhar, Judith
1998 Chinese medicine and the life of the mind: Are brains necessary? *North Carolina Medical Journal* 59: 188–190.

Fernando, Chitra
1996 *Idioms and Idiomaticity.* Oxford/New York: Oxford University Press.

Geeraerts, Dirk and Stefan Grondelaers
1995 Looking back at anger: Cultural traditions and metaphorical patterns. In *Language and the Cognitive Construal of the World*, John R. Taylor and Robert E. MacLaury (eds.), 153–179. Berlin/New York: Mouton de Gruyter.

Gibbs, Raymond W.
1994 *The Poetics of Mind: Figurative Thought, Language, and Understanding.* New York/Cambridge: Cambridge University Press.

Gibbs, Raymond W. and Gerard J. Steen (eds.)
1999 *Metaphor in Cognitive Linguistics*. Amsterdam/Philadelphia: John Benjamins.

Hansen, Chad
1992 *A Daoist Theory of Chinese Thought: A Philosophical Interpretation*. Oxford/New York: Oxford University Press.

H-YCD
1995 *Han Ying Cidian* [A Chinese-English Dictionary] (revised ed.). Beijing: Foreign Language Teaching and Research Press.

HYDCD
2000 *Hanyu Da Cidian* [Grand Dictionary of Chinese Language]. Shanghai: Grand Dictionary of Chinese Language Press.

Jäkel, Olaf
1995 The metaphorical conception of mind: "Mental activity is manipulation". In *Language and the Cognitive Construal of the World*, John R. Taylor and Robert E. MacLaury (eds.), 197–229. Berlin/New York: Mouton de Gruyter.

Ji, Xianlin (gen. ed.)
1995 *Chuanshi Cangshu* 《传世藏书》 [Collection of Classics over Centuries], Vols. 49, 50, 54, and 65. Hainan, China: Hainan International Journalism Publishing Center.

Kövecses, Zoltán
2000 *Metaphor and Emotion: Language, Culture, and Body in Human Feeling*. Cambridge/New York: Cambridge University Press.
2002 *Metaphor: A Practical Introduction*. Oxford/New York: Oxford University Press.
2005 *Metaphor in Culture: Universality and Variation*. Cambridge/New York: Cambridge University Press.

Lakoff, George
1999 The system of metaphors for mind and the conceptual system of analytic philosophy: A study of the metaphorical constraints on philosophical discourse. In *Cognition and Function in Language*, Barbara A. Fox, Dan Jurafsky and Laura A. Michaelis (eds.), 159–180. Stanford, CA: Center for the Study of Language and Information.

Lakoff, George and Mark Johnson
1980 *Metaphors We Live By*. Chicago: University of Chicago Press.
1999 *Philosophy in the Flesh: The Embodied Mind and Its Challenge to Western Thought*. New York: Basic Books.

Larre, Claude, and Elisabeth Rochat de la Vallée
1991 *The Heart in Ling Shu Chapter 8*. Cambridge, UK: Monkey Press.
1992 *Heart Master Triple Heater*. Cambridge, UK: Monkey Press.

1995 *Rooted in Spirit: The Heart of Chinese Medicine.* New York: Station Hill Press.
1996 *The Seven Emotions: Psychology and Health in Ancient China.* Cambridge, UK: Monkey Press.

LDCE
1978 *Longman Dictionary of Contemporary English.* Harlow/London: Longman.
2003 *Longman Dictionary of Contemporary English* (4th ed.). Harlow: Longman.

Lin, Min
2001 *Certainty as a Social Metaphor: The Social and Historical Production of Certainty in China and the West.* Westport, CT: Greenwood.

Niemeier, Susanne
This vol. To be in control: kind-hearted and cool-headed. The head-heart dichotomy in English.

OED
1989 *The Oxford English Dictionary* (2nd ed.). Oxford: Clarendon Press; New York: Oxford University Press.

Palmer, Gary
1996 *Toward a Theory of Cultural Linguistics.* Austin, TX: University of Texas Press.

Panther, Klaus-Uwe, and Günter Radden (eds.)
1999 *Metonymy in Language and Thought.* Amsterdam/Philadelphia: John Benjamins.

Parish, Steven M.
1991 The sacred mind: Newar cultural representations of mental life and the production of moral consciousness. *Ethos* 19: 313–351.

Sharifian, Farzad
2003 On cultural conceptualizations. *Journal of Cognition and Culture* 3: 187–207.

Shun, Kwong-loi
1997 *Mencius and Early Chinese Thought.* Stanford, CA: Stanford University Press.

Slingerland, Edward
2003 *Effortless Action: Wu-wei as Conceptual Metaphor and Spiritual Ideal in Early China.* Oxford/New York: Oxford University Press.

SOED
2002 *Shorter Oxford English Dictionary on Historical Principles* (5th ed.). Oxford/New York: Oxford University Press.

Wang, Qi, Luo Xijia, Li Yun and Liu Yanjiao
1997 *Zhongyi Zangxiang Xue* [Theory of Internal Organs in Chinese Medicine]. Beijing: People's Health Press.

Wierzbicka, Anna
1992 *Semantics, Culture, and Cognition: Universal Human Concepts in Culture-Specific Configuration.* New York/Oxford: Oxford University Press.
Wiseman, Nigel and Andrew Ellis
1996 *Fundamentals of Chinese Medicine.* Brookline, MA: Paradigm Publications.
Wiseman, Nigel and Ye Feng
1998 *A Practical Dictionary of Chinese Medicine* (2nd ed.). Brookline, MA: Paradigm Publications.
XDHYCD
1996 *Xiandai Hanyu Cidian* [Modern Chinese Dictionary] (revised ed.). Beijing: The Commercial Press.
Yearley, Lee H.
1990 *Mencius and Aquinas: Theories of Virtue and Conceptions of Courage.* Albany, NY: State University of New York Press.
Yu, Ning
1995 Metaphorical expressions of anger and happiness in English and Chinese. *Metaphor and Symbolic Activity* 10: 59–92.
1998 *The Contemporary Theory of Metaphor: A Perspective from Chinese.* Amsterdam/Philadelphia: John Benjamins.
2002 Body and emotion: Body parts in Chinese expression of emotion. In the special issue on "The Body in Description of Emotion: Cross-linguistic Studies", N. J. Enfield and Anna Wierzbicka (eds.). *Pragmatics and Cognition* 10: 333–358.
2003a Metaphor, body, and culture: The Chinese understanding of *gallbladder* and *courage*. *Metaphor and Symbol* 18: 13–31.
2003b Chinese metaphors of thinking. In the special issue on "Talking about Thinking across Languages", Gary B. Palmer, Cliff Goddard, and Penny Lee (eds.). *Cognitive Linguistics* 14: 141–165.
2004 The eyes for sight and mind. *Journal of Pragmatics* 36: 663–686.
2007a Heart and cognition in ancient Chinese philosophy. *Journal of Cognition and Culture* 7(1/2): 27–47.
2007b The Chinese conceptualization of the heart and its cultural context: Implications for second language learning. In *Applied Cultural Linguistics: Implications for Second Language Learning and Intercultural Communication*, Farzad Sharifian and Gary B. Palmer (eds.), 65–85. Amsterdam/Philadelphia: John Benjamins.
Zhang, Qiwen (gen. ed.)
2001 *A Practical Chinese-English Dictionary of Traditional Chinese Medicine.* Trans. by Sun Hengshan. Jinan, China: Shandong Science and Technology Press.

Zhang, Yu Huan and Ken Rose
2001 *A Brief History of Qi*. Brookline, MA: Paradigm Publications.
Zheng, Xunzuo, Zhao Zhentao, Zhang Wenting, and Zhou Dingzhi
1993 *Mengzi* [Mencius: A Chinese-English Bilingual Edition]. Trans. into modern Chinese by Zheng Xunzuo, and into English by Zhao Zhentao, Zhang Wenting, and Zhou Dingzhi. Shandong, China: Shandong Friendship Press.

The heart – What it means to the Japanese speakers

Yoshihiko Ikegami

Abstract

The paper proposes to discuss the traditional notion of the heart in Japanese and the functions popularly attributed to it. Two general points will first be pointed out:

(i) Thoughts and emotions tend to be conceptualized in Japanese as spontaneously arising rather than caused by some stimulus either internal or external (hence linguistically encoded in terms of either quasi-impersonal (e.g. Sad thought/feeling OCCURS TO me) or intransitive (e.g. I BECOME sad) constructions.

(ii) In relation to thoughts and emotions, body parts tend to be conceptualized as LOCI at which they arise rather than as ENTITIES which generate them.

With regard specifically to the heart, the native Japanese word in question is *kokoro*. The earliest attested examples of the word shows that the word was polysemous, being mediated by the metonymic association between the heart as a body organ and the functions associated with it – the latter covering the whole range of thinking, feeling and volition. (Notice that the mind/heart contrast did not exist in the native tradition.) In the later semantic development of the word, the word *kokoro* became more and more strongly associated with the workings of the heart rather than the heart as a body organ, the latter sense being taken over by the formal Chinese-derived compound, *shin-no-zo* ('storehouse of the heart') and finally *shin-zo* ('heart-storehouse'). The first sense, referring to the workings of the heart, on the other hand, managed to survive even after it became known anatomically that such workings were in fact to be attributed to the brains rather than the heart. At the same time, however, the idea of the heart as generator of thoughts, feelings, emotions and volitions has been partially eclipsed by that of the breast/chest as locus where such thoughts, etc. arise.

Keywords: body organs, conceptualization as entity versus as locus, emotion, evolution, heart, Japanese

1. Introduction: The Japanese “body-lexicon”

Phrases and idiomatic expressions involving words referring to body parts are called in Japanese ‘karada-kotoba’ or more formally ‘shintai-goi’ (either of the expressions equally translatable as ‘body-lexicon’) and we have a number of studies by native linguists, psychologists and ethnologists (e.g. Wada 1969, Miyaji 1979, Matsuki 1995, Kusumi 1996). These studies seem to indicate that while Japanese abounds in phrases and idiomatic expressions involving words referring to outward body parts (such as hands, feet, head, mouth and neck), as perhaps in any other languages, those involving words referring to internal body organs are comparatively few.[1] In the absence of precise statistical data on this point, I have to be content with impressionistic remarks for the moment, but it will be interesting to check whether this paucity of the words for internal organs is correlated with the predominantly agricultural (rather than nomadic) orientation of culture in the earliest stages of the country.[2]

2. Names for body-internal organs: native and borrowed

In the earliest linguistic records of Japanese, two names of body-internal organs are attested: *kokoro* and *kimo*. The two words are sometimes found to be used in juxtaposition. The first word, *kokoro*, is known to have specifically referred to the heart (as body organ); the second word, *kimo*, is presumed to have referred to the body-internal organs in general, but later tended to be identified specifically with the liver. In the early linguistic records, both words are also known to have been used to refer to mental/emotional states, suggesting that either the heart or the whole inside region of the breast/belly was associated with mental/emotional activities. With the introduction and gradual spread of the traditional Chinese medical doctrines, with their characteristic idea of 'five internal organs' (i.e. heart, lung, liver, spleen, kidney), these two words tended to be supplanted as terms for body-internal organs by the words borrowed from Chinese, *shinzo* (literally, 'heart-storehouse') and *kanzo* (literally, 'liver-storehouse'). These two borrowed words, however, have generally remained to this day only as technical terms. On the other hand, as words of native stock, *kokoro* and *kimo*, especially, the former, have provided by far the largest number of phrases and idiomatic expressions involving the names of body-internal organs. While *kokoro* has retained its original wide range of application, covering both mental and emotional activities, *kimo* has had its use narrowed down to a specific attribute, namely 'courage'/'bravery', suggesting that the liver came to be conceived of specifically as the seat of courage/bravery . Only a few of the idiomatic expressions involving *kimo* have survived to this day: *kimo o tsubusu* (literally, 'to crush the liver', meaning 'to be excessively disturbed', 'to be greatly alarmed') and *kimo o hiyasu* (literally, 'to cool the liver', meaning 'to be disturbed', 'to be frightened'), *kimo ga chiisai* (literally, 'to have a small liver', meaning 'to be cowardly/chicken-livered'). In the following, I am concentrating on the semantic development of the word *kokoro* (originally, 'the heart'), which has been most productive in generating a great number and variety of 'embodied' phrases and idiomatic expressions.

3. *Kokoro* and its related words with their metonymic substitutes

Etymologically, the word *kokoro* is presumed to be possibly related to the verb *koru* ('to become solid', 'to congeal'). The idea is that the heart and other internal organs were first created through congealment of some parts of a fluid which filled the body.[3] (Incidentally, the phonetic similarity to the Latin word *cor* is only accidental.) The earliest attested examples of the word (mostly in the eighth-century anthology of twenty volumes covering poems of the preceding three or four centuries) show that the word was already polysemous at that stage, supported by the metonymically mediated association between the heart as a body organ and the functions it was supposed to perform – the latter covering the whole range of thinking, feeling and volition, as it still does in modern Japanese (see section 3). Thus in archaic Japanese phrases like *kokoro araba* (literally, 'if (one) has a "heart"') and *kokoro shirazu* (literally, '(one) doesn't know the "heart"'), the word *kokoro* can mean 'a will (to do something)', 'a sympathetic feeling (toward someone)' or 'an idea, a thought'.

In its later semantic development, the word *kokoro* became more and more strongly associated with the workings of the heart rather than the heart as a body organ. The latter sense was taken over by the formal Chinese derived compound, *shin-no-zo* ('the storehouse of the heart') and finally by *shinzo* ('the heart-storehouse'). The first sense, referring to the workings of the heart, on the other hand, managed to survive even after it became known anatomically that such workings were in fact to be attributed to the brain rather than to the heart. The present-day conceptualization among Japanese speakers can perhaps be roughly summarized by saying that the brain is seen to be responsible for more intellectually oriented activities,[4] while the heart is responsible for more emotionally oriented activities. In fact, the newest arrival, the word *haato* borrowed from English, is used exclusively in association with the notion of romantic love.

It must be added that two other words, which refer not to body-internal organs but to outward body parts, have been in rivalry with *kokoro* ('the heart') and *kimo* ('the liver') as loci of emotions. They are *mune* ('chest/breast') and *hara* ('belly'). These refer to the parts of the body in which the heart and the liver are accommodated, so that what we have here is a common instance of metonymic transfer from the location to the entity associated with the location (or somewhat more specifically, from the container to what it contains). In fact, the words *kokoro* ('the heart') and *mune* ('chest/breast') are sometimes interchangeable in one and the same phrase: e.g. '*kokoro/mune ga odoru*' (literally, 'the heart/breast bounces',[5] meaning 'to have a buoyant feeling'), '*kokoro/mune ni ukabu*' (literally, 'to float in the heart/breast', meaning 'to occur to someone'). '*kokoro/mune o sadameru*' (literally, 'to set the heart/breast', meaning 'to make up one's mind'), '*kokoro/mune ga hiroi*' (literally, the heart/breast is wide', meaning 'to be tolerant, accommodating'. (The last two examples are attested and recorded in the dictionary, but the use of *mune* here is not as common as *kokoro* in the present-day speech.)

In contrast to *mune* ('breast/chest'), *hara* ('belly') is used as locus of a much more restricted range of emotions. It is interesting, however, that there are a few phrases in which *hara* is linterchangeable with *kimo*: e.g. '*kimo/hara o sueru*' (literally, 'to set the liver/the belly', meaning 'to be firmly determined').[6] Otherwise, *hara* tends to be conceived of as locus of non-normal and non-recommendable negative emotions, especially anger (cf. Matsuki 1995): e.g. '*hara ga tatsu*' (literally, 'the stomach rises', meaning 'to get angry'), '*hara o tateru*' (literally, 'to let the belly rise', meaning 'to get angry'), '*hara no mushi ga osamaranai*' (literally, 'the belly-worm can't be restrained', meaning 'One can hardly restrain oneself from getting angry'), '*hara ga kuroi*' (literally, 'the belly is black', meaning 'to be of deceptive/crooked character'). In fact, it is argued by some that the ritual of *hara-kiri* (literally, 'belly-cutting', being an honourable way of suicide by cutting one's belly open with one's own sword, formerly allowed only to respectable warriors) was intended to show and prove that the person in question had nothing evil in his belly. Moreover, the fact that the breast/chest and the belly are thought of as loci of love and anger, among others, respectively is arguably in accordance with a general principle, claimed by an anthropologist (Yamaguchi 1986: 56), that the upper half of the human body tends to be associated with 'higher' psychological functions and the lower half with 'lower' ones.

4. The polysemy of the word *kokoro*

Below is given a general sketch of the system of the polysemous senses associated with the word *kokoro*:[7]

[I] 'the heart as a body organ' (the sense now largely supplanted by the Chinese-borrowed word *shinzo*).

[II] 'the functions associated with the heart, broadly covering all of their intellectual, emotional and volitional aspects' (the sense derived from [I] through the metonymy: BODY PARTS ARE LOCI WHERE SOMETHING ARISES, or in a more formulaic representation, BODY PARTS FOR THEIR PERCEIVED FUNCTIONS) – now the commonest and prototypical sense of the word – certain specific aspects may be emphasized, depending on the context, such as (i) 'discretion', (ii) 'thoughtfulness', (iii) 'tender feeling'; (iv) '(not immediately apparent) true intention'; (v) 'ability to appreciate poetry'; etc.[8] Thus *kokoro no komotta motenashi* (literally, 'heart-involved hospitality': *kokoro* in the sense of (ii) or (iii)), *kokoro nai hitotachi* (literally, 'people without heart': *kokoro* in the sense of (i), (ii), (iii) and (v)), *kokoro shirazu* (literally, 'do not know the heart': *kokoro* in the sense of (iv)).

[III] 'an important aspect of something which is particularly relevant to it' (the sense derived from [I] and [II] combined through the metaphor: AN ENTITY MAY HAVE A HEART JUST AS A PERSON HAS A HEART (AS A FUNCTIONALLY IMPORTANT ORGAN), or in a more formulaic representation, THE VITAL ASPECT (OF SOMETHING) IS THE HEART (OF SOMETHING)), such as (i) 'an aspect of some entity which is to be appreciated', (ii) 'an essential aspect or the fundamental principle of something', (iii) 'a not immediately apparent point to be appreciated', (iv) 'the meaning conveyed by language', (v) 'the meaning of a poem to be appreciated', (vi) 'the reason supporting an assertion or the motivation leading to an action'. Thus *hana no kokoro* (literally, 'the flower's heart': *kokoro* in the sense of (i)), *cha no kokoro* (literally, 'the tea (ceremony)'s heart': *kokoro* in the sense of (ii)), *uta no kokoro* (literally, 'the poem's heart': *kokoro* in the sense of (v) or (iii)), *bunsho no kokoro* (literally, 'the (written) composition's heart': *kokoro* in the sense of (iv)), *kokoro shirazu* (literally, 'do not know the heart': *kokoro* in the sense of (vi)). Note that in [III], as in [II], most uses of the word are polysemous.

[IV] 'the center (of something)' now rare: e.g. *ike no kokoro* (literally, 'the heart of the pond': now obsolete) – the sense derived from [I] through the metaphor: SOMETHING QUALIFIES AS CENTER OF SOMETHING ELSE JUST AS THE HEART IS POSITIONALLY AND FUNCTIONALLY THE CENTER OF THE HUMAN BODY.[9]

5. Suggesting some further viewpoints to be taken into consideration in the research project, 'The Mind inside the Body'

5.1. Typological viewpoints

(i) Referring to body-internal organs vs. Referring to body-external parts:

The body-lexicon comprises two types of words: words referring to body-internal organs and words referring to body-external parts. The two types of body parts they refer to differ in the degree of perceptual accessibility. Body-external parts can be visually identified; moreover, some of these parts are known to be associated with particular functions so familiar to humans (like the eyes for seeing and the nose for breathing). Body-internal organs, by contrast, do not usually lend themselves to visual identification and hence their individual identities are not as readily confirmable. Among the body-internal organs, however, the heart enjoys rather unique status in this respect. Although not directly visible, it can make its existence felt by its beating, pounding and thumping, which can be heard. Hence a suggested hypothesis will be that it is probable that in encoding the workings of the mind in terms of the body lexicon languages will mobilize more words referring to body-external parts than words referring to body-internal organs and that among the words of the latter type, the word referring to the heart is uniquely (and probably, also universally) the most frequently accessed one across languages. Exceptions may arise if the language community is characterized by a strongly nomadically-oriented culture or if a certain medical doctrine based on expert knowledge has spread out and is somehow accepted as part of the folk knowledge by the members of the community (cf. note 2).

Thus, for the Japanese data, Wada (1969) cites altogether sixty-four body-part names with nearly two thousand compounds, phrases (not necessarily metaphorical) and idiomatic expressions involving these words. Only six of these sixty-four names refer to internal body organs and of these six, two are either a synonym or a written variant of two of the remaining four. We have thus four body-internal organs being referred to by six terms: *hai* ('lung'), *i*('stomach'), *fu*/*cho* ('bowels') and *kimo* ('liver') with two written variants *kan* ('liver') and *tan* ('liver'). Only two expressions are cited for *i* and three for *hai*. *Fu*/*cho* has eighteen expressions cited and *kimo* (with its two variants combined) forty-four. Unfortunately, *kokoro* is not included in the author's list, presumably because the word is understood in present-day Japanese as nearly exclusively referring to the functions of the heart and not to the heart as a body-organ (cf.section 3).

(ii) Metaphorical encoding with names for body-internal organs vs. Metonymic encoding with names for body-external parts:
The difference in perceptual accessibility between body-internal organs and body-external parts also have some consequences for the way in which their names are utilized in phrases encoding psychological processes. It is our common experience that certain private psychological processes are often correlated with certain expressions observable on some body-external parts. The latter serve metonymically as indices of the former. Thus in Japanese *me o hosomeru* (literally, 'narrow one's eyes') implies (unlike English) loving and affectionate feelings, *hana ga takai* (literally, '(one's) nose is tall') implies proud feelings, *kubi o chijimeru* (literally, 'shorten (one's) neck') implies fearfulness, and so forth. (Notice that the linguistic encoding does not necessarily describe an objectively occurring change observable on the body-external part in question. What is encoded can actually be fictive, as in '(one's) nose is tall' cited above.)[10]

Body-internal organs, which are usually not directly observable, do not seem to lend themselves as readily to indexical interpretation. Thus phrases in Japanese like *kokoro ga odoru* (literally, '(one's) heart bounces', meaning 'have a buoyant feeling') and *kimo o hiyasu* (literally, 'to chill (one's) liver', meaning 'to be extremely scared') can only be interpreted as referring at best to fictive events.[11] Hence they will more readily be interpreted as metaphors referring to certain psychological states (in the case of the examples above, a state of 'being full of hope and happiness' and a state of 'being extremely scared' respectively) than as metonymies referring indexically to some consequences of psychological states. The body-internal organs in these phrases will then be understood either as entities which generate certain psychological processes or as loci where certain psychological processes take place (see (iii) and (iv) below).

Notice, in this connection, that body-part names like *mune* ('breast/chest') and *hara* ('belly/stomach') occupy a rather ambiguous position. Primarily, they refer to body-external parts and are not names of body-internal organs. But the particular body-external parts they refer to are also supposed to serve as containers of certain body-internal organs (thus *mune* storing *kokoro* ('heart') in it and *hara* storing *kimo* ('liver') in it, among others). And since body-external parts are directly visible and hence cognitively more accessible, it will be no wonder if their names sometimes replace the name of the body-internal organs stored in them. Some relevant examples in Japanese are given in section 2.

The replacement of the name for a body-internal organ by a name for a body-external part may also affect the way the whole phrase is interpreted. Thus in the phrase, *kokoro ga odoru* (literally, '(one's) heart bounces'), the heart, as an entity, may be conceived of as 'generator' (i.e. something which generates happy feelings) or 'agent' in a somewhat metaphorical sense. In the phrase, *mune ga odoru* (literally, '(one's) breast/chest bounces'), the replacing terms, *mune* will be conceived of as a 'locus' rather than as an 'entity' (not to say, a 'generator' or an 'agent'), resulting in the interpretation that 'bouncing' takes place in, rather than is generated by, the breast/chest.[12]

(iii) Body-part conceptualized as 'entity' vs. Body-part conceptualized as 'locus':

An object can be construed not only as an 'entity' but also as a 'locus'. Thus the head is construed as an 'entity' in the phrase *cut off the he*ad, but in the phrase, *a fly on the head*, it is construed as a locus. In general, anything which is marked with a clearly delineated outline tends to be construed as an entity and if not, as a locus. Thus prototypically, an entity is associated with the notion of 'boundedness', while a locus can very well be associated with the notion of 'unboundedness'. The contrast is often reflected linguistically in the two series of spatial adjectives, one referring to three-dimensional entities (e.g. *big*) and the other referring to two dimensional expanses (e.g. *wide*).[13]

With regard to their relationships to the psychological processes they are associated with, body-parts can also be conceptualized either as entities or as loci. Body-internal organs, insofar as they are identifiable as such (whether in a real or in an imaginary way), will tend to be conceptualized as bounded entities. To the extent that they are not fully so identifiable, they will tend to be conceptualised as loci with more or less uncertain boundaries. Body-external parts, in contrast, will more readily be conceptualized as loci rather than as entities. The same conceptualization will apply to those cases in which the names for body-external parts are metonymically substituted for the names for body-internal organs (like *mune* ('breast/chest') for *kokoro* ('heart') and *hara* ('belly') for *kimo* ('liver'): see section 2). The breast/chest and the belly are here conceptualized as a sort of containers for the heart and for the liver respectively. Being contrasted with the entities they contain, however, they are readily conceptualized as loci in which those entities exist.

I would like to suggest in this connection that although anything can potentially be conceptualized either as an entity or as a locus, speakers of different languages may vary among themselves in preferring one type of conceptualization over the other. Thus there are cases in which, given one and the same object, the English and Japanese speakers seem to differ from each other in that the English speaker prefers conceptualization as entity while the Japanese speaker prefers conceptualization as locus. The contrast is quite systematically manifested in the dictionary definitions, as given below (cf. Ikegami (2000 [1998]: 226–31) on the cross-linguistic relativity of the notion of 'entity' and 'place'):

CREMATORIUM = 'a building in which the bodies of dead people are burned at a funeral ceremony'
KASOJO ('crematorium') = 'a place where the cremation takes place'
STADIUM = 'a building for sports, consisting of a field surrounded by rows of seats'
KYOGIJO ('stadium') = 'a place where people play sports'

It appears that the English speaker focuses on the constructed object *per se*, while the Japanese speaker conceptualizes the constructed object together with the characteristic activity people typically engage in *there*. A definition of the word *place* like 'an area, town, building, hotel or shop' (*Cambridge International Dictionary of English*) strikes the Japanese speaker as bizarre.[14]

Thus in quite a few Japanese phrases and idiomatic expressions involving the word *kokoro* ('heart') the referent is encoded as if it was an area where thoughts and emotions arise rather than as an entity which generates thoughts and emotions. The following examples are picked up from *Nihon Kokugo Daijiten*, 2nd ed. (2000-2002): cf.

(a) With nouns in apposition: *kokoro no ike/izumi/umi/kagami* (literally, 'the pond/spring (i.e. issuing water)/sea/mirror of the heart'), meaning 'the heart as compared to the pond/spring/sea/mirror'); with nouns indicating a location within the heart: *kokoro no oku/soko* (literally, 'the deep inside/bottom of the heart').

(b) with verbs of existence/locomotion/transportation: *kokoro ni ari/(ha)iru/ ukabu/osamu/kayou/todomu/nokoru/noru* (literally, 'to exist/enter/float/place/ reach/ keep/ remain/ ride in(to)/ on(to) the heart').
(c) with spatial adjectives: *kokoro ga hiroi/semai* (literally, 'the heart is wide/ narrow).[15]

(iv) Psychological processes construed as 'being caused to arise' vs. Psychological processes construed as 'arising spontaneously': There are languages, on the one hand, which encode the speaker's experience of psychological processes in terms of transitive verbs in the passive and languages, on the other, which encode them in terms of intransitive verbs. Thus there is a highly systematic contrast in this respect between English and Japanese: e.g. *be disappointed – shitsubosuru* (vi), *be perplexed – konwakusuru* (vi), *be satisfied – manzokusuru* (vi), *be surprised – odoroku* (vi). It can safely be assumed that the encoding of psychological processes in terms of transitive verbs in the passive is based on the construal of the processes as being caused by some stimulus/agent, while the encoding of them in terms of intransitive verbs is based on the construal of the same processes as arising spontaneously. I have no data on the distribution of these two contrastive features across languages, but the typological significance of the two orientations will be beyond doubt. Thus in talking not only about the affective / emotive process but also about the process of thinking, the speaker may prefer to construe and encode it in terms of the human agent engaging in an activity (i.e. 'I work out an idea') or in terms of the idea (as if) spontaneously arising (i.e. 'An idea occurs to me'). The contrast can ultimately be generalized as one between the two psychological tendencies which the Swiss linguist Charles Bally (successor to Ferdinand de Saussure) proposed back in 1920 and which he chose to call 'causal/transitif' and 'phénoméniste/intransitif' respectively (Bally 1920): see Appendix (1); cf. also Nakamura 1997, as quoted in Appendix (2). It will also serve as a significant pair of indices which typologizes languages into what I proposed to call 'SURU-teki gengo' ('DO-language') and 'NARU-teki gengo' ('BECOME-language') (Ikegami 1981, 1991).[16]

5.2. An evolutionary hypothesis

One of the intriguing books I encountered while preparing my paper for presentation at the ICLA conference in Seoul was *Naizo ga Umidasu*

Kokoro (Internal Organs Creating the Mind) by Katsunari Nishihara (Tokyo: Nihon Hoso Shuppan Kyokai, 2002). The author of the book is introduced as a well-known authority on the evolutionary process of the vertebrate animal and the designing of artificial organs. The book refers to a number of cases in which people who had a heart and lungs transplanted were observed to undergo a drastic change in personality after their successful operations: the transplantees took over the personalities of the donors from whom they received the organs. The author believes that this is a piece of evidence which suggests that the ultimate origin of the mind lay in the primitive internal organs. Specifically, the primitive coelenterate with only a single tube-like intestine was already sensitive to appetitive and procreative stimuli. Receiving such stimuli, which directly concerned the creature's basic needs for living, must have produced some sort of 'good' effects (primitive emotions) on the creature. It then responded by letting its own mobile muscles move in proper ways so as to take in any object functioning as stimulus-source. The author sees the origin of the mind in such primitive 'desires' and their associated dispositions created in the intestines.

To comment on the hypothesis proposed by the author is clearly beyond my professional knowledge. But if a hypothesis like this is indeed confirmed, it will certainly serve as a sure biological basis on which cognitive linguistic data on 'the mind inside the body' can be more securely sorted out and assessed.

Appendix

(1) Bally (1920: 261–262):

Deux tendances pyschologique opposées interviennent dans la perception des phénomènes et dans leur expression par le langage:

a) Tantôt l'esprit reste attaché de préférence à l'impression initiale et s'y absorbe ; tout ce qui se passe dans le monde extériur et chez le sujet pensant se présente sous la forme de faits purs et simples, même lorsque la cause peut en être atteinte spontanément ; et si cette cause est aperçue, elle reste indéterminée ; on ne lui attribue qu'une *activité interne, intransitive*, et non une action susceptible d'affecter un objet (un agi) et de produire un résultat. Cette attitude a un fondement logique seulement si la cause se dérobe à l'observation spontanée ; partout ailleurs elle repose sur une tendance subjective, une manière habituelle de voir les choses.

b) Tantôt au contraire la pensée se porte d'instinct vers la recherche de la cause et de l'effet ; le phénomène évoque l'idée d'un *agent* produisant une *action transitive*, propre à affecter un objet, un *agi* ; quand l'agent n'est pas offert par la réalité, on n'hésite pas à en imaginer un, et ce sujet imaginaire est tantôt le phénomène luimême, conçu comme une substance, tantôt le sujet pensant. Autrement dit : le phénomène, envisagé comme une action, n'est plus une représntation autonome, il devient *fonction* des êtres ou des choses (réels ou non) qu'il met en *relation*. Cette seconde attitude n'est donc logique que lorsque la cause ou l'effet, ou tous les deux, sont suggérés par la réalité ; dans tous les autres cas elle relève de pures habitudes de pensée.

Le premier mode d'aperception peut être appelé *phénoméniste* ou *impressionniste* (ce mot étant pris dans son sens strictement étymologique, sans arrière-goût artistique ou littéraire) ; la seconde tendance sera dite *causale* ou *transitive*. Toutes deux comportent naturellement une foule de degrés, de variétés et de combinaisons réciproques ; en outre elles se manifestent soit dans les formes individuelles de la pensée, soit dans ses formes collectives et sociales ; ou --- en termes linguistiques: soit dans des faits de parole et de style, soit dans la grammaire de la langue usuelle.

(2) Nakamura, Hajime (1997 [1964]: 350):

In the first place, we should notice that the Japanese are willing to accept the phenomenal world as Absolute because of their disposition to lay a greater emphasis upon intuitive sensible concrete events, rather than upon universals. This way of thinking with emphasis upon the fluid, arresting character of observed events regards the phenomenal world itself as Absolute and rejects the recognition of anything existing over and above the phenomenal world. What is widely known among post-Meiji philosophers in the last century as the "theory that the phenomenal is actually the real" has a deep root in Japanese tradition.

(3) Hartmann, Peter (1954: 115):

Der Mensch wird so in dem für sein Leben entscheidenden Kontakt mit der Umwelt nicht als ein selbständiges Wesen gesehen, das von sich aus von der Welt Besitz ergreift, sondern als ein von Vorgängen „affiziertes Objekt". Die in der Welt sich abspielenden Vorgänge erscheinen in der japanischen Weltsicht als die tragenden Elemente alles Geschehens; die Dinge, an denen sie sich abspielen, werden als Orte im weitesten Sinne gesehen: dem indogermanischen bewirkenden Tätersubjekt entspricht im Japaneschen ein Bereich, in den ein Vorgang gehört; es wird nicht einmal vermerkt, ob er in diesem Subjektsbereich seinen Ursprung hat, geschweige denn, daß dies „subjekt" ihn hervorgebracht habe. ... Statt des idg. Objektes finden wir im Japanischen einen vage ausgedrückten Bezug zur Umwelt, den ein vorgang beim Sich-Abspielen hat.

(4) The Expression of Emotions in terms of the Japanese Body-Lexicon (adapted from Fig. 2-1 (Kusumi 1996: 39) with the author's kind permission):

HAIR (ke, kami, hatsu)

dohatsu ten o tsuku (literally, '(one's) angry hair sticks into the sky') [ANGER]

BROWS (mayu)

mayu o shikameru ('knit (one's) brows') [DISPLEASURE]

mayu o hisomeru / yoseru ('knit (one's) brows') [DISPLEASURE]

mayu o hiraku (literally, 'open (one's) brows'; 'feel relieved') [PLEASURE]

HEAD (atama)

atama ni chi ga noboru ('the blood goes up to the head') [ANGER]

atama / tosaka ni kuru (literally, 'come to the head'; 'get angry') [ANGER]

atama kara yuge o dasu ('issue steam from the head') [ANGER]

atama kara mizu o abita yo ('as if one had water poured on (one's) head') [SURPRISE]

atama ga itai ('(one's) head is painful', 'have a pain in the head') [WORRY, AGONY]

atama o kakaeru ('hold (one's) head') [WORRY, AGONY]

atama o kaku ('scratch (one's) head') [EMBARRASSMENT]

EYE (me)

me o muku (literally, 'peel / bare (one's) eyes'; 'glare') [ANGER]

mekujira o tateru (literally, 'let (one's) eye-corners stand'; 'glare') [ANGER]

me ni kado o tateru (literally, 'let the corners stand in (one's) eyes'; 'glare') [ANGER]

me o sankaku ni suru (literally, 'make (one's) eyes triangular-shaped') [ANGER]

me ga suwaru ('(one's) eyes are set') [ANGER]

me (no tama) ga tobideru ('(one's) eye(-ball)s spring out') [SURPRISE]

me o shirokuro saseru ('(one's) eyes roll in black and white') [SURPRISE] me o miharu ('widen (one's) eye') [SURPRISE]

me o hosomeru ('narrow (one's) eyes') [PLEASURE]

me no kataki ni suru ('make an eye-enemy (of someone)') [HATE]

shiroi me de miru (literally, 'look at (someone) with cold eyes', 'look coldly upon (someone)') [DESPISE]

me mo aterarenai ('can't direct (one's) eye (to something)') [MISERY]

EAR (mimi)

mimi o kegasu (literally, 'dirty (someone's) ears'; 'tell (someone) something unpleasant') [DISPLEASURE]

mimi ni sawaru (literally, 'do something offensive to the ears') [DISPLEASURE]

mimi o ou (literally, 'cover the ears'; 'be unwilling to hear the news') [MISERY]

FACE (kao)

aosuji o tateru ('let blue veins stand out') [ANGER]

kesso o kaeru (literally, 'change the blood-physiognomy'; 'change colour') [ANGER]

manmen ni shu o sosogu (literally, 'pour vermillion over all the face') [ANGER]

kao kara hi ga deru (literally, 'fire issues from the face') [SHAME]

NOSE (hana)

hana ga takai ('(one's) nose is tall') [PRIDE]

hana o ugomekasu ('wriggle (one's) nose') [PRIDE]

hana ni kakeru ('hang (something) on (one's) nose') [BOASTING]

hana de ashirau ('treat (someone) with the nose'; cf. 'turn up (one's) nose at something') [COLD ATTITIDE, CONTEMPT]

ki de hana o kukuru ('rub (one's) nose with wood') [COLD ATTITIDE, CURTNESS]

hana ni tsuku ('stick to (one's) nose') [BEING ANNOYED, BECAUSE OF HAVING TOO MUCH]

kobana o fukuramaseru ('swell the nose-wings') [DISSATISFACTION]

CHEEK (hoho)

hoho o someru ('dye (one's) cheeks') [SHAME]

hoho o fukuramasu ('swell (one's) cheeks') [DISSATISFACTION]

TEETH (ha)

ha-gishiri o suru ('perform tooth-gnashing', 'gnash one's teeth') [ANGER]

sesshi yokuwan ('gnashing (one's) teeth and rolling up (one's) sleeves') [ANGER, CHAGRIN]

shiroi ha o miseru ('show (one's) white teeth') [LAUGH]

MOUTH (kuchi)

aita kuchi ga fusagaranai ('(one's) open mouth won't close') [SURPRISE]

kuchi hiki taru ('lower (one's) mouth(-corners)') [REGRET]

kuchi o togaraseru ('sharpen / pout (one's) mouth) [DISSATISFACTION]

kuchibiru ga hokorobiru ('(one's) lips come unsewn') [LAUGH]

TONGUE (shita)

shita o maku ('roll up (one's) tongue') [SURPRISE]

shita o dasu ('stick out (one's) tongue') [DISDAIN, SHAME]

JAW / CHIN (ago)

ago ga hazureru ('(one's) jaw is dislocated') [LAUGH]

ago ga kuichigau ('(one's) jaw misfits') [DISCOURAGEMENT]

ago o tareru ('let (one's) jaw down') [SUFFERING]

ago o naderu ('stroke (one's) chin') [SATISFACTION, PRIDE]

NECK (kubi)

kubi o chijimeru (literally, 'shorten (one's) neck', 'duck (one's) head') [FEAR]

kubi o nagakusuru ('lengthen (one's) neck') [EXPECTATION]

THROAT (nodo)

nodo yori ken o haku ('spit out a sword from the throat') [PAIN]

nodo yori chi o haku omoi ('blood-spitting thought') [PAIN]

BLOOD (chi)

chi ga waku ('the blood boils up') [COURAGE]

BREATH (iki)

iki o nomu (literally, 'swallow (one's) breath') [SURPRISE]

iki o tsuku ('take a breath') [RELIEF]

SHOULDER (kata)

kata de warau (literally, 'laugh with the shoulders') [DISDAIN, LAUGH]

kata-hiji o haru (literally, 'stick out (one's) shoulders-elbows') [BOASTFULNESS]

kata o furuwasu ('shake (one's) shoulders') [GRIEF]

kata o otosu ('lower (one's) shoulders') [DISAPPOINTMENT]

BREAST / CHEST (mune)

mune ga odoru ('(one's) chest dances / leaps') [DELIGHT]

mune ga suku ('(one's) chest empties') [DELIGHT]

mune ga ippaininaru ('(one's) chest becomes full') [DELIGHT, GRIEF]

mune o shimetsukeru ('(something) fastens (one's) chest') [GRIEF]

mune ga fusagaru ('(one's) chest becomes filled up') [GRIEF, DELIGHT]

mune saku ('chest-splitting') [GRIEF, REGRET]

mune tsubureru ('chest-crushing') [SURPRISE, GRIEF]

mune o kogasu ('burn (one's) chest') [LOVE]

mune ga sawagu ('(one's) chest is disturbed') [ANXIETY]

mune o hiyasu ('cool (one's) chest') [ANXIETY]

mune ga ita no yo ni naru ('(one's) chest turns into something like a board') [ANXIETY]

mune o nadeorosu ('stroke (one's) chest') [RELIEF]

mune o haru (literally, 'stick out (one's) chest') [PRIDE]

BELLY / STOMACH (hara)

hara no kawa o yojiru (literally, 'twist the skin of (one's) belly') [LAUGH]

hara ga tatsu (literally, '(one's) belly stands up') [ANGER]

hara ni suekaneru ('can't settle (something) in (one's) belly') [ANGER]

hara no mushi ga osamaranai ('the belly-worm can't keep itself quiet') [ANGER]

harawata ga niekurikaeru ('(one's) bowels boil up') [ANGER, RAGE]

dancho no omoi ('bowels-rending thought') [GRIEF]

hozo o kamu (''bite (one's) navel') [REGRET]

LIVER (kimo)

kimo o iru ('roast the liver') [ANGER]

kimo o nuku ('pull out the liver') [SURPRISE]
kimo o tsubusu ('crush the liver') [SURPRISE]
kimo o hiyasu ('cool the liver') [SURPRISE]
kimo o kudaku ('break the liver') [PAIN]
kimo o kesu ('vanish the liver') [ANXIETY]

WAIST (koshi)

koshi o nukasu (literally, 'pull out (one's) waist') [SURPRISE]
koshi o yoru ('twist (one's) waist') [LAUGH]

BUTTOCKS (shiri)

shiri ni hasamu ('put in the buttocks') [DISDAIN]

MI (body)

mi no ke mo yodatsu ('even the body-hair sticks up') [FEAR]

HAND / ARM (te)

te ga wananaku ('the hands tremble') [GRIEF]
te ni ase o nigiru (literally, 'clasp sweat in the hands') [ANXIETY]
te no mai ashi no fumu tokoro o shirazu (literally, 'the dancing hands don't know the places to step on') [DELIGHT]
te mo ashi mo denai (literally, 'neither the hands nor the legs can't reach out') [BEING DISTURBED]

FOOT / LEG (ashi)

ashi o sora ni suru ('turn the legs towards the sky') [SURPRISE]
ashi ga chi ni tsukanai ('(one's) feet won't reach the ground') [SURPRISE]

Notes

1. Cf. Wada (1969) referred to in section 4.1 (i).
2. Cf. Miyaji (1975:25), referring to an unnamed author. The idea will presumably be that people with nomadic culture will have far more opportunities to observe and deal with animal internal organs than agricultural people. Incidentally, only two names of the body-internal organs, *heart* and *liver* are included in the 100 word basic lexicon for glottochronology, first proposed by Morris Swadesh in the middle of the twentieth century (cf. Wang (1994: 1450)), side by side with

such body-part names as *belly, blood, breast, ear, eye, foot, hair, hand, head, heart, neck, nose*, and *tooth.*

3. The image, incidentally, is the same as the one found in the mythological account of the creation of the islands that constituted the country.
4. The word for the brain, *nou*, has remained a technical word to this day. Phrases and idiomatic expressions referring to the workings of the brain more commonly contain the word for the head, *atama*, rather than that for the brain, *nou*: e.g. 'atama ga yoi/warui' (literally, 'the head is good/bad, to have a good/bad head', meaning 'to be clear-headed/dull-headed'), 'atama no kaiten ga hayai' (literally, 'to be quick in the turn of the head', meaning 'to be (intellectually) smart/bright'). In modern Japanese, the verb *kangaeru* ('think') collocates with atama ('head'), while the verb *kanjiru* ('feel') collocates with *kokoro* ('heart'), thus *atama de kangaeru* (literally, 'think by the head') and *kokoro de kanjiru* ('feel by heart').
5. The image is similar to the one in Wordsworth's well-known lines, "My heart leaps up when I behold / A rainbow in the sky."
6. The word, *kimo*, can also alternate with the word for the heart, *kokoro*, in the phrase, 'kimo/kokoro ni meijiru' (literally, 'to impress on one's liver/heart', meaning 'to take to the heart).
7. The description given here is based on a number of academically acknowledged dictionaries of the Japanese language, including *Nihon Kokugo Daijiten*, 2nd edition, 13 volumes (2000–2002).
8. The definition of this prototypical sense of the word *kokoro* in a more popularly oriented dictionary will go as follows: '(something which is presumed to control) human mental and affective activities, involving intellect, knowledge, judgment, feeling, emotion, volition, etc.' The Japanese conceptualization in this respect is thus essentially the same as the Chinese one (cf. Yu 2003).
9. Logically, sense [III] could very well be derived from [I] through [IV] as well as from [I] through [II]. The reason why the latter, rather than the former, applies in Japanese is simply historical: for the word *kokoro*, sense [IV] has not been a common one, the same sense being more commonly encoded by the Chinese-borrowed term *shin*.
10. It has been suggested by an anonymous reviewer that the phrase is well motivated as it stands if it is associated with a situation in which "one hold one's head high, i.e. with one's face raised toward the sky." This is a possible interpretation. I have to add, however, that this is not exactly what modern Japanese speakers will think of. They associate the tall nose with the face of the fictional demonic human-like creature called *tengu*, physically big and strong in martial art, whose feature is a long protruding nose (which symbolizes the erected male sexual organ).
11. An anonymous reviewer's comment is that "'one's heart bounces' could be metonymic and indexical … [w]hen one is happy and excited, one's heart may

beat faster and harder ... the change in the heart beating can be heard by oneself." This is true and in fact the same point was mentioned when I discussed the ready substitution of the word *mune* for *kokoro* in section 2. But in Japanese the verb *odoru* may mean 'to dance' as well as 'to bounce' and the Japanese speaker's first response to the phrase, *kokoro ga odoru* will be to interprete it metaphorically, comparing it to a phrase like *watashi ga odoru* ('I dance.').

12. Thus the interpretation for the sentence *mune ga odoru* ('(one's) breast/chest bounces') is of the same semantic type (THE CONTAINER FOR WHAT IS CONTAINED) as the one given to an English sentence like *The kettle is boiling*. My impression, however, is that an interpretation in terms of such a 'looser' semantic relationship between the subject and the predicate is tolerated for a much wider range of cases in a topic-prominent language like Japanese than in a subject-prominent language like English (where the subject is prototypically associated with the semantic role of the agent).

    Kusumi (1996: 39) offers a chart showing how various feelings/emotions in Japanese are associated with various body-parts, both internal and external. In this chart, however, *kokoro* is not taken into account, presumably because the word is largely supplanted by the Chinese-borrowed word *shinzo* in present-day Japanese as a term referring to the body-internal organ.
13. What I have in mind here is the use of the word as in the phrase, *far and wide* (i.e. 'opening or spreading as much as possible'). This was apparently the archetypal image-schema associated with the word *wide*. The word "[n]ow ... tends to be restricted to applications in which actual mensuration from point to point is possible or contemplated ...." (*OED* ***wide*** a. II. 5. a.) For further information, see *OED* ***wide*** *a.* and ***wide*** *adv*. The earlier 'unbounded' sense seems to be much better preserved in German than in English.
14. To see a further possibility of reversal between 'agent' and 'locus', refer to the quotation from Hartmann (1954) given in Appendix (3). Hartmann points out in effect that the Indoeuropean notion of the agent ('Täter') performing an act corresponds to the Japanese notion of the locus ('Ort') where something takes place. It will be an interesting question to see whether all this had anything to do with the fact that Japanese nouns are not morphologically marked for the singular/plural contrast.
15. All this also indicates how far the meaning of the word *kokoro* has shifted metonymically from its original use of referring to the body-internal organ to the functions it is supposed to perform (cf. section 3). The Chinese-borrowed word for the heart as a body-internal organ, *shin* can combine with the size adjective *sho* ('small'), producing a compound, *shoshin* ('cowardliness'); the word *kokoro* cannot idiomatically collocate with the size adjective *chiisai* ('small').
16. 'BECOME-language' refers to the type of language whose speaker prefers engaging with a situation as a whole and construing it in terms of changing

from one phase to another (i.e. in terms of 'BECOMING'); 'DO-language' refers to the type of language whose speaker prefers focusing on some particular entity (prototypically, person) involved in the situation and construing it in terms of the focused entity (prototypically, person) performing an act (i.e. in terms of 'DOING'). Japanese is typically a BECOME-language, in contrast to English, which is eminently a DO-language.

References

Bally, Charles

1920 Impressionisme et grammaire. In *Mélanges d'histoire littéraire et de philologie offerts à M. Bernard Bouvier*, reprinted (1972), Genève: Slatkine Reprints.

Hartmann, Peter

1954 *Einige Grundzüge des japanischen Sprachbaues*, Heidelberg: Carl Winter.

Hoshino, Akira

1976 Shintaigoi ni yoru hyogen (Exressions in terms of body-part names). In *Nihongo no Goi to Hyougen* (Japanese Lexicon and Expressions), Takao Suzuki (ed.), 153–181. Tokyo: Taishukan.

Ikegami, Yoshihiko

1981 *SURU to NARU no Gengogaku: Gengo to Bunka no Taiporozi eno Shiron* (Linguistics of DOING and BECOMING: An Essay Towards a Typology of Language and Culture). Tokyo: Taishukan. (German translation (2007), Berlin: LIT Verlag)

1991 DO-language and BECOME-language: Two contrasting types of linguistic representation. In *The Empire of Signs: Semiotic Essays on Japanese Culture,* Yoshihiko Ikegami (ed.), 285-326. Amsterdam: John Benjamins.

1998 <MONO> to <TOKORO>: Sono tairitsu to hanten ('Entity' and 'Location': Their Contrast and Reversal). In *Tokyo Daigaku Kokugo Kenkyushitsu Soritsu Hyakushunen Kinen Kokugo Kenkyu Ronshu* (Papers in Japanese Linguistics in Commemoration of the 100th Anniversary of the Japanese Department of the University of Tokyo), 864–888. Tokyo: Kyukoshoin. (Also incorporated as Part Two of Yoshihiko Ikegami (2000) *Nihongoron eno Shotai* (Invitation to Japanese Linguistics), Tokyo: Kodansha and (2007) Nihongo to Nihongoron (Japanese and Japanese linguistics). Tokyo: Chikuma Shobo.

2003 How language is conceptualized and metaphorized in Japanese – An essay in linguistic ideology. In *Motivation in Language: Studies in*

Honor of Günter Radden, Herbert Cuyckens et al. (eds.), 259–271. Amsterdam: John Benjamins.

Imosaka, Naoyuki
1999 *Kansei no Kotoba o Kenkyusuru* (Studying the Language of Sensibilities). Tokyo: Shinyosha.

Kusumi, Takashi
1996 Kanjo-gainen to ninchi-moderu no kozo (Concepts of emotion and the structure of the cognitive model). In *Kanjo to Kodo/Ninchi/Seiri* (Emotions and Behavior/Cognition/Physiology), Shoji Tsuchida and Kazuhisa Takemura (eds.), 30–54. Tokyo: Seishin Shobo.

Matsuki, Keiko
1995 Metaphors of anger in Japanese. In *Language and the Cognitive Construal of the World,* John R. Taylor and Robert E. MacLaury (eds.), 137–151. Berlin: Mouton de Gruyter.

Miyaji, Atsuko
1979 Kango no teichaku – "Kokoro", "shin no zo", "shinzo" (Entrenchment of Chinese-derived words – *Kokoro, shin no zo, shinzo*). Ch.4 of the same author's *Shin-Shin-Goi no Shiteki Kenkyu* (A Historical Study of the Body/Mind Vocabulary), 80–100. Tokyo: Meiji Shoin.

Nakamura, Hajime
1997 [1965] *Ways of Thinking of Eastern People: India, China, Tibet, Japan.* London: Kegan Paul International.

Wada, Takashi
1969 Karada kotoba ko (An essay on expressions involving names of body parts). *Shiso no Kagaku* 94, 82–109.

Wang, W. S.-Y.
1994 Glottochronology, lexicostatistics, and other numerical methods. In *The Encyclopedia of Language and Linguistics*, Vol. 3, 1445–1450. Oxford: Pergamon Press.

Yamaguchi, Masao
1986 *Bunkajinruigaku no Shikaku* (From the Viewpoint of Cultural Anthropology). Tokyo: Iwanami.

Yu, Ning
2003 Chinese metaphor of thinking. *Cognitive Linguistics* 14: 141–165.

How to have a HEART in Japanese

Debra J. Occhi

Abstract

This paper investigates the Japanese cultural model comparable to that associated with "heart" in English from a cultural linguistic perspective. Thus, it investigates a range of naturalistic data from phonetic to discourse levels. The Japanese cultural model correlating to that of English HEART has as its conceptual core four aspects of HEART expressed in four terms. These terms are: the physical organ for heart (shinzô); the loanword *haato*, from English HEART, which is used with increasing frequency along with the symbol ♡; the spatial region mune 'chest'; and perhaps most interesting of all, the abstract term *kokoro*. Following the onomasiological survey of these expressions is a discussion of onomatopoeic *gitaigo* and sound-symbolic *giongo* terms associated with the heart, along with the physiological and psychological states they describe. As in English, reference to the heartbeat in Japanese metonymically indexes emotional experience. HEART as *kokoro* can also exhibit various levels of grammatical agency indicating an imaginal separation from the human body. Next, the paper examines references to the Japanese HEART as "the seat of the mind" and its imagined dimensions. Following this is an investigation of how gendered, figurative and poetic HEART expressions are taken up in a genre of popular song dubbed "the heart of the Japanese" known as *enka*. Finally, another look at HEART ideology is provided through the example of a recent children's textbook designed for use in morals classes to foster patriotism in Japanese youth.

Keywords: Japanese, cultural model, metaphor, sound symbolism, ideology, gender.

1. Introduction

This paper analyzes the contemporary Japanese concomitants of English HEART from a cultural linguistic perspective.[1] The data consist of naturally occurring expressions collected by the author since 1995 or cited from literature. The Japanese language is full of HEART expressions ranging from the onomotopoeic to the ideological. These HEART expressions reflect a *cultural model*, which is an observable component of human cognitive behavior defined thus: "Cultural models are presupposed, taken-for-granted models of the world that are widely shared (although not necessarily to the exclusion of other, alternative models) by the members of a society and that play an enormous role in their understanding of the world and their behavior in it" (Quinn & Holland 1987: 4).

Cultural models are composed of schemas (mental representations). Palmer (1996) stresses the importance of investigating linguistic components of cultural models starting at the level of the phoneme. His work, dubbed *cultural linguistics,* is innovative for using concepts of cognitive linguistics to reanalyze various linguistic/cultural phenomena commonly studied in linguistic anthropology. He concludes: "It is likely that all native knowledge of language and culture belongs to cultural schemas and that the living of culture and the speaking of language consist of schemas in action" (Palmer 1996: 63).

Cultural models are integral to human understanding and motivation. The Japanese cultural model correlating to that of English HEART has as its conceptual core four aspects of HEART expressed in four terms. These terms are: the physical organ for heart (*shinzô*); the loanword *haato*, from English HEART, which is used with increasing frequency; the spatial region *mune* 'chest;' and perhaps most interesting of all, the abstract term *kokoro.* Ikegami (this volume) presents etymologies for the key terms discussed here; this paper will focus on their use in contemporary expressions. Usage of these terms overlaps to some extent with that of other terms for body parts, such as *hara* 'stomach' (Matsuki 1995) as well as with the ethereal element attributed to living beings known as *ki* (McVeigh 1996). Onomatopoeiac and mimetic sound-symbolic terms (in Japanese, *giongo/ gitaigo*) also have roles to play in the Japanese cultural model.

This paper uses the cultural linguistic perspective (Palmer 1996) to investigate phenomena comparable to the conceptualization of humour/ faculties associated with English HEART. Data show that the Japanese cultural model also has components typically associated with MIND in English. However, for ease of comparison with other papers in the volume and with the paper that inspired it (Niemeier 1997) I will call the model "the Japanese HEART model." Cultural linguistics is useful in elucidating cultural models because of its roots in both cognitive and anthropological traditions as well as its scope from phonetic to discursive aspects of language (Palmer 1996). Defined by Palmer as "a synthesis of cognitive linguistics with Boasian linguistics, ethnosemantics, and the ethnography of speaking", cultural linguistics seeks to "embrace the goals of grasping the native point of view and of studying language use in its social and cultural context" (Palmer 1996: 35).

This paper follows the cultural linguistic tradition by incorporating natural instances of language use ranging from phonetics to discourse. These are discussed in contexts of occurrence in a search for a cultural model corresponding to and in comparison to English HEART.[2] Data are taken from popular dramas, music, and advertising, representing the 'culture industries' (McCreery 2000: 30) that are currently among Japan's most successful exports. Official sources of discourse also provide worthwhile data. Particularly useful among these is NHK (*Nippon Hôsô Kyoku*), the public broadcasting service of Japan, the national education agency *Monbukagakusho*, and other public agencies. Data from these official sources will show how the cultural model is used to explain and support state-supported ideology.

2. Onomasiological survey of Japanese HEART expressions

2.1. Four names for HEART in Japanese

Comparison to English HEART shows that Japanese HEART expressions have divided up the various domains of HEART into four major areas: *shinzô* (the physical organ known as the heart), *mune* (lit. 'chest'), *haato* (from English 'heart') and *kokoro* (abstractly, the human heart, human psychology; a mentality, the psyche; the mind; the emotions; a person's feelings [emotions]). These Japanese words thus share part of English HEART's semantic field.

In addition, Japanese also uses the terms *hara* (stomach) as well as a variety of body organs (esp. *kimo*, 'liver') and physical forces (such as *ki* 'spirit, mind, heart,' from Chinese *qi*) in the conceptualization and verbalization of emotional experience (Matsuki 1995, McVeigh 1996). In dictionaries, these terms and certain variations of them (e.g., *kimochi* 'feeling' lit. 'carrying *ki*') frequently appear in translation of each other, lending evidence for their status as nuanced participants in a rich lexical network describing emotion. These lexemes do not separate as neatly into categories of thought versus feeling as do their English concomitants (see Niemeier 1997).

Expressions involving Japanese HEART terms show similarities and differences between Japanese and English regarding metaphorical and sound-symbolic construals of physiological events and sensations associated with emotion. In referring to emotional expression, Japanese prefers the relatively abstract *kokoro* to the physiological *shinzô*, whose use is oriented towards medical discourse. Only a few metaphorical expressions stem from *shinzô*, such as *shinzô ga tuyoi* 'strong heart; gutsy, have nerve,' *shinzô ga yowai* 'weak heart; timid, shy,' *shinzô ni ke ga haite iru* 'have hair on one's heart; brazen, cocky'[3] (McVeigh 1996: 37). Native speakers report that these sayings are rather impolite and are thus seldom used. *Haato*, also written with the symbol ♡, is an English loan used to represent the heart shape in texts or the suit in playing cards. It is becoming more common and is taking on some of the semantic roles associated with the Japanese HEART cultural model including several innovative uses.

2.2. *Kokoro*

Kokoro is an indigenous Japanese term, written often in native *hiragana* script こころ or with the same Chinese character 心 pronounced *shin*, as in *shinzô* 'heart organ' 心臓 and many other complex lexemes that were also originally borrowed from Chinese. The various domains of experience associated with the etymology of *kokoro* are presented in Ikegami, this volume.

Though this chapter focuses on onomasiological rather than semasiological aspects of language (viz. Dirven & Verspoor 2004), dictionary definitions of *kokoro* are presented here to assist the non-Japanese speaker in understanding the mental and emotional faculties denoted by *kokoro* relative to those of HEART. The three main definitions of *kokoro* translated from the Japanese-Japanese *Kôjien* dictionary include:

1) 'a common name for a human organ, also carrying the meaning of spirit:' a) 'in opposition to the body [*karada*], the integral place of will, feeling, and knowledge; b) thoughtfulness, speculation; c) feeling [*kimochi*], [*kokoromochi*]; d) consideration, sympathy; e) the sensibility that understands mood; f) hope; g) special thoughts, feelings [*kokoromochi*] of betrayal or badness.'
2) '(metaphorically) a) flavor, effect, spice, look, appearance; b) circumstance, condition, matter, situation, affair; c) style, device, artifice, idea; d) meaning; e) reason, grounds for understanding a riddle; f) (specialist term from poetry review) contents, a poem's central theme, matter, idea.'
3) a)'heart [*shinzô*], chest [*mune*]; b) the center of a thing' (1995: 950–1).

In the Japanese-English dictionary published by *Kenkyûsha*, we find the following definitions for *kokoro* in context:

1) 人間の〜'human's heart:' the human heart; human psychology; a mentality; the psyche; the mind; the emotions; a person's feelings [emotions]; the way people feel; (literary) the mind [heart, mentality] of man/ 子供の〜'child's heart' a child's feelings [heart, emotions]; the way a child feels; the emotions [feelings] of a child; 子供のような〜'heart like a child' childlike feelings; the emotions of a child; innocence/ 庶民の〜 the feelings of ordinary people; ordinary people's psychology/ 日本人の〜 the Japanese psyche [mentality]; Japanese people's way of thinking; Japanese feelings [sentiment(s)]/ 伝統的な日本の〜the traditional mentality of the Japanese (people); traditional Japanese sentiment(s); the way the Japanese traditionally think (2003: 966).

It is clear through examining these definitions that in addition to its function as the locus of emotion, *kokoro* functions as the locus of experience that could (in English) be considered mental, e.g., as in references to 'will, knowledge, thoughtfulness' listed above. The scope of experience describable by reference to *kokoro* is also evident in the large number of morphologically complex expressions including it. For example, 64 expressions of the form [*kokoro* + particle + verb] and 351 direct collocations appeared in the monolingual Japanese dictionary *Kôjien* (1995: 950–956).

Links between *kokoro* and the chest region as well as the construal of *kokoro* as the locus of understanding are evident in JSL (Japanese Sign Language). The sign for *kokoro* is a circle made by the thumb and right index finger over the left side of the chest, where the heart organ would be located (Tani 2005: 165). Connected to the knowledge-related sense of *kokoro*, the verb *wakaru*[4] 'understand' is signed by passing the right hand downward over the chest (Tani 2005: 445). Its negative form, *wakaranai,* is signed by curling the right hand and stroking from the chest towards the right shoulder, away from the heart region, as though knowledge has not entered there.[5] These two signs are also used to interpret related terms for 'understanding/not understanding' such as *shiru/shiranai* and *rikai suru/rikai shinai* (Tani 2005: 445).

2.3 *Haato*

The English loanword *haato* is defined first as *shinzô* and second as *kokoro*, though its uses seem more aligned – and predictably so – towards *kokoro*. It is used in referring to the suit in Western playing cards known as 'hearts.' Its symbol ♡ has also been used in the casual letters young women share both as a topic boundary marker co-occurring with utterance-final particles such as *yo* 'I'm telling you' and *ne* 'don't you agree?' and is described as an emotive pictorial symbol (EPS) lending semantic nuance (Kataoka 2002: 232). The ♡ symbol is preprogrammed in the symbol sets used when sending text mail via cellular phones.

Haato, as well as other EPSs in Kataoka's data,[6] indexes a context of casualness and orality in epistolary writing, creating a "(quasi-) face-to-face interaction" (Kataoka 2003:12). His diachronic research shows that the increase in use of EPSs (particularly *haato* and teardrop/sweat shapes) has led to their semantic and pragmatic dilution, resulting in a near-grammaticalization over the past twenty-five years (Kataoka 2003: 18). Indeed, in the thank-you notes I received for a speech delivered to Japanese high schoolers on 16 September 2005, EPSs appeared in four letters, all written by girls. Among these, two employed ♡ in the following contexts:

(1)	*kyoo*	*wa*	*sensei*	*no*	*hanashi*	*ga*	*kikete*
	today	TOP	teacher	GEN	talk	SUBJ	hear-POT-CONT
	yokatta		*shi*	*tanoshikatta*		*desu*	♡
	good-PAST		and	fun-PAST		COP	EPS

'Today [I] was glad to be able to hear the teacher's [i.e. your] talk, and it was fun ♡'

(2) *Your speech is very interesting* ♡
I will go to Australia ♡
I am looking forward to visiting Australia in school excursion ♡
Thank you for speeching to us ♡♡♡

While the use of ♡ in the last lines of these letters may indicate affection towards the addressee, (2) clearly demonstrates the bleaching phenomena Kataoka describes as akin to grammaticalization.

Similarly, *haato* as a lexeme has enjoyed a recent broadening of usage from its original context of romantic love to that of social welfare. For example, a *haatofuru raibu* 'heartful live' music event is a charity concert, and the NHK *haatoten* 'heart exhibit' is a yearly showing of poetry written by persons with disabilities, conducted since 2002. Furthermore, the sticker currently designating automobile drivers with disabilities depicts a white four-leaf clover with heart-shaped leaves.

Haatobiru Hô 'heart building law,' is the popular nickname for the law concerning barrier-free buildings for persons of age or with disability (officially *kôreisha, shintaishôgaisha tô ga enkatsu ni riyô dekiru tokutei kenchikubutsu no kenchiku ni kansuru hôritsu,* 'Act on Buildings Accessible and Usable by the Elderly and Physically Disabled.' The nickname *haatobiru hô* 'heart buildings law' comes from the slogan under which this law was introduced: *haato no aru biru wo tsukurô* 'let's make buildings with heart,' where "buildings with heart" means buildings that are friendly (accessible) to people of all shapes, sizes and abilities. Buildings that meet the requirements of this law display a heart-shaped sign (Hanlon 2002). The use of heart-shaped signs as indicators of barrier-free buildings will undoubtedly reinforce association of the heart symbol with social welfare.

2.4 *Mune*

Mune 'chest' has three metonymically related major categories of meaning in the Japanese-English dictionary: 1) the breast; the chest, the thorax, the plastron (of a shirt); 2) [female] breasts; 3) the heart. Forty-four *mune* expressions[7] follow the third definition, with translations in which the English term 'heart' appears thirty-nine times, showing the close association of Japanese *mune* with English HEART. For example, the title of Elvis Presley's song "Whole Lotta Love" in Japanese is *Mune ippai no ai o* 'Heart [lit. 'chest'] Full of Love' (2003: 2534–5).

Though the abstract *kokoro* is the central term in the Japanese HEART cultural model, the physically based container *mune* 'chest' is also clearly and metonymically implicated, much more so than the technical *shinzô* 'heart organ.' This compares interestingly with Korean, as described in Yoon (this volume). The extent of *mune*'s involvement will emerge more clearly in the discussions below, which will describe sound-symbolic expressions associated with HEART.

3. Sounds of the HEART

Japanese has a vast array of onomatopoeic and mimetic[8] sound-symbolic resources known as *giongo*, categorized broadly (and fuzzily) into *giseigo* 'onomatopoeia' and *gitaigo* 'mimetics.' They are often associated with notions of naturalness and intranslatability (Asano 1978, Gomi 1989, Hasa-

Hasada 2001). Hamano's extensive study (1998) details the phonetic and sound-symbolic properties of this word class. She argues (as does Shibatani 1990) that sound symbolism pervades the entire language[9] and states that the extensiveness of *giongo* (often used adverbially) dovetails with the relative paucity and lack of differentiation of verbs in Japanese (1998: 1). *Gitaigo* 'mimetics,' which do not represent sound directly, include in their meaning contextualization in the sense of social scenario, described as "intuitive remembrance of an image" (Soga 1991). In Kita's words: "Japanese mimetics have a unique psychological effect. They evoke vivid 'images' of an experience, full of affect. This imagery is not only visual but can also be based on other perceptual modalities and physiological states. The meaning is felt, by native speakers, to be direct and real, as if one is at the scene" (Kita 1997: 386).

Kindaichi considers the entire subset of *gijyôgo* 'emotional *gitaigo'* as 'things that express the condition of the human heart' *ningen no kokoro no jyôtai wo arawasu yô na mono* (1974: 8). For reasons of space, however, this paper will only present the terms that pertain most specifically to the physical heart in its organ sense and its sensations. The most prevalent among these represent the sound of the heart itself: *dokin, dokiQ, doki doki*. These variations represent subtleties of a single loud thump of the heart (nasal stop), the heart skipping a beat (geminate stop), or a (reduplicative) continuative pounding of the heart. Other variants such as *dokin dokin* or *doQkin doQkin* are attested by literary example (Kakehi et al. 1996: 267–8). A popular children's song lyric shows possibilities of use for these terms:

(3)	*Doki doki*	*saseru*	*yo,*	*Dokin-chan*
	Heartbeat	cause	EMPH	heartbeat-DIM
	'Little Heartthrob makes hearts beat' (Izumi n.d.a.)			

This description of the bright red, selfish girlfriend of the 'bad guy' in the popular preschooler's Fuji TV animated cartoon *An-Pan Man* shows two natural instances of use. From this we learn that, as in English, quickened or noticeable palpitation can be brought on by the presence of another (e.g., *you make my heart pound*). *Dokin* 'one thump of the heart' is used with the diminutive naming suffix *chan* to form a nickname similar to "Little Heartthrob." Though *doki doki* and its variants represent the sound of the heart beating, emotional resonance is also implied, as the following examples will also show.

Because of their ability to evoke emotional resonances with the hearer's experience (Kita 1997) as well as with specific social scenarios (Occhi 1999), *giongo* and *gitaigo* are commonly used in advertisements. Another instance of induced *doki doki* palpitations was found in an advertisement for the eastern segment of Japan's railway system. A cartoon featuring two young women on a train introduced the *doki doki Tokyo* 'Heartpounding Tokyo' Japan Rail ad campaign of 1996. It appears that the use of *doki doki* here may not only indicate the pulse of the big city, but also the palpable excitement of visitors to Tokyo from the relatively peripheral Tohoku region in which the advertisement appeared.

Specific research on the grammar of reduplicative mimetics in various subject positions indicates that their strong ties to social scenarios override the possibility of a uniform usage rule. In comparison, Japanese has person restriction on use of emotional adjectives. That is, one cannot attribute adjectives describing an emotional state to non-first person subjects without evidential modification (Kuroda 1979, Teramura 1982, Higashi 1992). However, these rules do not apply as uniformly to *gitaigo* due to semantic differences between terms within the word class.[10] *Doki doki* enjoys a high degree of salience among *gitaigo*. Respondents to the open-ended questionnaire prompt used in Occhi (1999) named *doki doki* most frequently (45 times, N=75). This is especially interesting since the questionnaire focused strictly on *gitaigo* of emotion (described as *kanjô o arawasu gitaigo*) rather than onomatopoeia *per se*. Clearly *doki doki* was considered emotional by these respondents. Perhaps since it describes an internal physiological phenomenon, data also indicated that *doki doki* seems to require person restriction by more speakers than does the related term *waku waku* 'generalized excitement'. When asked to create sentences using the terms listed, respondents often used *doki doki* in two frames: for men, *joseitachi no mae de* 'in front of a woman' or more generally, *tesuto no mae ni* 'before a test.' These findings indicate that, as in English, reference to the heartbeat metonymically indexes emotional experience in Japanese as well (Occhi 1999).

Another *gitaigo* of the heart is *kyun*, which prototypically refers to the feeling in the heart of a young person, especially a girl, when falling in love for the first time. The vowel may be lengthened for emphasis, as when used in *manga* cartoons. A variant of *kyun* which collocates the sensation and its locus is *munekyun*, 'chest *kyun*.' When asked to describe *kyun*, a respondent held her hand before her chest and made a slow squeezing gesture. A dictionary defines *kyun* as 'the state of feeling emotional pain;' the accompanying sample sentence suggests love emotions in particular: *Okaasan mo anata to onaji yô ni gakusei jidai ni wa suki na hito ga ite ne, sono hito no koto o omou tabi ni mune ga kyun to natta mono yo* 'Your mother [i.e. I] also had someone she liked when she was in school, just like you, and every time I thought of him, I used to feel a pain in my heart' (Kakehi et. al. 1996: 738). Such memories are typically associated with the synaesthetic perceptual memory of 'sweet-sour,' said to be 'the taste of first love' for Japanese *amazuppai hatsukoi no aji,*[11] *vis-a-vis* English 'the bitter-sweet taste of first love').

There are other sound-symbolic resources associated with the heart that indicate emotional states. The *gitaigo*-like compound verb *zawatsuku* 'be noisy' is sound-symbolically related to the *gitaigo* term *zawa zawa* 'rustling' as well as to the verb *sawagu* 'be noisy.' This verb, used with *mune,* connotes a feeling of unease about a situation or a premonition of trouble, similar to the *funny feeling* that for English speakers is usually located in the pit of one's stomach and is not an expression of humor. *Zawatsuku* was used in a voiceover at the end of a 2003 NHK TV drama episode to describe the condition of the anxious titular heroine:

(4)	*Wakaba*	*wa*	*zawatsuku*	*mune*	*deshita*
	Wakaba	TOP	noise-attach	chest	COP-PAST
	'Wakaba had a funny feeling' [about what had happened]				

Not surprisingly, *gitaigo* of HEART can also represent states of romantic disappointment referred to in English as 'broken-hearted'. *Zuki zuki* 'throbbing pain' can be used in reference to the heartbeat. *Zuta zuta* describes even deeper heartbreak than that of *zuki zuki*. As the more intense term, *zuta zuta* can be also used to describe torn or ragged objects, but it differs in meaning from the English expression *my heart is torn* 'I have a difficult choice.'[12] Both describe *kokoro* (Gomi 1989: 85, 89).

Some HEART-related terms which are not strictly speaking part of the *gitaigo* category but are fuzzy members through sound symbolism may be used with either *kokoro* or *mune*. For example, *tokimeki* is the nominalized form of the verb *tokimeku* 'palpitate, beat fast.' It is considered similar enough to *doki doki* to appear with it in the same line item in a magazine reader poll[13] (Miyamoto 2005: 24). Like *kyun, tokimeki* also holds associations with young love. In use it is combined with the subject *mune*, as in the following: *Sakamichi no hodô de kanojo o machinagara, Fuyuki wa shônen no yô na mune no tokimeki o oboeta* 'Waiting for her on the sidewalk of the hilly road, Fuyuki remembered his heart's pounding, as when he was a young man' (Itsuki, in Nakamura 1979: 359).

As in the *doki doki* example above, romantic anticipation is expressed through reference to awareness of the heart beating. Though these expressions denote physiological states, there are also many HEART expressions in Japanese that denote psychological states without concomitant physical experiences.

4. Actions of the HEART

HEART in Japanese (as in English) is capable of actions and reactions exhibiting various levels of agency. As we see in (5), someone could be pulled into one's heart, or one's heart could be pulled towards another.

(5)	*kokoro*	*ni*	*hikareru*
	heart	DAT	pull-PASS

someone is pulled into the heart
the heart is pulled to someone
'romantic attraction'

In the NHK drama *Wakaba* already cited above, the young husband Masaya pledges that, though in the case of an earthquake he may have to physically leave his wife's side to check on the buildings he has designed, they will remain united:

(6)	*kokoro*	*wa*	*kanojo*	*kara*	*hanashimasen*
	heart	TOP	she	from	separate-POL-NEG

'[my] heart will not separate from her' (2005).

As shown in (5) and (6), *kokoro* expressions can indicate its imaginal separation from the physical body. *Kokoro* can also make decisions and harbor attitudes, framed in (7) as utterances by quotation marks:

(7)	*"X*	*ga*	*shitai!"*	*yori*	*mo*	*"Y*	*o*	*naosu"*	*no*
	'X	SUBJ	do-want'	than	more	'Y	OBJ	fix'	GEN
	kokoro								
	heart								

'[the] heart that [says] "[I'll] fix this!" rather than "[I] want to do that!"'
(Flipper 2002:44).

5. The HEART as the seat of the mind

Kokoro, as mentioned above, is the most abstract of the Japanese HEART lexemes, often translated as 'spirit' or 'mind' as well. This section shows *kokoro* in relationship to other indigenous terms in order to more fully differentiate the Japanese HEART cultural model.

It appears that though the *kokoro* HEART is a part of the body and can metonymically represent the whole, it can be contrasted to the body in a subtle way, as examples above have shown. Kuwayama (2004) discusses Japanese *wakaru* 'understand' as "a full-bodied experience" of Japanese understanding[14] relative to the Cartesian dualism (Damasio 1994) inherent in much Western thinking (and immortalized in phrases such as 'mind over body'). For Kuwayama, and seemingly even more so for the folklorist Yanagita Kunio, understanding relies in part on *kokoro* (2004: 82). Kuwayama quotes Lebra's contrast of *kokoro* as 'the [em]bodied mind' with *karada* (body) 'the minded body' (2004: 153). These indicate substantial conceptual overlap. Yet phrases in daily use such as *kokoro mo karada mo* 'heart as well as body' indicate some differentiation between the two, shown in reverse order in (8):

(8)	*karada*	*mo*	*kokoro*	*mo*	*motto*	*uruô*
	body	and	heart	and	more	moisten-HORT

'let's moisten the heart as well as the body more'
(2004 Coca-Cola ad)

In recent popular health literature, *kokoro* was described as being prone to *kôri* 'stiffening, freezing.' This ailment is commonly attributed to stress and usually attacks the shoulders of Japanese as *katakôri* 'stiff shoulders.' The ailment *kokoro no kôri* was depicted in a feedback cycle with the body, thus placing heart and body in mutually reinforcing juxtaposition. The body attacked by *kôri* develops poor circulation, a lowered metabolism, and misalignment resulting in chills, swelling, constipation, and a tendency to gain weight. The stiff, frozen heart brings on a lack of laughter, tears, and other feelings, resulting in a lack of overall health, derangement of the autonomous nervous system, and nervous irritation. (Yamaguchi 2007: 18) This leads to another cycle of increasing stress, in which the sufferer is hard to get on with, never smiles, and is quickly discouraged. Depression can result (Yamamoto 2007: 20–21). The physical and mental activities recommended for curing this problem were described in large type on the magazine's cover as *kokoro no detokksu* 'detoxification of the heart." Though framed as novel through use of the recent loan *detokksu*, this concept and the ailment *kokoro no kôri* hearken back to indigenous notions of *kokoro* pollution (Lebra 1976:161). Elsewhere, *kokoro* has also been implicated in reconfiguring depression as a medical illness – the common cold – curable through the use of pharmaceuticals:

(9) *kokoro* *no* *kaze*
heart GEN cold[15]
'depression' (replacing the common term utsubyô)
(1999 Glaxo-Smith-Kline ad)[16]

In a discussion of *zazen* 'seated meditation,' Rohlen describes the teacher's exhortation that sitting up straight would "bring one's spirit [*kokoro*] and body together in harmony" (1986: 311). This vignette appeared in a broader discussion of *seishin kyôiku* 'spiritual training.' In sumo wrestling, the winner of a bout writes the character for *kokoro* with his hand over the money envelope he is about to receive from the referee, indexing the sport's spiritual connections. A radio public service announcement requesting blood donors seems to indicate that the Japanese HEART's maturation may be accomplished – through bodily action – by this act of charity: *Kenketsu wa kokoro no seijinshiki* 'blood donation is the heart's coming-of-age ceremony' (radio PSA 2005). Coming-of-age day is a public holiday in Japan celebrating those who have reached adulthood, i.e., twenty years of age.

Furthermore, in the cultural subgroup of religious psychotherapy known as *Gedatsukai*, or Salvation Cult, Lebra cites notions of Shinto purity and Buddhist emptiness as motivators for expurgation, the final stage of self-reconstruction. "The ultimate moral value in Japanese culture is represented by the pure heart, *magokoro* (true, sincere heart), *muga* (ego-lessness), or *mushin* (mindlessness)" (1986: 364). She considered the cult's practices and beliefs as hyperbolized compared to those of typical Japanese people. Note that the "heart" here is not the unmodified *kokoro* but specifically the "true heart" of *magokoro.*

6. The HEART's dimensions

Notions of "true heart" bring to mind ideals such as those Niemeier discussed regarding the depth of HEART in English (1997). Japanese, too, conceptualizes the heart as a container. One's sincerity can also be indicated by reference to the 'heart's depths,' *kokoro no oku*. The 'bottom of one's heart,' *kokoro no soko*, is where real intention lies. To have a 'narrow heart,' *kokorohosoi*, is to be helpless, uncertain, and lonely. 'Crowding the heart' – *kokoro o komete* – is a way to indicate the fullness of one's sentiment. *Kokoro* can also be the source of speech; a common ending to formal, publicly made emotive statements is *kokoro kara môshiagemasu* 'to offer up words from the heart.'

7. HEART in lyrical context

Kokoro, like other emotion-laden terms across languages, is used in specific ideological contexts. That is, it appears in songs and textbooks as part of an implicit cognitive cultural model for the ideal Japanese society. A particular genre of song known as *enka* popular among adults in the post-WWII era is often described as *nihonjin no kokoro*, 'the heart/soul of the Japanese people.' *Kokoro* often appears in its lyrics (as well as *mune*, but not *shinzô* or *haato*), along with a coherent set of metaphoric and other tropes. [17] In *enka* images of nature are used figuratively to describe human gender roles, romantic love, and lament. Some Japanese claim to hate *enka*, although – or perhaps because – it represents a particular cultural model for heterosexual romance that is supported by Japan's contemporary political-economic framework (Occhi 2006). Examining this genre can show us more about *kokoro* and about emotional aspects of the model that *enka* supports. Since *enka* seldom uses *gitaigo* expressions yet is rich in metaphor and other tropes, it also shows how HEART-related expressions are manifested throughout various resources of the Japanese language.

The closest English genres of song to which *enka* can be compared are American country or blues. In *enka* the heart (and other body parts) are capable of many imagistic exploits. Examining these lyrics can resolve the apparent paradox of *kokoro*'s inclusion in and separation from the body. In *enka* the use of nature imagery as metaphoric for humans renders impossible the separation of lyrics into categories of dialogue versus narrative (Occhi 2000). For example, by saying *nami no kokoro tsumetai ya* 'the heart of an ocean wave is cold' the singer describes her feeling of bleak failure in seeking the man who left her behind. In the song *Aishu Minato* 'Harbor of Sorrow' the singer gazes into the ocean and sees the painful memory of her beloved which burns in her heart:

(10)

ima	*mo*	*kokoro*	*ni*	*sumitsuite*
now	too	heart	DAT	settle-CONT

kurai	*namima*	*ni*	*omokage*	*ga*
dark	wave-interval	DAT	visage	SUBJ

kiete	*ukande*	*mata*	*kiete*	*yuku*
die out-CONT	float-CONT	again	die out-CONT	go

miren	*setsunaku*	*moenokoru*
lingering attachment	unending	burn-remain

'Even now living on in my heart the visage in the dark waves (i.e., of the beloved) floats in and out unending heartbreak burns on'

Enka is filled with descriptions of physiological and psychological sensations couched in terms of natural phenomena such as these. Coldness and pain, especially stinging pain, are common sensations of the woman's body as described in these songs. In *enka* the woman is often portrayed metaphorically as a flower, beaten by rain and wind. This feminized flower should be protected. Such imagery indexes traditional Japanese cultural ideology about gender roles in heterosexual partnerships (Shibamoto Smith 1999). Hearts (*kokoro*), specifically, are scattered, seeking, deciding (i.e., not to remember the beloved) or being thrown away. Other body parts are also productively used in women's songs to index the male beloved. He is referred to by mention of his mouth, shoulders, back, and chest (*mune*) and strongly associated with water as well. These findings are borne out through quantitative as well as qualitative examination (Occhi 2000).

Enka's main audience is middle-aged and older. Its discourse of lament plays on memories of youthful heartbreak, reinforcing the idealized stability of married midlife couples. *Enka's* emphasis on imagery of coldness and winter contrasts neatly with songs about and for youth, which have summery themes. Among these, the following title – nostalgic yet far from the *enka* style – is a classic often sung by schoolchildren:

(11)

Watashi	*no*	*kokoro*	*wa*	*natsu*	*moyô*
me	GEN	heart	TOP	summer	pattern

'My heart is in summer mode' (Inoue 1990)

Another source for Japanese HEART ideology is the recently developed (and rather controversial) set of children's textbooks,[18] *Kokoro no Nôto*, '[the] Heart's Notebook' (*Monbukagakusho* n.d.). Designed for use in morals classes to foster patriotism in Japanese youth, the texts describe nature imagery, festivals, and familial practices with a variety of *kokoro* expressions. Examples (12a) to (12g) are section titles taken from the third/fourth year reader (aimed at nine- and ten-year-old children). They contain notions of guidance for the *kokoro* of a proper citizen:

(12) a. *Kokoro o migaku ookiku sodateyo*
heart OBJ polish large-CONT rear-HORT
'Let's polish the heart and raise it up big'

b. *Kisetsu o kanjiru kokoro o migakô*
season OBJ feel heart OBJ polish-HORT
'Let's polish [our] hearts to feel the seasons'

c. *Omoiyari no kokoro o sagasô*
thoughtfulness GEN heart OBJ seek-HORT
'Let's seek the thoughtful heart'

d. *"ima yori yoku naritai" to iu kokoro o*
now than well become-want QUOT heart OBJ
motô
carry-HORT
'Let's have a heart that says it wants to become better than now'

e. *Jibun ni shôjiki ni nareba, kokoro wa*
self DAT honest DAT become-if heart TOP
totemo karuku naru
very light become
'If [one] is honest to oneself, [one's] heart becomes very light'

f. *Reigi – katachi o taisetsu ni shite*
etiquette form OBJ important DAT make-CONT
kokoro o kayowaseau
heart OBJ communicate
'Making the form of etiquette important communicates the heart'

g. *Watashitachi* *no* *kokoro* *o* *sodatete* *kureru*
my-PL GEN heart OBJ rear-CONT give
furusato
hometown
'The hometown that kindly nurtures our hearts' (n.d.)

The examples in this section show how readily *kokoro* is employed in promoting cultural ideology through various genres of popular expression. As the recent move towards neo-nationalism develops in Japan, one might expect that the use of *kokoro* in similar rhetoric may expand as well.

8. Concluding comments

This paper has shown how Japanese HEART-related expressions manifest in everyday language, ranging from the phonetic level of mimesis to discourse-level ideological statements. Though there is some minor degree of overlap with the English cultural model for HEART, the examples and discussion presented here may first of all have shown that differences prevail. Perhaps a closer comparison would be possible between Japanese and Korean (Yoon, this volume).

Specifically, the sound-symbolic system of Japanese provides for a realm of expressions largely absent in English. Some of these are connected with *kokoro* and others with the more generalized *mune.* Pervasive metaphoric transference between humans and natural phenomena found in Japanese expressions provides a vast arena for the expression of verbal beauty. This 'sloppy selfhood' is also employed in ideological assertions, shown here in data taken from popular songs and children's educational material, which contribute to the cultural model.

Furthermore, though Japanese ideology is largely non-Cartesian, expressions evidencing HEART-body differentiation and agency manifest as well. While the more ancient terms *kokoro* and *mune* endure much of the workload within this model, we see in the Japanese data that the loans *shinzô* and most recently *haato* have taken on roles of their own. This provides a useful study of how a cultural model can incorporate words that originate in other languages, here expanding notions associated with HEART in Japanese.

Notes

1. With thanks to the members of Kaoru Horie's cognitive-functional linguistics seminar course at Tohoku University in 1995, Gary B. Palmer, Farzad Sharifian, and Janet S. Shibamoto Smith, as well as to my consultants and the reviewers of this paper.
2. Some of these data were gathered in the course of other research projects on Japanese I have conducted since 1992 (Occhi 1994, 1999, 2000).
3. The first and third of these terms would correspond to English 'gutsy' while the second brings 'lily-livered' to mind.
4. The character used for writing *wakaru*, however, is associated with notions of separation or division rather than with the heart *per se*.
5. When signing *wakaranai,* the head also tilts to the right and the eyebrows are lowered.
6. Data were taken from letters written by Japanese women from various parts of Japan 16–25 years of age.
7. This number is large in comparison to seven and four instances respectively for the first two definitions, with no 'heart' in either set.
8. In using the term "mimetic" I follow Hamano's usage; she states that the majority of words in this category "are sound-symbolic rather than onomatopoeic; they symbolize manners or psychological conditions" (1998:2).
9. An understanding of sound symbolism in Japanese yields superior insight into the Japanese mimetic system compared to that found in other explanatory mechanisms (e.g., NSM used in Hasada 2001; also see Yoon this volume).
10. For instance, *doki doki* is an internal sensation detected by the self, but *tobo tobo* 'trudging' is an external assessment of another's behavior.
11. Data supporting this common synaesthetic association were seen on the label of a cherry flavored soft drink circa 2001.
12. A closer translation of 'my heart is torn' is *kokoronokori* 'heart remains.'
13. That is, *Q: anata ga renai ni motomeru mono wa? Tokimeki, dokidoki kan* 35.8% 'Q: What thing do you seek in love? … heartpounding feeling 35.8%.'
14. This interpretation resonates with the reference to embodiment found in the JSL sign for *wakaru* discussed above.
15. Here 'cold' refers to the illness which is commonly translated as 'cold/flu,' not to temperature.
16. Thanks to Anne McLellan Howard for pointing out this example.
17. *Kokoro* expressions appear commonly in Japanese popular music, often with metaphoric nature imagery; however, the following section focuses on *enka* because of its ideological designation as '*the* heart/soul of Japanese people.'
18. Though Japanese textbooks are often written by outside agents and approved by the Ministry of Education, this series was developed in-house.

References

Anonymous
2002 Select the Board. *Bodyboarding Flipper* 16: 44.

Asano, Tsuruko
1978 *Giongo/Gitaigo Dictionary.* Tokyo: Tsunegawa Shoten.

Damasio, Antonio
1994 *Descartes' Error: Emotion, Reason, and the Human Brain.* New York: Penguin.

Dirven, René & Marjolijn Verspoor eds.
2004 *Cognitive Exploration of Language and Linguistics,* 2nd ed. Amsterdam: John Benjamins.

Gomi, Taro
1989 An Illustrated Dictionary of Japanese Onomatopoeia. Tokyo: Japan Times Press.

Hamano, Shoko Saito
1998 *The Sound-Symbolic System of Japanese.* Stanford, CA: CSLI Publications and Tokyo: Kurosio Shuppan.

Hasada, Rie
2001 "Meanings of Japanese sound-symbolic emotion words." In Jean Harkins and Anna Wierzbicka (ed.) *Emotions in Crosslinguistic Perspective.* Berlin: Mouton de Gruyter, 221–258.

Hanlon, Yuno
2002 *haato biru hô* [heart building law.]
http://www.jekai.org/entries/aa/00/np/aa00np59.htm.
Accessed 11 November 2005.

Higashi, Hiroko
1992 "Person Limitations in Sentences With Emotional Adjectives." In Tajima Gando & Niwa Kazuya (eds.) *Japanese Theoretical Research 3*. Osaka: Izumi Shoin, 45–68.

Ikegami, Yoshihiko
this vol. "The heart – What it means to Japanese speakers."

Inoue, Yosui
1990 *"Shônen Jidai"* [The Era of Youth]. Tokyo: Fire Music Publishers.

Izumi, Taku
N.d.a. *An-pan man to yukai na nakamatachi BGM* [Background music: *An-pan Man* and his lively buddies]. Tokyo: JASRAC.

Kakehi, Hisao, Tamori, Ikuhiro, and Lawrence Schourup
1996 *Dictionary of Iconic Expressions in Japanese.* Berlin: Mouton de Gruyter.

Kataoka, Kuniyoshi
2003 "Form and function of emotive pictorial signs in casual letter writing." *Written Language & Literacy* 6(1):1–29.
2002 "Emotion, Textual Awareness, and Graphemic Indexicality." In Kuniyoshi Kataoka & Sachiko Ide (eds.) *Culture, Interaction, and Language.* Tokyo: Hituzitisyobo, 215–242.
Kataoka, Kuniyoshi & Sachiko Ide (eds.)
2002 *Culture, Interaction, and Language.* Tokyo: Hituzitisyobo, 215–242.
Kindaichi, Haruo
1978 "An Outline of *Giongo & Gitaigo.*" *Giongo/Gitaigo Dictionary.* Tokyo: Tsunegawa Shoten, 4–25.
Kita, Sotaro
1997 "Two-Dimensional semantic analysis of Japanese Mimetics." *Linguistics* 35: 379–415.
Kuroda, S.-Y.
1979 "Where epistemology, style and grammar meet." In *The (W)hole of the Doughnut: Syntax and its Boundaries.* Ghent: E. Story-Scientifica, 185–203.
Kuwayama, Takumi
2004 *Native Anthropology: The Japanese Challenge to Western Hegemony.* Melbourne: Trans Pacific Press.
Lebra, Takie Sugiyama
1976 *Japanese Patterns of Behavior.* Honolulu: University of Hawaii Press.
1986 "Self-Reconstruction in Japanese Religious Psychotherapy." In *Japanese Culture and Behavior: Selected Readings.* Takie Sugiyama Lebra and William P. Lebra, (eds.) Honolulu: University of Hawaii Press, 354–368.
Matsuki, Keiko
1995 "Metaphors of anger in Japanese." In J.R. Taylor and R.E. Maclaury (eds.) *Language and the Cognitive Construal of the World.* Berlin: Mouton de Gruyter, 137–157.
McCreery, John
2000 *Japanese Consumer Behavior: From Worker Bees to Wary Shoppers.* Honolulu: University of Hawaii Press.
McVeigh, Brian
1996 "Standing stomachs, clamoring chests and cooling livers: Metaphors in the psychological lexicon of Japanese" *Journal of Pragmatics* 26:1: 25–50.
Miyamoto, Eriko
2005 *'Hontô no renai' wa doko ni aru?* [Where is "true love"?]. *Nikkei Woman* 251:12: 18–29.

Moeran, Brian
1986 "Individual, Group and *Seishin*: Japan's Internal Cultural Debate". In *Japanese Culture and Behavior: Selected Readings.* Takie Sugiyama Lebra and William P. Lebra, (eds.) Honolulu: University of Hawaii Press, 62–79.

Monbukagakusho
n.d. *Kokoro no Nôto* [(the) Heart's Notebook]. Tokyo: Monbukagakusho.

Nakamura, Akira
1979 *Kanjô hyôgen jiten* [Dictionary of Emotion Expressions]. Tokyo: Tokyo Dô Shuppan.

Niemeier, Suzanne
1997 "To have one's heart in the right place – metaphorical and metonymic evidence for the folk model of the heart as the site of emotions in English" in *Human Contact Through Language and Linguistics*, Birjit Smieja and Meike Tasch, (eds.) Frankfurt: Peter Lang, 87–106.

Occhi, Debra J.
1992 "*An-Pan Man*: language and culture in a Japanese children's cartoon." MA thesis, Department of Anthropology, UNLV.
1999 "Sounds of the Heart and Mind: Mimetics of Emotional States in Japanese" in *Languages of Sentiment: Pragmatic and Conceptual Approaches to Cultural Constructions of Emotional Substrates*, Gary B. Palmer and Debra J. Occhi, (eds.) Amsterdam: John Benjamins. pp. 151–170.
2000 "*Namida, sake,* and love: emotional expressions and Japanese *enka* music." PhD dissertation, Department of Anthropology, University of California, Davis.
2006 "Heartbreak's destination: Tohoku in the poetic discourse of *enka*" in *Wearing Cultural Styles In Japan: Concepts of Tradition and Modernity in Practice*, Christopher Thompson and John Traphagan, (eds.) New York: SUNY Press. Pp. 151–170.

Palmer, Gary B.
1996 *Toward a Theory of Cultural Linguistics.* Austin: University of Texas Press.

Quinn, Naomi and Dorothy Holland
1987 "Culture and Cognition" in *Cultural Models in Language and Thought,* Dorothy Holland and Naomi Quinn, (eds.) Cambridge: Cambridge University Press. Pp. 3–42.

Rohlen, Thomas P.
1986 "Spiritual Training in a Japanese Bank." In *Japanese Culture and Behavior: Selected Readings.* Takie Sugiyama Lebra and William P. Lebra, (eds.) Honolulu: University of Hawaii Press. Pp. 307–338.

Shibamoto Smith, Janet S.
1999 "From *Hiren* to *Happî-Endo:* Romantic Expression in the Japanese Love Story" in *Languages of Sentiment: Pragmatic and Conceptual Approaches to Cultural Constructions of Emotional Substrates*, Gary B. Palmer and Debra J. Occhi, (eds.) Amsterdam: John Benjamins. Pp. 131–150.

Shibatani, Masayoshi
1990 *The Languages of Japan.* Cambridge: Cambridge University Press.

Shinmura, Izuru, ed.
1995 *Kôjien*, fifth ed. Tokyo: Iwanami Shoten.

Soga, Tetsuo
1991 *Giongo & Gitaigo Reader.* Tokyo: Kogaku Shoten.

Tani, Chiharu
2005 *DVD de oboeru Shûwa Jiten* [Japanese sign language dictionary to learn by DVD]. Tokyo: Ikeda Shoten.

Teramura, Hideo
1982 *Syntax and Meaning in Japanese, Vol. 1.* Tokyo: Kurosio.

Watanabe, Toshirô, Edmund R. Skrzypczak, and Paul Snowden, (eds.)
2003 Kenkyusha's New Japanese-English Dictionary, fifth ed. Tokyo: Kenkyusha.

Yamaguchi, Hajime
2007 *Ira ira, utsu utsu, genki ga denai... sono genin wa 'kokoro no kôri'* [Irritation, depression, lack of health – the cause is freezing/stiffening of the heart]. *Nikkei Health* 12:2: 18–19.

Yamamoto, Haruyoshi
2007 *'shizen no nagare' ni sakarau kara sutoresu ga tamaru* [Stress builds up from "going against the natural flow"]. *Nikkei Health* 12:2: 20–21.

Yoon, Kyung-Joo
this vol. "The Korean cultural model of *heart*: An indigenous perspective."

The Korean conceptualization of *heart*: An indigenous perspective

Kyung-Joo Yoon

Abstract

The relationship between language, body, and culture is one of the current interests in cognitive science. It is widely agreed that the conceptuaslisations of body parts across languages and cultures may shed light on human cognition in general. This contribution attempts to establish the Korean *cultural model* (cf. Quinn and Holland 1987) of *heart*. In Korean there are three distinctive concepts corresponding to the English concept of *heart*: *simcang*, *kasum*, and *maum*. These words are frequently used in daily conversation as well as in literature. Therefore, knowing the meanings of these concepts is crucial in understanding the Korean view on human faculties. In order to illustrate the Korean perspective on them, I will describe the meanings of these concepts by using the Natural Semantic Metalanguage theory. This approach uses the 'reductive paraphrase' (Goddard 1998) method, which decomposes complex concepts into a combination of simple and intuitively intelligible meanings. By using only these concepts that are claimed to be universal, I tried to avoid the danger of ethnocentrism so that the meanings of cultural concepts can be comprehended with the same understanding by both insiders and outsiders of the Korean culture. The lexical semantic analysis of the three Korean concepts illustrates the Korean culture-specific way of conceptualising human faculties related to the English concept of *heart*. This chapter shows the possibility of understanding Korean ethnopsychology thorough the culture-specific concepts, and at the same time provides evidence of cross-cultural and cross-linguistic variability in the conceptualization of human faculties and body parts.

Keywords: Korean cultural model of heart, Natural Semantic Metalanguage Theory, reductive paraphrase, Cross-cultural variability

1. Introduction

It is widely accepted that different cultures develop different cultural pychology and that language reflects this variation. The aim of the present paper is to establish the Korean cultural model of *heart* from an indigenous perspective. In Korean there is no one exact semantic equivalent to the English concept of *heart*. Instead, there are three translational equivalents: *simcang* 'heart', *kasum* 'heart/chest', and *maum* 'heart/mind'. These three lexical items can all be translated as *heart* in English, despite the semantic differences between them. However, to my knowledge, there has been no lexical semantic analysis in relation to these concepts in any previous literature. My primary intention in this paper is, therefore, to describe the meanings of the three concepts, *simcang*, *kasum*, and *maum*, in an exhaustive way on the basis of linguistic evidence. All these concepts are frequently used in Korean every-day conversations, as well as in various forms of literature, in relation to emotion and cognition, which suggests that understanding them is of central importance in understanding Korean ways of conceptualising the human faculties of emotion and cognition.

In order to make sense for cultural outsiders of how Koreans view human faculties (both physical and psychological), semantic analysis has to be done in a non-language specific and non-culture specific way. Otherwise, non-Koreans will not be able to understand the indigenous perspective as it is. An analytical tool for this kind of semantic analysis is, therefore, indispensable. Without an appropriate and systematic analytical tool, the meaning of cultural concepts will not be adequately comprehended by the outsider. This paper adopts the Natural Semantic Metalanguage (NSM) approach because this framework enables an attempt to escape from ethnocentrism by providing a culture independent metalanguage. What I mean by 'culture independent' is that this metalanguage is built on universal human concepts attested by a great number of cross-linguistic studies (see Goddard and Wierzbicka 1994, 2002; Goddard 1997, 2002 for details). This approach has been found to be a practical tool for semantic analysis, particularly for culturally salient concepts, since it is not associated with any specific language and culture (cf. e.g., Wierzbicka 1999; Goddard 1996a, 1996b, 1997, 2001; Goddard and Wierzbicka 1995; Harkins and Wierzbicka 2001; Hasada 2000; Ye 2000).[1]

The NSM approach is interested in capturing the necessary semantic components of the defined concept and paraphrasing them in simple language that is self-explanatory. The present inventory of NSM theory consists of around 60 concepts or 'semantic primitives' which have been proposed after "a great deal of trial-and-error experimentation in diverse areas of semantic analysis" (Goddard 1997: 3). The concepts of this inventory are presented both in English and Korean terms in the Appendix in order to show that definitions or 'explications' can readily be transposed into Korean without any change of meaning[2]. Although in this paper the explications are presented only in English due to space limitations, they can be transposed into Korean and in principle into any other natural language, if the set of semantic primitives has been agreed upon.

In order to analyse the meanings of the three concepts, collected data were examined first in their original contexts to establish preliminary explications based on the linguistic use of native speakers. Formulating hypotheses on the basis of linguistic evidence is essential for semantic analysis using the NSM approach. As a second step, the examples were tested by being substituted for the defined expressions. All the explications in this paper are proposed following this process. Nonetheless, these explications are left open to further discussion or modification, as it is possible that counterexamples may be found. The aim is for explications to be open to improvement and to be compatible with all usages.

The data collected for my investigation were taken from various sources: Korean corpora, novels, drama scenarios, popular songs, and newspaper articles, and were originally written in Korean. Not all of them could be cited in the present paper due to space limits. Examples used in this paper are only a part of all those that were examined and were taken mainly from the Korea Advanced Institute of Science and Technology (KAIST) corpus. Although they do not represent a wide variety of sources, the examples cited all reflect contemporary usage of the analysed concepts among speakers of the Korean language in South Korea.

2. Translational equivalents of heart in Korean: *simchang*, *maum*, and *kasum*

Among the three translational equivalents of *heart* in Korean, the word *simcang* is mostly used to refer to an internal body organ. The two other concepts, *kasum* and *maum*, have wider meanings than *simcang*, referring not only to a body part but also to psychological entities. While the concept of *simcang* is Sino-Korean borrowed from a Chinese character, the two other concepts, *kasum* and *maum*, are purely Korean. According to the etymology of these words, the meanings of *kasum* and *maum* have changed since the 15th century. The first meaning of *maum* is similar to that of the contemporary use. However, it had the second meaning of the internal body organ, *simcang* 'heart'. The use of *maum* can be found in *Wuelinsekpo* (1459) as shown below:

masam (the old form of *maum*):

(1) The mother could not get rid of her *masam* [*maum*] of wanting to eat …
(*Wuelinsekpo- cwung* 23: 89 in *Pyocwun Kwuke Taycacen* 1999)

(2) If there is a hell, they will eat the *masam* [*simcang*] of sinners …
(*Wuelinsekpo- cwung* 23: 75 in *Pyocwun Kwuke Taycacen* 1999)

The translations of the old form of *masam* are given in square brackets. It indicates that the word *masam* expressed two different meanings: psychological faculty and physical organ. Due to the lack of data in the historical Korean linguistics, it is not clear when the second meaning of *masam* disappeared. However, the Sino-Korean word *simcang* may have replaced the second meaning of *masam* in the course of centuries. The contemporary use of *maum* never refers to the physical organ.

The same source suggests that the word *kasam* (old form of *kasum*) exclusively denoted the physical body part, chest, in the 15th century. However, the meaning of *kasam* may have gradually extended during the last few centuries. As a result, the contemporary use of *kasum* seems to refer to either the body part or the psychological faculty. However, it is interesting to note the second example of the use of *kasum* below, which suggests that the *masam* was conceptualized to be located in *kasam* already from that period.

kasam (the old form of *kasum*)

(1) If there was a hell, their *kasam* and back would have been speared …
(*Wuelinsekpo: cwung* 21: 80 in *Pyocwun Kwuke Taycacen* 1999)

(2) At first, *masam* that was contained in *kasam* …
(*Dwusienhay: cho* 9: 17 in *Pyocwun Kwuke Taycacen* 1999)

According to the data shown above, *kasam* (*kasum*) was a body part, referring to the chest area, where *masam* (*maum*) was located. Moreover, the concept of *masam* used to express the meaning of an internal body organ, *simcang* 'heart'. The two concepts have gone through significant semantic changes during the last six centuries. It is not clear when the word *simcang* first appeared in the Korean lexicon. Presumably this word would not have been under a dramatic semantic change unlike the two other purely Korean words.

In contemporary Korean, all three words are frequently and widely used. Koreans refer to them when they express their emotion, intention, cognition, and behaviour in everyday life. Before proceeding to the semantic analysis of each concept, dictionary definitions of the three concepts are presented below. English-Korean and Korean-English bilingual dictionary entries show that the meanings of the three Korean words overlap and that there is no single corresponding counterpart of *heart* in Korean.

heart: 1. *simcang*
2. *kasum* (*sok* [inside])
3. *maum* (*sok* [inside])
(The Random House Dictionary of the English Language 1991)

Among the sixteen entries for heart given in the English-Korean dictionary, the three above are the most frequent translational alternatives in Korean. By comparison, a Korean-English dictionary gives the following translations. It would be interesting to see whether these translations coincide with those of English-Korean dictionaries.

simcang: 1. the heart
2. nerve

kasum: 1. the breast; the chest; the thorax, the bust; the bosom
2. the heart
3. *maum*; one's heart; one's mind; feelings

maum: 1. mind; spirit; mentality; an idea; a thought
2. heart; feeling mood;
3. sincerity; wholeheartedness
4. consideration; sympathy
5. attention
6. will; intention; inclination

(Dong-A 1996)

Apart from what is chosen above, these bilingual dictionaries reveal that, while the word *simcang* has only one or two counterparts in English, the two other words have different meanings from English heart. They also show that the word *simcang* has a narrower range of meanings than the two other words. In fact, this word is relatively straightforward in meaning, referring to an internal bodily organ just as the first entry of any English dictionary definition of heart. Therefore, it is less puzzling from an outsider's point of view.

The two other words, *kasum* and *maum*, however, are translated into various different words in English in addition to the word heart: for instance, breast, chest and bosom for *kasum* and mind, spirit and mentality for *maum*. This suggests that these two words are polysemous and at the same time they lack counterparts in English and vice versa. However, not only are the two words, *kasum* and *maum*, unclear in their meaning, they are defined in a circular way, i.e., via each other, i.e., *kasum* via *maum* and vice versa, in both bilingual (Korean-English) and in monolingual (Korean-Korean) dictionaries. Consequently, people can be misled into assuming that they are synonymous. More problematically, the suggested English equivalents are sometimes heart, mind, or spirit, which are all very distinct and Anglo culture-specific concepts.

The examples given below taken from a collection of translations of modern Korean literature show the variability of translation according to the context.[3]

(1) 눈물로 적시고 또 적시어도
속절없이 식어가는 네 흰 가슴 [*kasum*] 이
저 꽃으로 문지르면 더워 오리야
Your pale **breast** [*kasum*] grows colder and colder,
though I bathe it with tears, to no avail:
will it gain warmth if I rub it with this flower?)
(K → E, So 1996: 94–95)

(2) 윙윙그리는 불벌의 떼를
꿀과 함께 나는 가슴 [*kasum*]으로 먹었노라
… but I drew honey into my **heart** [*kasum*]
with whining wild bee swarms
(K → E: So 1996: 54–55)

(3) 아아 지금도 살아서 내가슴 [*kasum*] 에
굽이친다
지리산이여
지리산이여
You still alive, reverberate in my **mind** [*kasum*]
Oh! Chiri Mountain,
Chiri Mountain
(K → E: Kim 1998: 53)

(4) 번개와 같이 떨어지는 물방울은
취할 순간조차 마음 [*maum*]에 주지않고
나타와 안정을 뒤집어 놓은 듯이
높이도 폭도 없이
떨어진다
The water drops, falling like lightning,
drops
without height or breadth
as if confounding sloth and rest,
not granting the **mind** [*maum*] a moment's rapture
(K → E, Brother Anthony 2002: 22–23)

(5) 처음 내 마음 [*maum*]은
수천만 마리
노고지리 우는 날의 아지랑이 같었읍니다…
바닷물이 적은 여울을 마시듯이
당신은 그를 데려가고
그 훠-ㄴ 한 내마음 [*maum*]에
At first my **heart** [*maum*]
Was like the haze on days when myriads of skylarks sing …
You took him away again
like the ocean swallowing tiny streams,
and in my bright empty **heart** [*maum*]
(K → E, So 1996: 156–157)

The word *kasum* is translated as 'breast' in (1), as 'heart' in (2), and as 'mind' in (3), whereas the word *maum* is translated as 'mind' in (4), but as 'heart' in (5). Looking at these examples, it is puzzling to comprehend the range of the given concepts in Korean. Unless they are semantically unpacked in terms of simple and universal concepts, their meanings will remain unclear to cultural outsiders. This is what the present analysis intends to achieve for each of the three terms.

3. *Simcang*

The word *simcang* is particularly widely used in medical contexts with reference to the internal bodily organ that keeps up the circulation of the blood. Most collocations of the word *simcang* are related to technical terminology as shown in the examples below:

- *simcang susul*: heart surgery
- *simcang mapi*: heart attack
- *simcang paktong*: heart beating
- *simcang swuchwuk:* heart systole
- *simcang isik*: heart transfer

The meaning of *simcang* looks straightforward, although it is also used metaphorically to express 'something in the middle' or 'important' in a similar manner as in English *in the heart of the city*. The metaphor seems to suggest that there are conceptual links between *simcang* and the significance of something. The word *simcang* can be used to refer to an internal organ of creatures other than humans, too. However, the meaning of a non-human organ will not be dealt with in this paper, hence no polysemy has been proposed here. On the basis of the examination of the collected data, I posit the following explication for *simcang*.

simcang
(a) a part of a person's body
(b) people think about this part like this:
  "it is inside the upper part of the body (there are some other parts)[4]
(c) when a person feels something because this person thinks something
(d) this person can feel something in this part because of this"
(e) sometimes a person can think like this:
(f) "something is happening in this part
(g) I can hear it
(h) if something very bad happens to this part of a person,
  this person cannot live anymore"

Components (a) to (d) capture the physical characteristics of *simcang* seen from the viewpoint of common cultural knowledge, rather than from scientific expertise. Component (b) indicates that *simcang* is one of many other internal organs. The morphological structure supports this aspect. In Korean many other internal organs share the same morpheme, *-cang* 'organ' with *simcang*, e.g., *kancang* 'liver', *wycang* 'stomach', *taycang* 'great intestine', and *socang* 'small intestine', so on. It is well known among people that *simcang* is inside the chest area.

(6) 도천백은 무의식중에 왼손을 끌어올려 심장 부근을 가렸다.
Tochenpayk unconsciously covered the area where his heart [*simcang*] was with his left hand. (MT: K → E, KAIST)

Component (d) indicates people's awareness of any physiological change in this part. For instance, when a surprising incident happens, people immediately seem to feel something abnormal as shown in (7).

(7) 따르릉." 정신이 아득해지고 있었다. 가슴이 뛰고 심장 부위가 옥죄어 아파 오고 얼굴이 화끈거리며 손발에 땀이 났다.
"Tring Tring". He (she) was losing his mind [*cengsin*]. His heart [*kasum*] was beating, and he felt pain in his heart [*simcang*]. His face was flushed and he was sweating in hands and feet.
(MT: K → E, KAIST)

Components (e) to (g) capture the most obvious and prevailing characteristic of *simcang*. The various expressions referring to the heartbeat, such as *simcangpaktong* 'heartbeat', *kotongsori* 'sound of heatbeat', and *simcang ttuynun soli* 'sound of heartbeat' seem to suggest that this aspect is the most salient in distinguishing *simcang* from all other internal organs.

(8) 나는 밥을 씹다 말고 화면을 노려보았다. 심장 고동이 다시 커지고 있음을 스스로 느낄 만큼 흥분되었다.
I stopped eating and stared at the screen. I was excited so much that I felt the sound of my heartbeat was growing. (MT: K → E, KAIST)

(9) 내 심장의 두근거림과 그의 심장 고동 소리를 들을 수 있을 정도였네.
I could hear my heartbeat and the sound of his heartbeat.
(MT: K → E, KAIST)

The last component indicates the general assumption of people that *simcang* plays a critical role in the human body. It is not clear at this stage whether this is influenced by scientific knowledge or conceptualised as an invariable semantic component of this concept in the mind of speakers.

However, people seem to be aware that this part is directly linked with matters of life or death.

(10) 여인의 코에 손을 대 보았다. 숨이 느껴지지 않았다. 당황한 서춘근은 여인의 심장에 귀를 갖다대었다. 역시 아무런 고동이 느껴지지 않았다.
He tried to reach his hand to her nose. He could not feel her breath. Se Chwunkun agitated and put his ear to her heart [*simcang*]. Again, there was no sound or movement. (MT: K → E, KAIST)

(11) '대한제국 만세''황제폐하 만세'를 힘차게 불렀다. 그리고 다시 앉아 권총의 총구를 자기 심장에 대고 방아쇠를 당겼다. 단 한 발의 권총 탄환으로 그는 피를 흘리면서 숨을 거두었다.
He loudly acclaimed hurrah 'Kingdom of Tayhan'. Then he sat down and triggered the gun aiming at his heart [*simcang*]. Bleeding, he died from that bullet. (MT: K → E, KAIST)

Common metaphors such as (12) can be seen as a further support for this metaphoric use of *simcang*.

(12) 살아오는 동안 이 천혜적인 자리를 탐하는 이민족의 침입도 끊이지 않았다. 그것은 한국이 동서의 심장부이기 때문이다. 한국의 위치는 중국·소련·일본 삼국을 분리도 연결도 시키고 있다.
There have been a number of invasions to obtain geographical advantages of this country. That is because Korea is at the central area [*simcang*] for East and West. Korea can either link or cut three countries, China, Russia and Japan, by its location.
(MT: K → E, KAIST)

4. *Kasum*

When Koreans refer to their feelings the word *kasum* seems to be one of the most frequently used words. Ordinary use of the word *kasum* indicates that this word is polysemous, expressing different concepts which are interrelated. For instance, *kasum* in a context as *kasum kiphun kos* 'deep inside the *kasum* [heart]' refers to the psychological faculty of a person, while this concept in such expressions as *kasumey kkochul talko* 'put a flower on the *kasum* [chest]' concerns the physical body part, although sometimes the two meanings are ambiguous if there is no contextual clue. The two meanings of *kasum* will be dealt with in turn: $kasum_{(1)}$ as a body part term, and $kasum_{(2)}$ as an intangible emotional faculty of a person.

Firstly, $kasum_{(1)}$ is used to refer to a body part, roughly *chest*. Although this part can be referred to not only in the human body but also in other creatures[5], only it's meaning of the human body part will be dealt with here. The word *kasum* is used ambiguously to refer to the chest area either inside or outside. When referring to the part of a woman's body, this word is often translated as 'breast', although in relation to both men and women it is more frequently translated as *chest*, *heart*, or *bosom*. Since this word is polysemous between the physical body part term and the emotional faculty, there are other devices that clarify the intended meaning in natural discourse. Collocations and contexts can also give clues as to the correct interpretation of particular examples, e.g., *kasum twulley* (girth), and *kasum sen* (bust line), or *kasum ppye* (chest bone), or *kasumkkey* (chest area). From an English point of view, it can be described, roughly, as 'chest viewed as a location of various internal organs including *simcang* 'heart''. On the basis of the data examined, I propose the following explication for $kasum_{(1)}$.

$kasum_{(1)}$
(a) a part of a person's body
(b) people think about it like this:
(c) "it is in the middle of the upper part of the body
(d) this part of a woman's body is not like this part of a man's body
(e) a person can feel some things in this part when this person thinks about something"
(f) when a person feels some things because they think, this person can think like this:
(g) "something is happening in this part of my body now,
(h) I can hear something because of it"

Component (a) indicates that $kasum_{(1)}$ is a part of the human body. Component (b) intends to capture the conceptual scenario the Korean people have in mind in relation to this part. It tries to show the cultural view, which is also applicable to component (f). The location of $kasum_{(1)}$ around the chest area is captured in component (c). Note that this component is phrased 'in the middle of upper part' not 'inside the upper part' as is used in the explication of *simcang*. The use of the semantic primitive INSIDE is significant in distinguishing the two different concepts of *simcang* and $kasum_{(1)}$. Although both are body parts, *simcang* is inside the body while $kasum_{(1)}$ is not. The concept of *kasum*$_{(1)}$ appears to include both the inside and outside of that particular part of the body. People use this word to refer to the breast or chest, as well as to various physical organs, such as the heart, lungs, and throat. The difference in this part, between male and female has to be included in the meaning of the word, since it is used to refer to the 'breast' (perhaps euphemistically, and since there are other words that refer only to this body part: *yupang*) as shown in example (13).

(13) 가슴은 이러이러한 모습이어야 한다는 사회의 통념에 따라, 가슴을 고치는 방법에는 세 가지가 있다. 첫째, 큰 가슴을 작게 줄일 수 있고, 둘째는 작은 가슴을 크게 할 수 있다.
There are three ways of fixing the ***kasum*** according to social views on the shape of the ***kasum***. First, it is possible to reduce the size of the ***kasum***, second, it is also possible to enlarge the size of the ***kasum***. (MT: K → E, KAIST)

The obvious visible difference between male and female is captured in component (d). One may argue that the word *kasum*$_{(1)}$ is polysemous and that it can mean also "a woman's breast"[6]. Component (e) indicates that people can feel some kind of physiological sensation in this part when their body reacts either to nice surprises or to negative emotions as shown in (14). Koreans use various words such as *thek thek* (*makhita*) (roughly, clogged), *chelleng* (*naylyeancta*) (roughly, collapsed), *aphu-* (sick), *celi-* (roughly, be numbed), *ali-* (roughly, tingling), or *ttakap-* (roughly, sting) in order to describe negative sensations and *ttwy-* (roughly, pound) to denote surprise in combination with *kasum*. Some other body parts can be referred to when people express their emotions: *kan* 'liver' for fearful surprise, and *hepha* 'lung' for feeling giggly, and *pay* 'tummy' for 'jealousy'. However, among the different body parts, *kasum*$_{(1)}$ is the most frequently used, and accordingly this aspect is significant enough to be included as part of its meaning.

(14) 가슴이 쿡쿡 저려오는 것 같았다.
My ***kasum*** seems to be sore (kwukkwuk). (MT: K → E, KAIST)

Components (e) to (g) account for the fact that people conceptualise what they experience in *kasum*$_{(1)}$ as 'something happening' by using various mimetic words. Self-awareness of bodily changes is expressed in psychomimetic words, such as *khwungkhwang*, *wulleng wulleng*, *twukun twukun*, *pelleng pelleng* (beating hard or pulsating hard), which express strong movement occurring in the *kasum*$_{(1)}$ (presumably the 'heart' although *kasum*$_{(1)}$ includes other organs such as lungs or throat). It appears that people notice any change in this part immediately. The first noticeable reaction of the *kasum*$_{(1)}$ is beating, or throbbing hard at either positive or negative emotions. Then people express that they hear some sounds in their *kasum*$_{(1)}$ as the consequence of the movement, as shown in example (15) where the sound is symbolised by the drum (*bwuk*). This aspect is accounted for in component (h). However, it can well be argued that this component is not necessary and rather belongs to the meaning of *simcang*. It is true that when people refer to sounds from their *kasum* they use this word as a synonym of *simcang*. At the same time people choose to use *kasum* over *simcang*, which suggest that the two concepts are still distinctive in the mind of Korean speakers. Taking account of this fact, I am inclined to include component (h) because this aspect appears to be highly characteristic not only for the concept of *simcang* but also for the concept of *kasum*.

(15) 창대는 백여사가 무슨 말을 하는지 알아들을 수가 없었다. 울렁거리는 가슴의 북소리만 들릴 뿐이었다.
Changtay could not understand what Mrs.Payk was talking about. He could merely hear the sound of a drum from his beating ***kasum***.
(MT: K → E, KAIST)

It is true that *kasum*$_{(1)}$ can be used interchangeably with *simcang* in some contexts without carrying any noticeable meaning difference, as in (15).

Turning to the second meaning of *kasum*$_{(2)}$, one can immediately notice that there is a symmetrical use of *kasum*$_{(2)}$ with *meli* (head or brain), whereas *maum* is used symmetrically with *mom* (body) (I will return to this point in the next section), in such expressions as *nayngchelhan meli ttattushan kasum* (lit. 'cool brain, warm *kasum*', roughly brain and *kasum*$_{(2)}$ function for logic and affection, respectively). This seems to suggest that the main function of *kasum*$_{(2)}$ is feeling, while that of *meli* is thinking. Although *kasum*$_{(2)}$ and *maum* are used interchangeably in many contexts, there are instances where *kasum*$_{(2)}$ is not likely to be replaced by *maum*. The following book titles (published recently in Korea) show contexts where *kasum*$_{(2)}$ prevails over *maum*.

- *nelohaye wunun kasumi issta*: There is a *kasum* which is crying because of you

- *hangsang kasum ttellinun cheumipnita*: When one starts something new, one's *kasum* always trembles
- *silin kasumeyto kkochun phiko*: A flower blossoms in a chilled *kasum*

I propose the following explication after examining the collected data.

kasum$_{(2)}$
(a) a part of a person
(b) people think about it like this:
(c) "it is inside *kasum*$_{(1)}$
(d) people can never see this part of a person
(e) because of this part, a person is not like other living things
(f) when something happens to a person,
this person can feel some things in this part at the same time
(g) when something very good happens to a person,
this person can feel something very good in this part
(h) when something very bad happens to a person,
this person can feel something very bad in this part"
(i) when something very good happens to a person,
this person can think like this:
(j) "it is like something happens to my *kasum*$_{(1)}$"
(k) when something very bad happens to a person,
this person can think the same
(l) this person can feel these things for a long time
(m) because of this part, a person can feel good things towards other people

Component (a) indicates that the term *kasum*$_{(2)}$ refers to a part of a person, though not in the same way as *kasum*$_{(1)}$ does. People think that the location of this faculty is in the chest area, conceptually inside the *kasum*$_{(1)}$, although they know that it is not a physical organ as *simcang* is. When people refer to *kasum*$_{(2)}$, using metaphoric expressions such as *kasumuy sangche iculkil epsney* 'I cannot forget the scar left in my *kasum*', they think that the *kasum* is part of a person though it is an invisible and intangible psychological faculty as captured in components (c) and (d). This faculty is exclusively human and does not refer to any other creatures in the world as captured in component (e).

The primary function of *kasum*$_{(2)}$ appears to be that of experiencing various feelings. Moreover, this part is able to feel emotions immediately and instantly. It is not necessarily influenced by conscious thought from the *meli* or *maum* at the time of the event, as shown in (16).

(16) 흑인들의 상당수는 레이건의 인종차별 정책이 어떻든 그 순간 가슴 뭉클한 감동을 느끼기 마련이었다.
Regardless of Reagan's racial segregation policy, a majority of black Americans felt that their ***kasum*** was filled with emotion at that moment . (MT: K → E, KAIST)

Component (f) reflects that *kasum*$_{(2)}$ feels emotions instantly. Besides, the word *kasum*$_{(2)}$ is also frequently used to refer to extreme feelings. In the examples below people experience intense feelings in their *kasum*$_{(2)}$ rather than in their *maum*:

(17) 죽은 사람 생각하믄 가슴이 아퍼서 못 살겄다.
I cannot live anymore because of the pain in my ***kasum*** whenever I think about the dead one. (MT: K → E, KAIST)

(18) 금강산과 더불어 이 설악산이 하늘이 만든 명산이라는 말은 들었지만, 가슴이 벅찰 정도로 아름다울 줄은 몰랐다.
Although I have heard that Seolak mountain is as beautiful as *Kumkang* mountain, that is a masterpiece of God, I didn't expect that it was so beautiful that my ***kasum*** felt overwhelmed.
(MT: K → E, KAIST)

It can be argued that it is not clear whether the pain in *kasum* in example (17) is physical or psychological. Both can be true while thinking about a dead person. However, painful feeling in such context often refers to emotion rather than bodily reaction. This expression is an idiomatic way of stating one's extreme psychological pain.

One can choose to use *maum* in place of *kasum*$_{(2)}$ in the same examples (17) and (18) above. However, a subtle difference in meaning may result, implying less intense feelings of the experiencer. The word *kasum*$_{(2)}$ is not likely to be replaced by *maum* when a physical process is invoked to describe what the person feels like. Predicates such as *mwuneci-* (collapse), *pwuseci-* (break down), *naylyeanc-* (sink), *makhi-* (be choked, be clogged), *cciceci-* (tear apart) are also typically used in combination with *kasum*$_{(2)}$, although the combination of these words with *maum* is not unacceptable. All these collocational characteristics of *kasum*$_{(2)}$ suggest that Koreans express intensive feelings via *kasum*$_{(2)}$ rather than *maum*, although *maum* functions to describe various emotions as well. The function of *kasum*$_{(2)}$ in feeling extreme emotions is captured in components (g) and (h). Similar to the use of *kasum*$_{(1)}$, *kasum*$_{(2)}$ is commonly used in combination with psychomimetic or psycho-onomatopoetic expressions as shown in (19) with which *maum* does not tend to co-occur.

(19) 가슴이 부글부글 끓는 것을 애써 감추고 서울역으로 차를 몰았다.
I hid the fact that my ***kasum*** was boiling (*pwukul pwukul* :sound of bubbling), and drove towards the Seoul train station.
(MT: K → E, KAIST)

When the word *kasum*$_{(2)}$ is used in a context such as a mother's loss of her only child, it is usually difficult to draw a clear distinction as to where the acute pain occurs, in *kasum*$_{(1)}$ or *kasum*$_{(2)}$. It can be both, since a grieving mother may feel a powerful negative feeling in her body as well as in her inner psychological world. Koreans strongly believe in the close link between the body and what is happening in *kasum*$_{(2)}$ (as much as in *maum*). This close relationship between the psychological faculty and the human body appears to be a significant aspect that has to be captured as a part of the meaning. This aspect is reflected in the components (i) and (j). One can argue that it is still not clear whether the concept of *kasum* in examples (17) to (19) is strictly psychological or not. According to the explication above, the psychological effects in *kasum*$_{(2)}$ cause physiological reactions in *kasum*$_{(1)}$ as indicated in (j). Thus *kasum* used in these examples can be seen as the physical organ. However, the body part *kasum*$_{(1)}$ is not compatible with such expressions as *pekcha-* 'overwhelmed', or *pwukul pwukul kkulh-* 'boiling', although *aphu-* 'ache' can be used for both the body part and the psychological faculty. The question as to whether *kasum* in Korean has the meaning of body part and the use of this concept as a reference of various emotions can simply be seen as a metonymy or a metaphor has not been fully answered in the literature. However, the collocations of *kasum* including those that are shown in examples (17) and (18) suggest that the concept of *kasum* has developed another meaning beyond that of the body part.

Another characteristic aspect of *kasum*$_{(2)}$ is the storing of memories of events or words. This feature is shown in the examples below where *kasum*$_{(2)}$ is treated as a place for keeping or preserving memories. When extreme feelings are experienced in one's *kasum*$_{(2)}$ they can remain there for a long period.

(20) 목숨을 구해준 도대협의 은혜 이 화모 평생 가슴 속에 담아두리다.
I will keep the favour you did me in saving my life in my ***kasum*** for the rest of my life. (MT: K → E, KAIST)

In the same vein, there is an expression that is common when one loses one's child. '*casikun kasumey mwutnunta*' (One buries one's own child in one's *kasum*). This expression means perhaps that one can never forget one's own child no matter how much time passes after the death. The examples presented above suggest that people generally believe that once something gets into one's $kasum_{(2)}$, regardless of one's will, the memory of the feeling at the moment the $kasum_{(2)}$ reacted to it can be preserved as it is and does not disappear for a long time. This is possible presumably because $kasum_{(2)}$ does not have the specific function of thinking, and consequently the thing kept in $kasum_{(2)}$ can remain without being influenced by any ideas or thoughts. This is captured in component (l).

Unlike *maum*, the word $kasum_{(2)}$ tends not to be used in combination with value terms, although there are expressions such as '*kasumi ttattushan salam*' (a person with a warm *kasum*), which denotes a good personality that is compassionate and sympathetic toward others. There are no expressions like 'good *kasum*' or 'bad *kasum*', although there are expressions such as '*kasumi chakapta*' (*kasum* is cold: cold blooded) and '*kasumi ttattushata*' (*kasum* is warm). The expression '*kasumi epsnun seysang*', (a society without *kasum*) is used as a book title (Kim, 1993) to refer to modern Korean society where one experiences all kinds of inhumane treatment. The meaning is not 'a society without any feeling' but 'a society without compassion, affection, and sympathy'. This element of meaning is reflected in component (m).

5. *Maum*

The term *maum* is very frequently used by Koreans[7], thus indicating the central nature of this concept in Korean life. The importance of this concept is also reflected in various combinations of this word with other words, such as predicates and attributives, and also with a number of set phrases. The following are some of the established metaphorical expressions[8].

1. *maumi kata* (lit. *maum* goes): tend to be attracted
2. *maumul kkulta* (lit. pull one's *maum*): cause someone to be attracted

3. *maumi nata* (lit. *maum* comes up): want to do something
4. *maumi naykhita* (lit. *maum* becomes agreeable): inclined to do something
5. *maumul ssuta* (lit. use *maum* for someone/something): pay attention to someone/something

The concept of *maum* is critical to understand the specifically Korean conceptualization of one's mental faculties since this concept is quite distinct from the English concepts of the heart and the mind, though it is closer to the latter. The following is the proposed explication built on linguistic evidence (see Yoon 2003 for details).

maum
(a) one of the two parts of a person (one part is the body, this is the other part)
(b) people think about it like this:
(c) "it is somewhere inside the middle of the upper part of a person's body
(d) people cannot see this part
(e) because of this part, a person is not like other living things
(f) because of this part, a person can want to do many things
(g) a person can feel some things in this part when this person thinks about aomething
(h) when a person thinks about something, it is good if this person thinks with this part[9] (like one can do something with a part of one's body)
(i) when a person does something, it is good if this person thinks at the same time in this part like this: 'I want to do this thing well'
(j) if this part of a person is good this person wants to do good things
(k) if this part of a person is bad this person wants to do bad things"

The concept of *maum* is commonly used symmetrically with *mom* (body). The customary contrast between *maum* and *mom* in colloquial everyday use, as well as in the literature of various fields including psychology, suggests that they are considered to be the two main parts of a person. The concept *maum* is used to describe various psychological activities that are not normally experienced in the body. The binary contrast of *mom* (body) and *maum* suggests that Koreans think that *maum* is one of the two parts of a human being. The following are some selected book titles that have been published in Korea in the last decade (1990–2002)[10].

- *thunthunhan mom kenkanghan maum*: The healthy body and the healthy *maum*
- *momkwa maumuy sayngmwulhak:* The biology of body and maum
- *momuy poyak maumuy poyak*: Healthy medicine for body and *maum*

Moreover, *maum* exists exclusively in human beings and not in other living creatures. Koreans believe that while *mom* (body) belongs to the physiological, material world, *maum* is invisible and intangible and exists as a mental faculty. However, *maum* is conceptualised, just like English *soul* used to be, as being located somewhere inside one's chest area. All these aspects are captured in components (a) to (e).

Among various functions that *maum* has, it can desire something as shown in (21). There are a number of fixed expressions, such as '*maumi kancelhata*' (*maum* is eager to), '*maumul mekta*' (lit. eat *maum*: decide to do something) '*maumi naykhita*" (lit. *maum* becomes agreeable: inclined to do something), and '*maumi nata*' (lit. *maum* comes up: wants to do something).

(21) 우리 국민들은 오랫동안 마음속 깊이 변화와 개혁을 열망해 왔습니다.
We people of this nation have had a strong desire for change and revolution deep inside our ***maum***. (MT: K → E, KAIST)

Second, *maum* seems to be -what is commonly called- the seat of various emotions, as shown in examples (22). The concept of *maum* is not used with physiological sensation terms, as shown in (23) although there are metaphoric expressions such as *maumi ssulita* (*maum* aches), *maumi chwupta* (cold in the *maum*) or *maumi aphuta* (sick at *maum*).

(22) 허영심, 수치심, 질투심, 욕심 등 여러 가지 감정이 마음속에 자리잡고 있다.
Various emotions including vanity, a sense of shame, jealousy, and greed and others are placed inside the ***maum***. (MT: K → E, KAIST)

(23) **maum-i kalye-we-se* ...
maum-NM itch-and then
For ***maum*** is itch ...

The word *maum* is often used to describe people's emotional states, mostly those feelings that can be described as 'good' or 'bad' (e.g., 'feel joy or feel fear in one's *maum*'), although there is a feeling that is not specified by these terms as shown in (24). The Korean emotion term *ttelita* 'tremble' does not denote 'good' or 'bad' itself.

(24) 손을 잡을까 눈을 감아보라 할까
너무도 떨리는 내마음 조금이라도 내게 다가와 그대 이밤이 가기전에
Should I hold your hands? Should I ask you to close your eyes? My ***maum*** is trembling so. Come closer to me while this night is young.
(MT: K → E, First Kiss,
http://login.bugsmusic.co.kr/home_login.htm,
last accessed Sep 20th. 2002)

In addition to being the seat of emotions, *maum* has another function. One can also think in this part about something and/or someone, as shown in example (25). This is the most distinct aspect of *maum* compared to the English concept of the *heart*, although in the literature the most frequent translation of *maum* is the *heart*.

(25) 어떻게든 제가 모셔 오도록 하겠습니다. 스님은 마음속 깊이 생각한 바가 있어 진주 속가로 내려갔다.
I'll do my best to bring him here. The monk went back to his family at *Cincwu* since he had a thought deep in his ***maum***.
(MT: K → E, KAIST)

However, the *maum* is not the primary organ for thinking. There is another body part to which all the conceptual or intellectual functions are ascribed: the *meli* (brain; lit. head). Therefore, one can think not because of *maum* but can think in *maum* if one wants to think sincerely without any self-interest, and as it were morally. Thinking only with the *meli* (head) is not as good as thinking in collaboration with the *maum*, since the *meli* can calculate things according to self-interest. It is unclear whether or not this suggests that the *maum* is a fundamental good part of a person that enables a person to be altruistic. More examples are required in order to determine whether this aspect is a part of the meaning of *maum*.

Furthermore, one can do many things putting one's *maum* in one's activities, like English *put your soul into your work*, which means doing something, concentrating all one's psychological energy and activating one's ability to feel and think, as shown in example (25)

(25) 칼질은 힘으로 하는 것이 아니라 마음으로 하는 거다.
The action of cutting has to be done with ***maum***, not with your physical strength (MT: K → E, KAIST)

One's *maum* is often described with value terms such as good or bad, beautiful or ugly, decent or indecent, and kind or unkind, as seen in the various collocations below:

- 고마운 마음 (an appreciative *maum*): with appreciation
- 이기적이고 초조한 마음 (a self-centred and distracted *maum*): self-centredly and agitatedly
- 따스한 마음 (a warm *maum*): warm-heartedly
- 아름다운 마음 (a beautiful *maum*): kind-heartedly or tender-heartedly
- 좋은 마음 (a good *maum*): good will
- 나쁜 마음 (a bad *maum*): bad intention

When one's *maum* is good, one tends to do good things for others, and when it's bad, one would tend to behave selfishly as shown in components (j) and (k).

There are two points worth mentioning in relation to components (h) and (i). One may argue that these are not part of the meaning of *maum* but rather of other expressions involving *maum* and various actions or thinking. There are expressions such as *maumulo semkita* or *sengchalhata* (serve someone with *maum* or examine one's conscience with *maum*) that are already metaphoric on their own. One may take the view that these expressions have to be explicated separately from *maum*, and that a component like (h) or (i) should be included in the explications of these other concepts, rather than in that of *maum* itself. This possibility is worth exploring in more depth in further study. Explications in relation to the concept of *maum* can be elaborated on the basis of comprehensive studies with an extensive range of examples. However, at this stage, I am inclined to include the two components (h) and (i) as a part of the meaning of *maum* on the basis of the examples that I examined as well as of the consistent combinatorial pattern between *maum* and words for thinking or doing in Korean.

Another aspect that could be improved is the phrasing of component (h). In terms of grammar of the metalanguage, the valency option 'think with' is not proposed as universal. This means that this phrase cannot be transposed into all languages. However, there seems to be an analogy between 'think with this part' and 'do something with a part of one's body'. Therefore, the phrase can be interpreted taking that relationship into account for the time being. Alternatively, the component could be phrased as follows:

when a person thinks about something, it is good if this person thinks like this:
  "sometimes when I have to do something, I do this with a part of my body
  I want to do the same when I think about something
  I want to think about things with this part"

Despite the problem of phrasing 'think with', this component reflects the Korean point of view in relation to *maum*. Koreans use *maum* a great deal in relation to 'think' and it seems to be important to capture that aspect in the explication. This expression, however, could be improved with better phrasing since, as mentioned above, the valency option of THINK does not include think with (as opposed to do something with) in the current NSM.

6. Conclusion

In this paper I have attempted to establish the Korean cultural model of the human *heart* or heart-like faculties on the basis of a lexical semantic analysis of the three different concepts: *simcang*, *kasum*, and *maum*. These concepts are paraphrased into semantic primitives proposed in the theory of Natural Semantic Metalanguage. This framework consists of a set of lexical items and their syntactic properties (see Wierzbicka 1996; Goddard and Wierzbicka 1994, 2002) which are based on cross-linguistic investigations. By using the NSM approach, this paper aims to illustrate the cognitive structure of the given concepts in such a way that everyone should be able to see them through the same window as native speakers of Korean. Regarding the accuracy of the posited explications, any component is open to verification and may be put in dispute based on counter examples. If it is possible to dispute these explications, this is significant as it means that the explications are intelligible and have the potential to be improved by further research. All the proposed analyses are submitted as a first attempt and a reference point for further investigation.

According to the current semantic analyses of the given concepts, the English concept of *heart* is neither objective nor universal. The conceptual difference between English and Korean in relation to notions like *heart* is revealed as a result of semantic analysis based on various examples taken from natural usages in contemporary Korean society. The three Korean concepts, *simcang*, *kasum*, and *maum*, are of critical importance for understanding Korean way of viewing or its 'ethnopsychology', as it is usually called in NSM approaches. They reflect the Korean cultural conceptualization of the human faculties related to emotion and cognition.

The concept of *simcang* is used to refer to the bodily organ responsible for blood circulation. The Korean way of conceptualising this part, including its physiological characteristics, is described in the explication. The explication reflects what ordinary people think about this part as revealed in everyday use of this word. It is found that people perceive *simcang* as an internal organ that plays a critical role in human life. This word has its metaphoric use as 'something important and significant', reflected in such expressions as *tosiuy simcang* 'heart of the city (centre of the city)' or *hoysauy simcangpu* 'the heart of the company'. However, this use can be seen as a metaphoric use rather than as a second meaning of *simcang*. Accordingly, there is no polysemy proposed for this concept.

On the other hand, the concept of *kasum* was found to be polysemous: a body part term and the psychological faculty of a person. The meaning of *kasum*$_{(1)}$, appears to share some semantic components with the concept of *simcang*. In fact they appear to be used interchangeably in some contexts though they are semantically different. One can say that *simcang* is in *kasum*$_{(1)}$, but not vice versa, which suggests that *kasum*$_{(1)}$ has a relatively wider meaning than *simcang*. People's awareness of the movement and the subsequent sound of *simcang* are commonly found in both concepts (because *simcang* is one of many internal organs located in *kasum*$_{(1)}$). Both concepts refer to physical body parts, as distinct from *kasum*$_{(2)}$ and *maum*, which do not.

The two concepts, *kasum*$_{(2)}$ and *maum*, do not stand for bodily organs but for psychological and/or mental faculties which are not visible or tangible at all. Despite semantic differences they are translated into English by the same word, *heart*, owing to their function related to emotions. This aspect of their meanings allows them to be used interchangeably in some contexts. However, while the concept of *kasum*$_{(2)}$ is seen as the seat of emotions by Korean people, *maum* is perceived as the faculty of all kinds of psychological activity and is used as a metaphor for a person.

In terms of the semantic difference between them, apart from the fact that *kasum* is a polysemous word while *maum* is not, *maum* can think while *kasum*$_{(2)}$ does not have this function. People freely use the concept of *maum* in combination with cognitive predicates such as *sayngkakha-* 'think', *al-* 'know', *myengsangha-* 'meditate' and so on. Furthermore, *maum* plays an essential part in carrying out important tasks. It is, generally, positively valued for people to act with their *maum*. There is significant linguistic evidence that reflects this aspect. On the other hand, $kasum_{(2)}$, rather than *maum*, carries out the function of experiencing extreme emotions, though *maum* can also feel various kinds of emotion. This is suggested by the prevailing use of $kasum_{(2)}$ rather than *maum* in association with physical sensation terms such as *sili-* or *ali-* (*kasum/*maum silin iyaki*: a story which makes the *kasum* ache). These expressions are often used to refer to extreme emotional states[11].

Finally, evaluative concepts are not used in combination with $kasum_{(2)}$ while they are with *maum*. For instance, there are no expressions like *cohun/nappun kasum* (good/bad *kasum*) or *chakhan/miwun kasum* (virtuous/ugly *kasum*). This can be related to the Korean view of emotional states as something that cannot be controlled by the human will. Consequently, what happens in one's $kasum_{(2)}$ is a reaction to an antecedent, which should not be judged in terms of positive or negative values. The semantic overlap and the difference between these concepts explain precisely why they can serve as translational equivalents in some contexts but not in others.

In conclusion, the three Korean concepts related to the English concept of *heart* have been found to be highly culture-specific. The explications illustrate the Korean way of conceptualising human faculties either as body parts or as psychological entities, highlighting the cognitive content of the given concepts. As an analytical tool, the NSM method was found to be useful in explicating this kind of cultural concept since it does not rely on English and its language-specific categories. There have been several attempts to explicate heart-like concepts in different languages, though they are not presented in this paper: *duša* and *heart* by Wierzbicka 1992, *hati* by Goddard 2001, *kokoro* by Hasada 2000. A wider range of cross-linguistic investigations may contribute to a better understanding of ethnopsychology focused on the same part of the human body.

Acknowledgments

I am grateful to Anna Wierzbicka for her insight and comments on an earlier version of this paper. Also I would like to thank René Dirven, Yu Ning, and reviewers for their valuable comments.

Appendix

Table 1 The proposed Korean exponents of universal semantic primes (Yoon, 2003)

Substantives	*na*/I, *ne*/YOU, *nwukwu*/SOMEONE, *salamtul*/PEOPLE, *mwues*/SOMETHING/THING, *mom*/BODY
Determiners	*i*/THIS, *ttokkath-*/THE SAME, *talu-*/OTHER
Quantifiers	*han*/ONE, *twu*/TWO, *myech/etten -tul*/SOME, *motun*/ALL, *manh-*/MUCH/MANY
Evaluators	*coh-*/GOOD, *nappu-*/BAD
Descriptors	*khu-*/BIG, *cak-*/SMALL
Mental predicates	*sayngkakha-*/THINK, *al-*/KNOW, *wenha-*(V + *ko·siph-*/ WANT, *nukki-*/FEEL, *po-*/SEE, *tut-*/HEAR
Speech	*malha-*/SAY, *mal*/WORDS, *sasil*/TRUE
Actions, events and movement	*ha-*/DO, *ilena-(sayngki-)*/HAPPEN, *wumciki-*/MOVE
Existence and possession:	*iss-*/THERE IS, *kac-*/HAVE
Life and death	*sal-*/LIVE, *cwuk-*/DIE
Time	*ttay (encey)*/WHEN/TIME, *cikum*/NOW, *cen*/BEFORE, *hwu*/AFTER, *olay(-tongan)*/A LONG TIME, *camkkan (-tongan)*/A SHORT TIME, *elma tongan*/FOR SOME TIME
Space	*kos (eti)*/WHERE/PLACE, *yeki*/HERE, *wi*/ABOVE, *alay*/BELOW, *mel-*/FAR, *kakkap-*/NEAR, *ccok*/SIDE, *an*/INSIDE
Logical concepts	*an* (V+*ci·anh*)/NOT, *ama (u)l kes i-*/MAYBE, *(u)l·swu·(ka)·iss-*/CAN, *ttaymwun(ey)*/BECAUSE, *(u)myen*/IF
Intensifier, augmentor	*acwu*/VERY, *te*/MORE
Taxonomy, partonomy	*conglyu(-uy)*/KIND OF, *pwupuwn(-uy)*/PART OF
Similarity:	*kath-*/LIKE

The NSM model has changed a lot since it was first advanced in the early 1970s. In Anna Wierzbicka's 1972 book *Semantic Primitives*, only 14 semantic primitives were proposed and in her 1980 book *Lingua Mentalis*, the inventory was not much bigger. Over the 1980s and 1990s, however, the number of proposed primes was expanded greatly, reaching a current total of 60 or so. The same period also saw the development of some important new ideas about the syntax of the semantic metalanguage. Each member of the inventory has been selected using various theoretical bases and following cross-linguistic investigations. The current inventory, therefore, has been established on the basis of empirical findings (cf. http://www.une.edu.au/arts/LCL/disciplines/linguistics/nsmpage.htm)

Typographical conventions

The translations of Korean and English are represented either by K → E or E → K to indicate that the text is translated from Korean to English and vice versa, respectively. When translations of original texts are available, they are indicated with the year and page number immediately after this (e.g., K → E, Shin 1999: 29). The names which appear are either the original authors or translators.

The square brackets (i.e., [ ]) that are used in translated texts from Korean to English indicate the original Korean word or expression. Except where other names are indicated, all translations are mine. This is indicated by MT: K → E (My translation from Korean to English).

Notes

1. This study can be seen as an exploration in cross-linguistic semantic comparison. If heart-related concepts in different languages are analysed using the same analytical tool (i.e., the NSM framework), conceptual differences and similarities between different languages and cultures can be compared with great precision, semantic component by semantic component.
2. The Korean natural semantic metalanguage was constructed by testing the lexicon and syntax of the proposed NSM theory (see Yoon 2003 for details). The results of the tests generally support the NSM hypotheses. All the proposed semantic primitives are lexicalised in Korean, and their combinatorics are along the same lines as in the proposed universal grammar.
3. Translations of the examples in this paper are mine. Some of them may sound somewhat clumsy and unnatural in English because I tried to capture the Korean way of speaking in the translations. At the same time, I had to transfer the meanings into English in such a way that they make sense to English readers. The translations of book titles and collocations are particularly awkward maybe because they are more language specific.

4. There are several non-NSM primitives used in the explication such as ‘middle’ or ‘upper’ (and ‘man’ and ‘woman’ in the explication of *kasum*$_{(1)}$). They are semantic molecules that are used only for the purposes of the reader’s convenience. Their meanings are not decomposed because it is beyond the scope of this research. According to Goddard (1998a: 254) “There shouldn’t be anything surprising about this … One can presume that having recurrent semantic complexes packaged into word-like ‘chunks’ makes it easier for people to acquire and manipulate the huge amounts of semantic information involved. Presenting explications at the intermediate level … rather than trying to resolve them all the way down to the level of primitives, allows us to ‘see’ the recurrent chunks or semantic molecules appearing in the form of simple (though not primitive) words.”
5. 호랑가시나무에 얽힌 전설에는 '가슴 붉은새'의 이야기가 있습니다.
   There is a story about ‘the bird with a red feathered kasum’ in the legend of the *Holangkasi* tree. (MT: K-E, KAIST)
6. The question of whether or not this aspect of *kasum*$_{(1)}$ needs to be elaborated positing polysemy will be explored in depth in a further study based on more examples.
7. According to the KAIST corpus (http://csfive.kaist.ac.kr/kcp/), the frequency of *maum* is as follows: [count=8674, ratio=0.000638] total voca:165998; total word:13605457; ratio=count/total words.
8. In the dictionary *Kumsengphan Kwuke Taysacen* (1991: 919–920), there are 30 idiomatic expressions which are in combination with all the different predicates.
9. Koreans believe that they can think in their *maum*, as reflected in such expressions as *maumulo ihayhata* ‘understand things with the *maum*’. This feature of *maum* appears to be similar to the Chinese concept of *xin* (Yu 2003).
10. Although the total number of books that deal with the same subject is unknown, there are around 27 book titles containing *mom* and *maum* (body and maum) in the catalogue of Korea’s national library alone. One may presume that this can be an influence of Western dualism on the Korean view, but this dualism can be found in old records according to *Pyocwun Kwuke Taysacen* 1999.
11. One often refers to bodily symptoms when experiencing intense emotions.

References

Abelson, R.
1981 Psychological status of the script concept. *American Psychologist* 36 (7): 715–729.

Alford, C. F.
2000 *Korean values in the age of globalization*. Seoul: Tosechwulphan Green Bee.

Amberber, M.
2002 Testing emotional universals in Amharic. In *Emotions in Crosslinguistic Perspective*, Jean. Harkins and Anna Wierzbicka (eds.), 35–67. Berlin/New York: Mouton de Gruyter.

Brother Anthony of Taizé (with collaboration with Young-Moo Kim)
2002 *Variations / Three Korean Poets by S.Y Kim, K.N Shin, and S.Y. Lee*. Cornell University Press.

Chang, Suk-Jin
1996 *Korean*. Amsterdam: John Benjamins.

Choi, Il-To
2001 *Chamulo socwunghakiey cokumssik nohacwuki*. Seoul: Jung-Ang M & B.

D'Andrade, Roy G
1995 *The development of cognitive anthropology*. Cambridge; New York: Cambridge University Press.

Dirven, René
2002 Metonymy and metaphor: Different mental strategies of conceptualization. In *Metaphor and Metonymy in Comparison and Contrast*, René Dirven and Ralf Pörings (eds.), 75–111, Berlin/New York: Mouton de Gruyter.

Goddard, Cliff
1996a The "social emotions" of Malay (Bahasa Melayu). *Ethos* 24(3): 426–464.
1996b Cross-linguistic research on metaphor. *Language and Communication* 16 (2): 145–151.
1997 Contrastive semantics and cultural psychology: 'Surprise' in Malay and English. *Culture and Psychology* 3 (2): 153–181.
1998 *Semantic Analysis: A Practical Introduction*. Oxford: Oxford University Press.
2001 *Hati*: A key word in the Malay vocabulary of emotion. In *Emotions in Crosslinguistic Perspective*, Jean Harkins and Anna Wierzbicka (eds.), 167–195. Berlin/New York: Mouton de Gruyter.

2002 Towards a systematic table of semantic elements. Paper presented at *The workshop of metalanguage, scripts, explications: New developments in NSM theory*, Australian National University.
1997 Studies in the syntax of universal semantic primitives. *Language Sciences* 19 (3).

Goddard, Cliff and Anna Wierzbicka
1995 Key words, culture and cognition. *Philosophica* 55: 37–67.
1997 Discourse and culture. In *Discourse as Social Interaction*, T. A. van Dijk (ed.), 231–257. London: Sage Publications.

Goddard, Cliff and Anna Wierzbicka (eds.)
1994 *Semantic and Lexical Universals-Theory and Empirical Findings*. Amsterdam: John Benjamins.
2002 *Meaning and Universal Grammar-Theory and Empirical Findings I.II*. Amsterdam: John Benjamins.

Hasada, Rie
1994 The Semantic Aspect of Onomatopoeia: Focusing on Japanese Psychomimes. M.A diss., School of Language Studies, Australian National University.
2000 An exploratory study of expression of emotions in Japanese: Towards a semantic interpretation. Ph.D. diss., School of Language Studies, Australian National University.

Kim, Bong Jin
1995 *Namcaka ihonul kyelsimhalttay*. Seoul: Saymwunan.

Kim, Chiha
1998 *Heart's Agony: Selected Poems*. New York: White Pine Press.

Kim, Won-Woo
1993 *Kasumepsnun seysang [Society without kasum]*. Seoul: Seykyeysa.

Lakoff, George, and Mark Johnson
1980 *Metaphors We Live By*. Chicago: University of Chicago Press.

Lee, Yik-Seob, Sang-Eok Lee, and Wan Chae
1997 *Hankwukuy Ene [The language of Korea]*. Seoul: Shinku Munwha Sa.

Lyons, John
1977 *Semantics*. Cambridge: Cambridge University Press.

Niemeire, Susanne
this vol. *To be in control: kind-hearted and cool-headed: The head-heart dichotomy in English.*
1997 To have one's heart in the right place – metaphorical and metonymic evidence for the cultural model of the heart as the site of emotions in Englsh. In *Human Contact through Language and Linguistics*, B. Smieja and M. Tasch (eds.), 87–106. Frankfurt: Peter Lang.

Quinn, Naomi and Dorothy Holland. (eds.)
1987 *Cultural Models in Language and Thought.* Cambridge: Cambridge University Press.

Sapir, E.
1951 *Selected Writings of Edward Sapir in Language, Culture, and Personality.* Berkeley: University of California Press.

So, Chong-Ju
1996 *The Early Lyrics 1941–1960.* Seoul: DapGae Books.

Sohn, Ho-Min
1994 *Korean-Descriptive Grammar.* London: T.J Press (Padstow) Ltd.

Wierzbicka, Anna
1992a *Semantics, Culture, and Cognition: Universal human concepts in culture-specific configurations.* New York: Oxford University Press.
1992b Japanese key words and core cultural values. *Language in Society* 20: 333–385.
1993 A conceptual basis for cultural psychology. *Ethos* 21(2): 205–231.
1996 *Semantics: Primes and Universals.* New York: Oxford University Press.
1997 *Understanding Cultures Through Their Key Words: English, Russian, Polish, German, Japanese.* New York: Oxford University Press.
1999 *Emotions across Languages and Cultures: Diversity and universals: Studies in emotion and social interaction.* Cambridge: University Press.

Ye, Zhengdao
2000 The language of emotions in Chinese: A study based on Hong Lou Meng, M.A diss., School of Language Studies, Australian National University.

Yoon, Kyung-Joo
2003 Constructing Korean Natural Semantic Metalanguage. Ph.D. diss., School of Language Studies, Australian National University.

Yu, Ning
2003 Chinese metaphors of thinking. *Cognitive Linguistics* 14: 141–165.
this vol. The Chinese heart as the central faculty of cognition

Internet sources

http://kr.kordic.yahoo.com (accessed on December 10, 2004)
http://clid.yonsei.ac.kr:8000/dic/default.htm (accessed on December 9, 2004)
http://csfive.kaist.ac.kr/kcp (accessed on March 20, 2005)

Dictionaries

Dong-A's Prime Korean-English Dictionary (2nd ed.)
1996 Seoul: Dong-A Publishing & Printing Co.
Kumsengphan Kwuke Taysacen [Kumseng Korean Big Dictionary]
1991 Seoul: Kumseng.
Kwuke Taysacen [Korean Language Big Dictionary]
1965 Seoul: Mincwungsekwan.
Pyocwun Kwuke Taysacen [Standard Korean Big Dictionary]
1999 Seoul: National Academy of the Korean Language.
Saywulimal Khun Sacen [A new Korean big dictionary]
1974 Seoul: Seoul Sinmwunsa.
Si-sa Elite Korean-English Dictionary
1998 YMB Sisa yong-o sa.
The Random House Dictionary of the English Language
1991 Si-sa yong-o sa.

D. Dualistic heart/head-centering conceptualizations

Conceptualizations of *del* 'heart-stomach' in Persian

Farzad Sharifian

Abstract

This chapter explores conceptualizations that are reflected in the use of Persian expressions that include the body-part term *del* 'heart-stomach'. This body-part term provides a conceptual base for profiling a relatively large number of conceptualizations in Persian. By examining a number of expressions from Modern Persian the chapter shows how *del* is conceptualized as being either equal to, or the container of, emotions, desires, patience, courage, compassion, and also thoughts and memories. A set of expressions in Persian also reflects conceptualizations of *del* as the centre of personality traits, character, and mood. Overall, it appears that the body part *del* provides a conceptual foundation for speakers of Persian to represent their cognitive, emotional, socio-cultural, and linguistic experiences. The observed conceptualizations of *del* are very likely to have originated in cultural models derived from Iranian Traditional Medicine and Persian spiritual belief systems such as the Sufi worldview, which has influenced Persian cultural cognition through Sufi literature. The linguistic expressions discussed in this chapter provide a clear example of how language can serve as a 'memory bank' for cultural conceptualizations that have prevailed at different stages in the history of a speech community.

Keywords: Persian (Farsi), cultural model, embodiment, cultural conceptualizations, emotion, metaphor.

1. Introduction: The body in human conceptualization

The role of the body in human conceptualization has received much attention in recent years (e.g., Johnson 1987; Lakoff and Johnson 1999). The thesis of such works "stresses the continuity and motivating character of the relationship between pre- or non-linguistic bodily experience, and cognition; and seeks deep explanatory principle in human neurobiology" (Sinha and Jensen de López 2000: 18). A number of cognitive linguists have also explored the role of the body as a source domain in metaphorical mappings of different kinds (e.g., Kövecses 2000). The majority of studies of this kind view the conceptualization of the body to somehow interact with culture (e.g., Csordas 1994; Gibbs 1999; Maalej 2004; Yu 2002, 2003). Gibbs (1999) maintains that "embodied metaphor arises not from within the body alone, and is then represented in the minds of the individuals, but emerges from bodily interactions that are to a large extent defined by the cultural world". Yu (2003: 29) describes the relationship between body, culture, and metaphor as follows: "Conceptual metaphors are usually derived from bodily experiences; cultural models, however, filter bodily experiences for specific target domains of conceptual metaphors; and cultural models themselves are very often structured by conceptual metaphors."

The relationship between metaphors and cultural models is also emphasized by Cienki (1999), Emanatian (1999), Kövecses (1999, 2005), and Quinn (1987). In general, there seems to be a consensus that the notion of "body" largely lends itself to *cultural conceptualizations* (Sharifian 2003a), which are conceptual structures such as *categories*, *schemas*, and *metaphors* that deploy at the cultural level of cognition (Sharifian 2008). The embodiment thesis needs to place more emphasis on how people's experience of the human body is a function of more general cultural models that embody cultural groups' philosophies, cosmovisions, etc.

This chapter aims at contributing to an understanding of how body-part terms are the input for abstract conceptualizations of emotions, feelings, intellectual and/or spiritual faculties by analyzing the use of the body-part term *del* 'heart-stomach' in Persian, which occupies a central place in Persian language and culture. The chapter also explores the cultural models that appear to have motivated such conceptualizations. The following section provides brief background information about Persian and examines conceptualizations that are reflected in the use of expressions that include the word *del*.

2. Conceptualizations of *del* in Persian

Persian (also known as Farsi) is an Indo-European (more specifically, Indo-Iranian) language that is mainly spoken in Iran and some parts of Afghanistan and Tajikistan. Persian is also spoken by communities of Iranians and Afghanis living in different parts of the world. Historically, the development of Persian has been described in terms of three distinct phases: Old Persian, Middle Persian, and Modern Persian. Modern Persian has retained a simplified syntax from Middle Persian and is written in Persian-Arabic script. The lexicon of Modern Persian includes borrowings from other languages such as Arabic, Turkish, French and English.

The semantic system of Persian is a repository for the history and rich cultural traditions of Persian people (Asdjodi 2001; Beeman 2001; Sharifian 2004). Persian acts as an archive that reflects the thought patterns of those who have lived as Persians and who have spoken Persian throughout the history of its existence.

The word *del* is translated in the *Aryanpur Persian-English Dictionary* (Aryanpur Kashani 1984) as "heart, stomach, abdomen, belly, guts, mind, courage, patience, middle". When used literally in contemporary Persian, it refers to the area of the body below the chest and above the pelvis, roughly similar to the area described in English by *abdomen.* One may talk about a pain or a heightened sensation of food in the oesophagus as *ghazaa*[1] *sar-e-del-am moondeh* (food top-of *del*-my has.remained[2]). Also one may say, *del-am dard mikoneh* (*del-my* pain does) meaning one feels a pain somewhere in one's *del,* such as in the stomach or kidney. In its figurative usage, including its usage in literary texts, and also in some ethnomedical texts, the word *del* refers to the heart. For example, many Persian poems clearly reflect that the organ that is conceptualized is the blood-pumping organ. Consider the following from Eghbal Lahouri:[3]

(1) *Del az zogh-e tapesh del bood likan.*
del from zeal-of beating *del* was yet
'*Del* earned its eminence due to its love of beating, otherwise.'

(2) *Cho yek-dam az tapesh oftaad gel shod.*
as a-moment from beating fell mud became
'If it ceased to beat for a second it became mud.'

Persian also has the word *qalb,* which is a borrowing from Arabic, referring to the heart. The word *qalb* has been used to refer to the physical heart and has also increasingly been conceptualized as the seat of emotions, reflected in sentences such as *to dar qalb-e mani* 'you are in my heart' meaning 'I love you'.

A number of idiomatic expressions in Persian reflect particular image-schemas that underlie the conceptualizations of *del.* For example, the expression *shesh gooshe-ye del*[4] (six corners-of *del*) suggests the involvement of a six-sided container image-schema. The expression may be used in relation to willingness and satisfaction, as in (3) below:

(3)	*Ageh*	*shesh*	*gooshe-ye*	*del-et*	*raazi-e*	*boro.*
	if	six	corner-of	*del*-your	happy-is	go

'If you are really willing, you can go.'

Conceptualization of *del* as a container is also reflected in expressions which suggest that *del* can be emptied or filled (e.g., *khaali kardan-e del* [emptying-of *del*] 'to vent one's negative feelings, especially anger'). *Del* may also be conceptualized as having a bottom and a top, referring to sincerity. A sincere remark is said to be made from the bottom of the *del* (*az tah-e del* [from bottom-of *del*]) and an unfounded remark may be described as being from the top of the *del* (*az sar-e del* [from top-of *del*]). Moreover, there are expressions that suggest *del* can be narrowed and tightened. A very productive verb in Persian, *deltangi kardan* (*del*.narrrow do), which refers to 'missing something or someone', associates the feeling of missing with tightening or narrowing of *del*. In fact the English sentence *I miss him/her* translates into Persian as follows:

(4)	*Del-am*	*baraayash*	*tang*	*shodeh.*
	del-my	for.him/her	narrow/tight	has.become

'I miss him/her.'

One's *del* may also be described as 'thin' (*delnaazok* [*del*+thin]). This expression is close to 'having a tender heart' in English. *Del* may also be described as 'collapsed' (*del-am rikht* [*del*-my collapsed]), which refers to being shocked or startled. A shocking incident may be described as 'scratching the *del*' (*del-kharaash* [*del*+scratch]) and one may describe their horrifying experience in terms of their *del* being pulled out of its place, as in the following sentence:

(5)	*Del-am*	*az*	*jaash*	*kandeh*	*shod.*
	del-my	from	its.place	pulled.out	became

'I was horrified.'

The word *del* can also be used in Persian to refer to the middle or inner part of something. The middle of the night, for example, might be referred to as the *del-e shab* (*del*-of night), a conceptualization that is parallel to the English *the heart of the night*.

3. *Del* in psychological, intellectual, and person-bound concepts

3.1. *Del* as the seat of emotions, feelings and desires

A number of Persian verbs profile the schema of *del* as the seat of love. Among these are *del-baakhtan* (*del*+losing), meaning 'falling in love', and *delbari kardan* (*del*+winning do) and *del-robaayi kardan* (*del*+stealing do), which roughly refer to the act of charming a person. A lovelorn or lovesick person, or someone deeply in love, may be described as having a tired *del* (*del-khasteh* [*del*+tired]) or having given in their *del* (*del-daadeh* [*del*+given]). A lover in Persian may also be described as *delbar* (*del*+winner), which reflects conceptualization of a lover as a person who has won, or has taken, someone's *del*. Emotional attachment may also be attributed to one's *del* being tied to someone or something, as in the following sentence.

(6)	*Vey*	*be*	*moosighi*	*del-bastegie*	*faraavan*	*daasht.*
	he/she	to	music	*del*-tied	a.lot	had
	'He/She was very fond of music.'					

Similarly, emotional detachment may be described as cutting one's *del* from something or someone (*delboridan* [*del*+cutting] or pulling out one's *del* from something [*delkandan* [*del*+pulling.out]. Something charming and alluring may be described as being able to deceive one's *del* (*delfarib* [*del*+deceive]). Various kinds of endearment may also be conceptualized in Persian in association with *del*. Someone's much loved family members and relatives, such as a son or a daughter, may be referred to as being dear to one's *del*, as in the following sentence.

(7)	*Rezaa*	*pesar-am*	*aziz-e*	*del-am-e.*
	Reza	son-my	dear-of	*del*-my-is
	'Reza is my beloved son/I love my son Reza so much.'			

Also to be able to attract someone's love and attention may be described as being able to place oneself in someone's *del*, as in the following example.

(8)	*Khodesho*[5]	*to*	*del-e*	*baabaash*	*jaa*	*kardeh.*
	himself/herself	in	*del*-of	father	place	has.done
	'He/she has made his/her dad love him/her.'					

On the other hand, losing someone's favour and love may be described as 'going out of their *del*':

(9) *az* *del* *biroon* *raftan*
from *del* out going
'fall into disfavour'

It thus appears that *del* is conceptualized as a place in which only loved ones may reside. The following Persian saying equates residence in *del* with being seen.

(10) *Az* *del* *beravad* *har* *aankeh* *az* *dideh* *beraft.*
from *del* would.go anyone that from vision goes
'One who is not seen will not be missed/will be forgotten.'

The above saying is close to the English idiom *out of sight, out of mind.* It can be seen that in the Persian saying remembering is attributed to *del* while it is attributed solely to the "mind" in the English saying. It appears that the Persian version conceptualizes the act of forgetting as involving a loss of affection too. It is to be noted that both memory and affection in Persian may be conceptualized in association with *del.*

Besides love, *del* also profiles schemas that are associated with a whole range of feelings and emotions. For instance, *beh del gereftan* (to *del* take) refers to taking something personally and getting upset about it. *Del-am az dastesh poreh* (*del*-my from his/her.hand[6] is.full) means 'I am very upset with him/her'. It can be seen that being upset is conceptualized as 'having a full *del*' and as mentioned earlier, venting one's anger may be described as 'emptying one's *del*' (*khaali kardan-e del*). Another associated conceptualization is that of 'having a dirty *del*' (*del-cherkin* [*del*+dirty]), which refers to being slightly upset or unsatisfied with a state of affairs or a person.

Other conceptualizations of *del* in association with emotions include 'breaking someone's *del*' (*shekastan-e del),* similar to the English expression *breaking someone's heart.* Also, being able to charm someone or regain someone's favour may be described as 'gaining someone's *del*' (*del-e kasi raa be dast aavardan* [*del*-of someone DO.marker to hand bring]), which is also similar to the English expression *gaining someone's heart.*

Del is also used in expressions that are associated with compassion and mercy. One may talk of their *del* 'coming to mercy', as in the following:

(11) *Del-am* *baraash* *beh* *rahm* *oomad.*
del-my for.him/her to mercy came
'I felt pity for him.'

A set of Persian expressions capture the conceptualization of sympathy and being filled with pity as the 'burning of *del*', as in (12) below:

(12) *Del-am* *baraash* *sookht.*
del-my for.him/her burnt
'I felt sorry for him/her.'

The word *del* may also be used in expressions that reflect 'being offended and irritated', as in the following:

(13) *Az* *harf-e* *man* *del-khor* *shod-i?*
from word-of mine *del*-eater became-you?
'Did you get offended by what I said?'

It can be seen in the above example that being offended and upset is conceptualized as being a '*del*-eater' and someone can make another person a '*del*-eater', that is, make them upset. It is as though when being upset one would start eating one's *del.* A higher degree of upset may be described as being *del-khoon* (*del*+blood), as in (14) below:

(14) *Del-am* *az* *dastesh* *khoon-e.*
del-my from his/her.hand blood-is
'I am very upset with him/her.'

The above conceptualizations may also be expressed as *del-e por-khooni az dast-esh daaram* (*del*.ART full-blood.ART from hand-his/her I.have). It seems that being upset and hurt here is conceptualized as having caused a wound in one's *del,* filling it up with blood. Conversely, the act of consoling someone in Persian may be described as 'seeking someone's *del*' [*del-jooyi kardan* (*del*+seek do)]. Another expression that involves *del* and blood is *khoon-e del khordan* (blood-of *del* drink/eat). This expression refers to selfless efforts that people may make as well as sufferings that they may go through in achieving a goal or carrying out a task, such as raising a child. This is exemplified in the following example.

(15) *Khoon-e* *del* *khord-am* *taa* *in* *bacheh* *ro* *bozorg*
blood-of *del* drank/ate-I till this child DO.marker big
kard-am.
did-I
'I went through a lot to raise this child.'

In this sentence, the sufferings appear to be conceptualized as having caused wounds to one's *del*. The noteworthy point here is that going through the hardship is conceptualized as 'drinking one's blood', an expression that reflects the association between the *del* and eating/drinking channels, possibly the stomach.

A number of expressions in Persian profile a schema that associates *del* with anxiety. One may talk about 'turbulence in one's *del*' in association with anxiety, a conceptualization that has a physical relevance. That is, the fact that anxiety can lead to an upset stomach might have been the basis for conceptualizing anxiety as various types of turbulence in one's *del*, as in the following example:

(16) *Too* *del-am* *shoor* *oftaadeh.*
in *del*-my disturbance has.fallen
'I am very worried.'

Another noteworthy conceptualization of *del* in Persian is as the 'agent for desires and cravings'. Persian speakers usually attribute their likes and dislikes, as well as their cravings, to their *del*. Consider the following examples:

(17) *Del-am* *bastani* *mikhaad.*
del-my ice.cream wants
'I fancy ice-cream.'

(18) *Del-am* *parmizaneh* *baraa* *yek* *sikh* *kabaab.*
del-my wings[7] for one skewer kebab
'I long for a kebab.'

(19) *Del-am* *ne-mikhaad* *baa* *kasi* *harf* *bezanam.*
del-my doesn't-want with anyone talk do
'I don't want to talk to anyone.'

In this context, Persian speakers may describe their hopes and desires as *moraaad-e del* (wish-of *del*) and may also talk of something desirable as *del-khaah* (*del*+wish). It should be noted that the desires and wishes that can be attributed to *del* are of various types, ranging from physical, such as a craving for some kind of food, to psychological, such as a wish to see somebody or a wish to achieve some goals. Closely related here is the use of *del-neshin* (*del*+seat) to refer to something pleasing, as in (20) below.

(20)	*Sedaay-e*	*kheili*	*del-neshin-i*	*daar-in*
	voice-ART	very	*del*-sit-ART	have-you
	'You have a very pleasant/nice voice.'			

Thus, for something to be pleasing to one's senses it should be able to 'sit' in one's *del.* This conceptualization appears to be close to the English *Your voice soothes my soul.* The above-mentioned expressions suggest that in Persian the intake of all human senses is somehow conceptualized as affecting one's *del*, in one way or another.

Del may also be conceptualized as 'cold' or 'warm', suggesting absence or presence of assurance and encouragement. That is, cold *del* (*del-sard* [*del*+cold]) may be glossed as 'discouraged' or 'disappointed' while warm *del* (*del-garm* [*del*+warm]) may be translated as 'assured', 'confident', or 'motivated'. Note that *warm-hearted* and *cold-hearted* in English refer to friendliness and not encouragement and assurance.

3.2. *Del* as the centre of thoughts and memories

As mentioned above, *Aryanpur Persian-English Dictionary* considers 'mind' to be an English equivalent of *del.* This in fact arises from the schema that associates *del* with reason and mental activities. One may talk about the 'secret of *del*' (*raaz-e del* [secret-of *del*]) or 'word of *del*' (*harf-e del* [talk/word-of *del*]), meaning one's genuine thought or desire. Both these expressions may refer to thoughts that one may want to keep in confidence. Having one's *del* on the tip of one's tongue (*del-esh sar-e zaboon-esh-e* [*del*-his/her tip-of tongue-his/her-is]) refers to the inability to keep secrets. One may also ask someone to keep a secret in his *del*, as in the following example:

(21) *Chiz-i ke mikhaam behet beg-am too del-et*
Thing-ART that want.I to.you say-I in *del*-yours
bezaar/negah-daar va be kesi na-goo khaaheshan.
put/keep and to anyone don't-say please
'Please keep what I am going to tell you as a secret.'

The act of 'placing in the *del*', mentioned above, may also be expressed as 'hiding in the *del*' or 'keeping in the *del*'. In these expressions, *del*, as a container of some sort for the safekeeping of secrets, is conceptualized as carrying a "mental" function. An idiomatic expression in Persian describes honest people as those whose tongue and *del* are the same:

(22) *Del o*[8] *zaboon-esh yeki-e*
del and tongue-his/hers one-is
'He/she is honest.'

In this idiom, one's tongue being the same as one's *del* refers to saying honestly what one thinks. The expression clearly reflects conceptualization of *del* as the "seat of thinking". Also note that the compound noun *del-aagaahi* (*del*+awareness) in Persian refers to 'consciousness'. *Del* is also associated with inspiration. The expression *beh del-e kasi oftaadan* (to *del*.of someone fall), literally referring to ideas falling in someone's *del*, may be glossed in English as 'someone being inspired'. Another expression that suggests a similar conceptualization of being inspired is *del-am behem migeh* (*del*-my to.me says), in which *del* is conceptualized as having the ability to talk. This expression is used to refer to people's inner feelings, particularly about something that is happening or going to happen.

It should be mentioned here that in Modern Persian, thinking is predominately associated with *zehn*, which is roughly similar to the conceptualization of the mind in English. For example, the notion of 'occurring to someone' is expressed as *be zehn aamadan* (to mind come) 'to come to mind'. *Zehn* is mainly associated with a person's brain.

3.3. *Del* as the centre of personality traits, character, and mood

Del in Persian also profiles a schema associated with mood. Being gloomy might be described as having a 'dull' or a 'nebulous *del*' (*del gerefteh*) and cheering up may be described as 'opening of the *del*' (e.g., *del-am baaz shod* [*del*-my open became]). With regards to more stable temperament, someone who is usually cheerless may be referred to as having a 'dead *del*' (*del-mordeh* [*del*-dead]) and someone who is usually lively and cheerful may be described as having a 'living *del*' (*del-zendeh* [*del*-alive]).

A set of phrasal verbs in Persian profiles the cultural conceptualization of "*del* is patience". For example the expression *del-am aab shod* (*del*-my melt became), refers to the exhaustion of patience, which may also be expressed as the finishing of one's *del*, as in the following example:

(23) *Zood* *begoo* *chi* *shodeh,* *del-am* *tamoom* *shod.*
quickly tell what has.happened, *del*-my finish became
'Come on! Tell me what happened, I can't wait to hear it.'

Also note that the adjective *del-gondeh* (*del*-big) is used to refer to someone with excessive patience. On the other hand, someone who is usually very impatient is described as having a small *del* (*del-kouchik).* Another expression that also reflects this conceptualization is *del too del-am nist* (*del* in *del*-my is.not) roughly meaning 'I can't wait'.

It is to be noted here that conceptualizations such as "*del* is patience" may be regarded as metaphors involving mapping from one domain onto another. While it is agreed that conceptual mapping is among the fundamental conceptual processes, it may not always be possible to characterize such processing on the basis of lexical appearance. At some stage, speakers of Persian, for example, might have conceptualized their *del* to be the real seat of emotions, without any mapping involved. It is only when later generations attribute emotions to other parts such as the brain that it makes sense to view expressions of "*del* as the seat of emotion" as being metaphorical. Even those later generations may simply use the expression as a whole to mean something without explicitly thinking about the body parts that are involved. Also sometimes viewing a conceptualization as metaphor reflects an English ethnocentric viewpoint toward the conceptualization involved. For example, as I have maintained elsewhere (Sharifian 2003b), in the context of Australia while "crying rain" might be considered to be a metaphor from an Anglo-Australian perspective, from an Aboriginal perspective the event may be conceptualized as the real crying of Ancestors.

A number of idiomatic expressions in Persian profile the schema of "*del* as courage and bravery". The expression *del-daashtan* (*del*-having) refers to being courageous. By contrast, *del na-daashtan* (*del* not-having) means lack of courage. One may refer to someone as having a lot of *del* and courage (*kheili del o jor'at daareh* [a.lot.of *del* and courage has.he/she]). Courage in Persian is also associated with *jegar* 'liver'. Speakers of Persian may use both these body parts in one expression to refer to courage, as in the following example:

(24) *Rezaa kheili del o jegar daareh.*
Reza a.lot *del* and liver has
'Reza is very brave.'

Another expression which reflects the conceptualization of *del* as associated with courage and risk is *del be daryaa zadan* (*del* to sea hit), which refers to taking a risk. Different cultures may conceptualize courage in association with different body parts. In Australian English, "guts" is usually associated with "courage" whereas in Chinese "gallbladder" is conceptualized to be the seat of courage (Yu 2003).

The word *del* is also frequently used in Persian speakers' narrative on illness. For example, one may refer to "mental stress" as *del-shooreh daashtan* 'having a turbulent *del*'. In fact, the majority of psychological conditions are often attributed to *del,* or *qalb,* in Persian.

In summary, the analysis of the expressions in which the word *del* is used shows that *del* provides a semantic base for a relatively large number of concepts. This base appears to include conceptualization of *del* either as being equal to, or the container of, emotions, desires, patience, courage, compassion, and also thoughts and memories. Some Persian expressions reflect the conceptualization of *del* as something that can be tightened, opened, wounded, emptied, filled, burned, collapsed, pulled out, broken or stolen.

4. Cultural models behind the conceptualizations of *del*

The observed conceptualizations of *del* in Persian are very likely to have originated from several sources, including Iranian Traditional Medicine, and spiritual belief systems such as the Sufi worldview. Iranian Traditional Medicine was influenced to a great extent by the Greek medical theory of four humours. The theory of four humours has been very influential in Persian culture and medicine through the work of Abu Ali Sina (Avicenna). According to Iranian Traditional Medicine, the body parts "heart" and "liver" are among the "chief" organs that govern all other organs, and the heart in particular occupies a significant place in the human body, in that it is the seat and place of origin of *ravaan* 'soul/psyche'.

In Avicenna's version of the theory of four humours, human personality is tied to human temperaments, and a natural predominance of a given humour in the body provides specific characteristics of body and behaviour. For instance, yellow bile, which was believed to be secreted from the liver, is associated with courage. This is most likely the origin of the cotemporary Persian expression of *jegar daashtan* 'to have liver', which refers to being courageous, similar to that of *del daashtan* 'to have *del*'. Also, in Persian, patience is associated with the heart (i.e., *del-koochik* 'small heart' = lacking patience), which seems to be associated with Avicenna's (1984: 521) attribution of patience to having a 'cold' heart. For Avicenna there is also a link between the heart and understanding. For example, being slow in understanding is attributed to having a 'cold' heart.

Another possible source of the conceptualizations of internal body organs in Persian is the Sufi spiritual belief system. Ahmadi and Ahmadi (1998: 40) maintain that the Iranian version of Sufism, which was rooted in the Iranian traditional worldview, has had a remarkable influence on the Iranian psyche and Persian literature. The Persian conceptual system seems to have been very much influenced by Sufism through the work of Sufi poets such as Rumi, Hafiz, and Sa'di, whose works are widely recited by Persian speakers. O'Shea (2000: 87) calls Iran "a nation of poets" and observes that "poetry has an importance in Iranian society that is probably not replicated in any other country".

In Sufism, *del* refers to the spiritual 'heart' not the physical heart. In the Sufis' pathway, *del* is a spiritual stage, or domain between the material nature and the innermost consciousness, where only God exists (Nurbakhsh 1992: 3). Material nature is viewed as a biological heritage which develops into *nafs*[9] 'ego, self'. In Sufism, all the "earthly", negative human tendencies, including wants and desires, are attributed to *nafs*. A Sufi's mission in life is then to battle with this *nafs*, for example by suppressing wants and desires, towards perfection. Nurbakhsh (1992: 71) observes that "when the *nafs* has reached the ultimate degree of perfection, it passes on to the level of the heart". But it is also possible for the heart to be dominated by the *nafs* and that is when it is described as being "darkened by the tarnish of the *nafs*' multiplicity, taking on its hue" (Nurbakhsh 1992: 71). *Del* is situated between *nafs* and *ruh* 'spirit', which is spiritually a higher realm than the heart. Nurbakhsh observes that in Sufi tradition, *nafs* is the seat of desires, *del* is the seat of knowledge, and *ruh* is the seat of love, in its spiritual and divine sense.

From Nurbakhsh's discussion it appears that the word *del* has been used metaphorically to refer to the spiritual heart. He observes that

> The physical heart is in a constant state of alteration, regulating the changes between arterial and subtle blood and venal or gross blood. The spiritual heart is also in a constant alteration, here rotating back and forth between the subtle influences of the spirit and the gross influences of the *nafs* … The physical heart sends blood to the arteries and receives the gross blood of the veins; it is essential to the process of purifying the human body. Likewise, the spiritual heart receives the gross traits of the *nafs* and purifies them with the help of the spirit, ... (Nurbakhsh 1992: 75)

It is clear that the spiritual stage of purification of the human psyche is called 'the heart' in Sufism due to its resemblance to the functions of the physical heart. In general the way in which the heart is conceptualized in Sufim shows a degree of similarity to the conceptualizations of *del* that were presented earlier in this chapter. For example, in Sufism the heart is the site of knowledge and reason, which is consistent with the conceptualizations of *del* explored in section 3.2 of this chapter. However, some of the conceptualizations that have been noted in the use of *del* expressions rather suggest a "misinterpretation" of Sufi concepts, as discussed by Nurbakhsh. Nurbakhsh (1992: 79) refers to this as follows:

> Ordinary people tend to confuse the heart with the *nafs*. When someone says "it is my heart's desire that …"[10] it is not the heart that is desiring; the desire comes from the *nafs*. When one says, "This appeals" or "does not appeal to my heart,"[11] it is not the heart but the *nafs* which accepts or rejects that thing; the heart is above acceptance or rejection.

It should however be kept in mind that Sufism has various versions and the role of the heart has not been identically conceptualized in all of these. For example, in one version, the heart is mainly the seat of emotions, while in other versions the heart is the "house" of God (e.g., Chittick 1989). On the other hand, the ways in which Sufi poets have used the concept of *del* have not been entirely consistent with the account provided above. Sufi poets in general seem to have conceptualized *del* as the seat of not only reason but also desires, wishes, and emotions. A prominent concept in Sufi poetry is *qam-e del* 'grief-of *del*', which reflects the conceptualization of *del* as the seat of emotions. An interesting conceptualization of *del* as the site of reason in Sufi poetry is revealed when the act of thinking is described as *baa del goftan* (with *del* say) 'say to *del*'.

A thorough treatment of Sufism and Sufi poetry falls beyond the scope of this chapter but it seems plausible to attribute the origin of some of the conceptualizations of *del* to cultural models that have been developed out of the influence of Sufi tradition on Persian literature. As an example, the attribution of something that a person fancies to one's *del* (e.g., *del-am x mikhaad* [*del*-my x wants] 'I fancy x') seems to have its roots in the Sufi's view of wants and desires as negative human traits. That is, such negative traits should be attributed to *part* of a person rather than the individual as a whole. In a sense, saying 'My *del* wants x' instead of 'I want x' is an attempt to distance oneself from the desire or want and somehow blame it on the *del*. A similar phenomenon seems to be very frequent in Sufi poetry, which has also influenced non-Sufi poets. That is, the poet often attributes "non-Godly" desires and wants to their *del* and may even condemn their *del* for that. The following is an example of a poem by Baba Taher in which he addresses his *del* in a complaining tone: [12]

(25)

Megar[13]	*shir-o*	*pelang-i*	*ey*	*del!*	*ey*	*del!* –	*beh*	*mo*
question.marker	lion-and	leopard-be	oh	*del*	oh	*del*	with	me

daayem	*beh*	*jang-i*	*ey*	*del!*	*ey*	*del!*
always	in	fight-be	oh	*del*	oh	*del*

(Addressed to *del*) 'Are you a lion or a leopard, as you always fight with me?'

It can be seen here that the poet conceptualizes his *del* as a wild animal fighting with him, a conceptual mapping implying that the emotions that reside in the *del* always go against the will of reason. This suggests a conceptualization of EMOTION AS AN OPPONENT.

Finally, it is to be noted that the cultural models that have been briefly explored in this chapter in relation to the conceptualizations of *del* in Persian should not be viewed as being totally distinct from each other. For example, Avicenna and Sufism do not represent two distinct philosophies of the human nature, but in fact there is ample evidence that Avicenna both influenced and was influenced by Sufism (e.g., Morewedge 1971, 1972). But this does not lessen the strength of the argument that conceptualizations of body are largely mediated by cultural models.

5. Conclusion

The analysis presented in this chapter reveals that *del* provides a conceptual base for profiling a relatively large number of conceptualizations in Persian. These include conceptualizations that regard *del* either as being equal to, or the container of, emotions, desires, patience, courage, compassion, and also thoughts and memories. Some Persian expressions reflect the conceptualization of *del* as something that can be tightened, opened, wounded, emptied, filled, burned, collapsed, pulled out, broken or stolen. Among the conceptual attributes of *del* is the ability to talk and even host people. These conceptualizations appear to provide a basis for the expression of a plethora of cultural meanings. Overall, the observations made in this chapter suggest that language serves as a "collective memory bank" (wa Thiong'o 1986) for cultural conceptualizations that have prevailed at different stages in the history of a speech community and these can leave their traces in current linguistic practice (see more in Sharifian to appear). In this sense language can be viewed as a primary mechanism for storing and communicating cultural conceptualizations.

Acknowledgments

I would like to thank Mrs Mansoureh Sharifzadeh for her generous help with the dictionary search required for this study. René Dirven and Ning Yu also deserve special thanks for their helpful comments on an earlier draft of this chapter.

Notes

1. In Persian transcriptions, the letter "a" symbolizes a low front vowel which is close to the sound of "a" in the word "cat". The "aa" sequence stands for a low back vowel which is close to the sound of "a" in the word "father".
2. The English glosses are semantically based and are attempts to provide the closest possible equivalence to the Persian expressions. It is acknowledged that even terms like 'emotion' and 'reason' capture culture-specific conceptualizations.
3. The equivalents given under each word in the examples are intended to serve as a semantic guide and are not to be taken as the exact morphosyntactic equivalents.

4. It is to be noted that there are certain cross-dialectal differences in the use of the expressions that include the word *del* in contemporary Persian and thus some Persian readers may find some expressions included in this chapter unfamiliar.
5. The ending *o* in expressions like *khodesho* is the spoken form of *raa*, which is a direct object marker postposition in Persian. For more information see Dabir-Moghaddam (1990).
6. The expression *az dastesh* (from his/her.hand) in Persian is a metonymical expression meaning 'from/with him/her', particularly in expressing negative feelings towards someone.
7. In Persian, the use of *par zadan* 'to wing' in conjunction with *rooh* 'soul' or *del* suggests a very strong desire and longing.
8. *O* here is a spoken form of *va* 'and'.
9. See http://islam-usa.com/im17.html for a discussion of the meaning of *al-nafs* in non-Sufi Islam.
10. "It is my heart's desire that ..." seems to be the translation of the expression *del-am mikhaad keh* 'my *del* wants that '
11. "This appeals to my heart" seems to be the translation of *beh del-am mishineh* (lit. 'it sits to my *del*')
12. Baba Taher's poetry is written in a rustic local dialect, which adds a great degree of beauty and genuineness to it. The translation to English, it is acknowledged, makes Baba Taher's poetry largely devoid of its charm, including its dialectal beauty.
13. The interrogative marker *megar* is pronounced *magar* in Standard Persian.

References

Ahmadi, Nader and Fereshteh Ahmadi
1998 *Iranian Islam: The Concept of the Individual.* MacMillan Press.

Aryanpur Kashani, Abbas, and Manouchehr Aryanpur Kashani
1984 *Aryanpur Persian-English Dictionary.* Tehran: Amir-Kabir Publications.

Asdjodi, Minoo
2001 A comparison between *ta'arof* in Persian and *limao* in Chinese. *International Journal of the Sociology of Language* 148 (1): 71–92.

Avicenna (Abu Ali Sina)
1984 *Ghanoon dar teb* [Canon of medicine], Abdolrahman Sharafkandi (trans.). Tehran: Soroush Publications.

Beeman, William O.
2001 Emotion and sincerity in Persian discourse: Accomplishing the representation of inner states. *International Journal of the Sociology of Language* 148 (1): 31–57.

Chittick, William
1989 *The Sufi Path of Knowledge*. Albany: State University of New York Press.
Cienki, Alan J.
1999 Models as profiles and bases. In *Metaphor in Cognitive Linguistics*, Raymond W. Gibbs Jr., and Gerard J. Steen (eds.), 189–203. Amsterdam: John Benjamins.
Csordas, Thomas J. (ed.)
1994 *Embodiment and Experience*. London: Cambridge University Press.
Dabir-Moghaddam, Mohammad
1990 On postposition *raa-* in Persian. *Journal of Iranian Linguistics* 13 (1): 2–60.
Emanatian, Michele
1999 Congruence by degree: On the relationship between metaphor and cultural models. In *Metaphor in Cognitive Linguistics*, Raymond W. Gibbs Jr., and Gerard J. Steen (eds.), 205–218. Amsterdam: John Benjamins.
Gibbs, Raymond W., Jr.
1999 Taking metaphor out of our heads and putting it into the cultural world. In *Metaphor in Cognitive Linguistics*, Raymond W. Gibbs Jr., and Gerard J. Steen (eds.), 145–166. Amsterdam: John Benjamins.
Johnson, Mark
1987 *The Body in the Mind: The Bodily Basis of Meaning, Imagination, and Reason*. Chicago: The University of Chicago Press.
Kövecses, Zoltán
1999 Metaphor: Does it constitute or reflect cultural models? In *Metaphor in Cognitive Linguistics*, Raymond W. Gibbs Jr., and Gerard J. Steen (eds.), 167–188. Amsterdam: John Benjamins.
2000 *Metaphor and Emotion: Language, Culture and Body in Human Feeling*. Cambridge: Cambridge University Press.
2005 *Metaphor in Culture: Universality and Variation*. Cambridge: Cambridge University Press.
Lakoff, George and Mark Johnson
1999 *Philosophy in the Flesh: The Embodied Mind and its Challenge to Western Thought*. New York: Basic Books.
Maalej, Zouhair
2004 Figurative language in anger expressions in Tunisian Arabic: An extended view of embodiment. *Metaphor and Symbol* 19 (1): 51–75.
Morewedge, Parviz
1971 The logic of Emanationism and Sufism in the philosophy of Ib Sina (Avicenna), Part I. *Journal of the American Oriental Society* 91 (4): 467–476.

1972 The logic of Emanationism and Sufism in the philosophy of Ib Sina (Avicenna), Part II. *Journal of the American Oriental Society* 92 (1): 1–18.

Nurbakhsh, Javad
1992 *The Psychology of Sufism (Del va Nafs).* London/New York: Khaniqahi-Nimatullahi Publications.

O'Shea, Maria
2000 *Culture Shock: Iran.* Portland, Oregon: Graphic Arts Publishing Company.

Quinn, Naomi
1987 Convergent evidence for a cultural model of American marriage. In *Cultural Models in Language and Thought*, Dorothy Holland, and Naomi Quinn (eds.), 173-192. Cambridge: Cambridge University Press.

Sharifian, Farzad
2003a On cultural conceptualisations. *Journal of Cognition and Culture* 3 (3): 187–207.
2003b Cultural conceptualisations in intercultural communication between Aboriginal and non-Aboriginal Australians. Paper presented at the 8th International Cognitive Linguistics Conference, University of La Rioja, Spain. July 20–25, 2003.
2004 Cultural schemas and intercultural communication: A study of Persian. In *Outer Limits: A Reader in Communication and Behavior across Cultures*, James Leigh, and Eric Loo (eds.), 119–130. Melbourne: Language Australia.
2008 Distributed, emergent cultural cognition, conceptualisation, and language. In *Body, Language, and Mind, Vol. 2: Sociocultural Situatedness*, Roslyn M. Frank, René Dirven, Tom Ziemke, and Enrique Bernárdez (eds.),109-136. Berlin/New York: Mouton de Gruyter.
To appear On collective cognition and language. In *Language* and *Social Cognition,* Hanna Pishwa (ed.), Berlin/New York: Mouton de Gruyter.

Sinha, Chris and Kristine Jensen de López
2000 Language, culture and the embodiment of spatial cognition. *Cognitive Linguistics* 11: 17–41.

wa Thiong'o, Ngugi
1986 *Decolonising the Mind: The Politics of Language in African Literature*. London: Heinemann.

Yu, Ning
2002 Body and emotion: Body parts in Chinese expression of emotion. *Pragmatics and Cognition* 10 (1): 341–367.
2003 Metaphor, body, and culture: The Chinese understanding of *gallbladder* and *courage*. *Metaphor and Symbol* 18 (1): 13–31.

Expressions concerning the heart (*libbā*) in Northeastern Neo-Aramaic in relation to a Classical Syriac model of the temperaments*

Daniel P. Wolk

Abstract

Northeastern Neo-Aramaic (NENA), the language spoken by Assyrians, seems to share key metonyms and metaphors of HEART in English. Examples can be found of both the HEART FOR THE PERSON metonym and THE HEART IS A CONTAINER OF EMOTIONS metaphor. However, in at least three important ways, *libbā* does not correspond precisely to "heart". First, its place among other organs of the body differs from English. It encompasses the stomach area as well, as in *dḇiqleh libbeh* ("his heart was gripped" ⇒ he lost his appetite). The *ʿaynā*, anatomically the eye, unlike in English, also is a CONTAINER of something resembling *emotions*. Whereas *libbā* contains those emotions related to attitudes toward others or moral character traits (akin to Aristotle's *hexeis*), *ʿaynā* contains desires of wanting or coveting.

Second, although NENA, like English, bears relics of the ancient Galenic medical model of the temperaments, the traces it has preserved differ from those in European languages. The Galenic model had much influence in Classical Syriac, the liturgical language of the Assyrians and an important part of their heritage. Unlike English, the modern language does not know the humors. However, another aspect of the ancient theory survives more markedly than in English, namely the HEART being a combination, or "mixture" (*muzāḡa*), of WET/DRY and HOT/COLD properties. So there are many expressions attributing one of these properties to a person, or describing a change in one them in him or her. But most remarkably, the sense of the heart as the source of movement in the body is expressed through idioms concerning the heart staying put (*libbā pyāšā*) and going (e.g. *beʾzālā*).

Third, other models come into play, such as the one underlying the highly productive metaphor BEING MORAL IS BEING CLEAN AND WHITE. A person with a "clean heart" (*libbā tammiz*) is above suspicion. A man who looks at a woman with a "dirty" one (*šiḫtānā*) has only lascivious desires.

In the course of the analysis, the paper cautions against overgeneralizing the meaning of "emotion". It suggests that for the purpose of comparative research, it is necessary to distinguish among the various components of emotions, such as feelings, affects, somatic changes, and perceptions. More importantly, it is necessary to distinguish among transitory, episodic states, which is what is normally meant by "emotion", and longer-term dispositions and character traits.

Keywords: Aramaic-Syriac, cultural models, post-metonymy, metaphor, heart, embodiment.

1. Introduction

Like the other ones in this volume, this paper focuses its attention on the tacit understandings and meanings underlying linguistic usages in one specific language. It does not concern a particular emotion, but rather the conditions and workings of a particular body organ, paralleling what is called in English the "heart". The language chosen is the one spoken by members of a Christian minority, namely Assyrians, whose ancestral lands are currently located in southeastern Turkey and adjacent regions of northern Iraq and Iran, and who now live in the latter two countries, as well as Syria, Lebanon, and an extensive diaspora in Europe, North America, and Australia. Speakers refer to the language simply as "Assyrian", but the current consensus among linguists is to call it Northeastern Neo-Aramaic (hereafter "NENA").[1]

The contemporary sociolinguistic situation of NENA is intricate and evolving, and it poses difficulties for researchers. Until World War I, the vast majority of Assyrians lived in villages. Many of these villages, especially in the Hakkari mountains in the Ottoman Empire (now southeastern Turkey), were isolated from each other and as a result exhibited much dialectal diversity. A majority of Assyrian NENA speakers took refuge in Iraq after World War I. To a certain degree, especially in the cities, speakers of different dialects mixed with each other and developed an urban "koinē" form. Native speakers who were consulted for this paper, including Oraham Yalda Oraham, are speakers of this "new language", but also have familiarity with "older" local dialects, such as the Lower Ṭyārē dialect in Oraham's case. With very few exceptions, it was not until the 19th century, under the influence of American Presbyterian, Anglican, and French Catholic missionaries, that vernacular forms of the language were committed to writing. Until today, only a minority of Assyrians are able to read the language.

Like other Semitic languages, NENA has an elaborate verb and noun morphology. Words are based on three or four root letters, or radicals, that undergo much interweaving to generate inflectional and derivational forms. The language shares more roots with Hebrew than with Arabic. Unlike other Semitic languages, NENA uses many copular forms in a variety of tenses and aspects, and the uses of active, passive, and, as some have argued, ergative forms (e.g. Chyet 1995), differ considerably from other Semitic languages. A great many words and idioms are cognate with those of neighboring languages, but vary dramatically from dialect to dialect. The Assyrians of Iran borrowed many words and calqued numerous expressions from Azerbaijani Turkish and Persian. In Ottoman territories, there are many Kurmanji Kurdish words. In southern dialects can be found many words borrowed from the local forms of Arabic.

Although the vernacular language seldom was written prior to the influx of Western missionary activity, deacons and priests were more or less versed in Classical Syriac, which continues to be used as the liturgical language in most Assyrian churches. The written form of this language flourished in the third to fifth centuries A.D. Subsequent writings, like those of Latin in Europe, tried to maintain earlier forms of the language that ceased being spoken. There is consensus among linguists that the contemporary spoken forms, including the Iraqi koinē, which is the main subject of this paper, is a rather different language from Classical Syriac, both lexically and grammatically. Nevertheless, from about the time that the koinē began to coalesce after World War I, there has been a growing movement to purge the language of Azerbaijani Turkish, Persian, Kurmanji, and Arabic words. Although the movement has had a significant impact on contemporary written forms, it has had only modest influence on how Assyrians speak in everyday life. In this paper, I have relied primarily on texts, as well as native speakers' testimony, that bear little influence of these contemporary trends to purify the language. But, as I will suggest, Classical Syriac does seem to have left its mark in cultural models that underpin the spoken language today.

Aside from applying cognitive linguistics to a language that has never been investigated within this paradigm before, this paper has specific theoretical goals: First, like others in this volume, it aims to call into question some of the strong claims that have recently been made about the "embodiment" of specific metaphorical structures. According to such claims, there are primary human metaphors that are universal on account of their entailment by the same experiences humans in all cultures share by virtue of inhabiting the same type of physical bodies. Close investigation of figurative uses, primarily metaphors and metonyms, in NENA concerning the "heart", as it relates to other body organs, will show that in spite of resemblances to those in English, such metaphors and metonyms are distinctive. Second, much light is shed on the tacit meanings of contemporary expressions through the "unearthing" of an archaic model.

The paper follows a simple strategy. In the first part, it shows that at first blush, many NENA usages connected with *libbā*, or 'heart', seem to be instances of the metaphor THE HEART IS A CONTAINER OF THE EMOTIONS. With some differences, it appears that the same emotions that are contained in the "heart" in English are also contained in the NENA *libbā*. Nevertheless, I show that this impression is overly simplistic, for the word "emotion" in English is a vague, inclusive one, and it is unwarranted to assume that the concept of it that English speakers hold corresponds exactly to a concept in NENA (or any other language). It is necessary both to decompose the concept of EMOTION into its components, and distinguish it from more perduring aspects of persons such as moods, temperaments, character traits, and the Aristotelian notion of *hexeis*, which resonates well with many NENA usages.

In keeping with other papers in this volume, it is important to avoid presuming that *libbā*, even when it clearly refers to bodily organs, corresponds exactly to the 'heart' in English. Rather, in many usages, it applies to a larger area that includes the stomach. Also, the emotions or desires contained in the *libbā* contrast in an interesting way with the desires contained in another body organ, the *ʿaynā* ('eye').

After completing this initial account of *libbā* in NENA, I turn to a sample of texts in Classical Syriac (abbreviated "CS"). From these texts, it is possible to outline some key features of an archaic model of the heart (also *lebba*)[2] that bears influence of Galenic medicine. In many ways this model resembles one in Persian described by Sharifian (this volume), as well as ones stemming from Galenic medicine in Western Europe. However, it is also colored by early Christian mysticism. The archaic Syriac model relates the basic elements of cold, dryness, heat, and wetness to human temperaments, literally "mixtures" (*muzāḡē*). It also puts great emphasis on the *libbā* being the source of healthy, perpetual movement that distributes light to the brain.

I then point out that NENA speakers are not ostensibly familiar with the archaic CS model. Nevertheless, paralleling the situation in contemporary European languages, there are many vestiges of the archaic model reflected in linguistic usages today. It is curious that hardly a trace of the theory of the humors and mixtures is preserved, but other aspects of the old model concerning heat and cold, light, and above all movement, are reflected in many conventional usages. It is also shown that current usages of *libbā* activate COURAGE.

After showing how vestiges of the archaic CS model are preserved in contemporary usages, I offer examples of other usages of *libbā* in the contemporary language that most likely have other sources. Concepts of BLACK VS. WHITE, with strong moral overtones, are often used in conjunction with the *libbā*. Most likely they cannot be attributed to archaic CS models. As one would expect, there is no single cognitive cultural model that accounts for the range of linguistic usages of *libbā* in NENA.

1.1. A note on the indications of tenses in NENA

In NENA, verb tenses are formed out of verbal stem forms in combination with copular forms and tense markers. Here and in the following, I use the following abbreviations: IMP = imperative stem; JUS = jussive stem; PFT = perfect stem; IN-PFT = inflected perfect stem; CONT = continuous stem;[3] COP = copula; PCOP = past copula;[4] PRST = present tense marker; FUT = future tense marker; and PST = past (preterite) tense marker. We are adapting the system developed by Hetzron as explicated by Fox (1997: 22ff.).

2. *Libbā* in relation to the EMOTIONS[5]

The generic metaphor THE BODY IS A CONTAINER OF THE EMOTIONS, which has been alleged to be, or treated as, a good candidate for universality, is filled out and elaborated in all kinds of ways in different cultures. Typically, various body parts and organs, including the heart, are conceived of as one type of container or another of emotions. For example, Yu has argued that in Chinese, the CHEST, BELLY, GALL BLADDER, BROWS, EYES, and HEART are all metaphorically conceptualized as CONTAINERS OF EMOTIONS, with the heart being the general container for all emotions (2002, for the heart, see 351ff.; this volume: 140–141). According to Niemeier, in English too, the HEART is metaphorized as a CONTAINER OF THE EMOTIONS (Niemeier 2000: 206–209). She argues additionally that in English there is a HEART FOR (A WHOLE) PERSON metonymy (2000: 199–202; this volume: 352), so that the metaphor THE HEART IS A CONTAINER OF THE EMOTIONS is an instantiation of THE PERSON IS A CONTAINER OF THE EMOTIONS.

With the help of Lakoff and Johnson's influential account of the SELF (1999: 267–289), the relationship between this relatively generic metaphor THE PERSON IS A CONTAINER OF THE EMOTIONS and the more specific one of THE HEART IS A CONTAINER OF THE EMOTIONS can be understood as part of an elaborate hierarchy of metaphors. Lakoff and Johnson think this hierarchy is universal, and although it seems to apply in English, the jury will be out for a long time yet before it is clear how universal it is. At each level in it, both the target-domain and source-domain concepts can also be expressed by the image schema CONTAINER OF THE EMOTIONS. First, at the most general level, the person is the usual source domain of the target domain of THE SUBJECT, which is "the locus of consciousness, subjective experience, reason, will, and our 'essence', everything that makes us uniquely what we are" (1999: 268) – thus THE SUBJECT IS A PERSON.

Second, the rest of our inner lives, besides the SUBJECT, including our bodily experience, our social roles, our histories, etc., is our SELF. This too can be expressed in terms of the person (THE SELF IS A PERSON) but also as a thing (THE SELF IS A THING).

Third, THE BODY is an aspect or an instance of THE SELF (1999: 271). Thus, although Lakoff and Johnson do not come out and say so, the relationship between the two is a member-of-category-for-category metonymy, BODY FOR SELF. With the substitution of BODY for SELF there results the metaphor THE BODY IS A CONTAINER OF THE EMOTIONS, exemplified by "His whole body was boiling with rage."

Finally, at the fourth level down in the hierarchy, not discussed by Lakoff, are the DIFFERENT (RELEVANT) PARTS OF THE BODY that also can be CONTAINERS OF EMOTIONS. Obviously part-whole metonymies are in play here. Aside from the HEART, the following body parts have been shown to be CONTAINERS FOR THE EMOTIONS in English: HEAD, EYES, BREAST, BOSOM, FACE, CHEST, BOWELS, STOMACH, SPLEEN, and THE SIDES (Peña Cervel 2003: 70–84). It should be pointed out that Peña Cervel finds the "soul" and the "mind" to be expressed as CONTAINERS OF EMOTIONS as well, finding them also to be parts of THE SELF (Peña Cervel 2003: 72–73, 78–79), but it is doubtful that they should be thought of as THE BODY aspect of THE SELF.

NENA at first blush also seems to make use of the metaphor THE HEART IS A CONTAINER OF THE EMOTIONS and the metonym HEART FOR PERSON. Closer scrutiny, however, plants doubt as to just how closely NENA usages correspond to these purported generic concepts. That is, as impressive as accounts of such hierarchizations of metaphor seem to be, they are inadequate to the task of finding the tacit schemas underpinning figurative usages of language. On the one hand, they brush over the subtleties of such concepts as "EMOTION" and on the other, they pay insufficient attention to cultural models that have explanatory power for those who tacitly subscribe to them.

2.1. THE HEART IS A CONTAINER FOR THE EMOTIONS

Without question there are many metaphorical usages in NENA that appear to be, as in English and many other languages (including Chinese and Zulu), instances of THE HEART IS A CONTAINER OF THE EMOTIONS:

(1) a. *ʾIn lē hāwī hwā leh snêtā b-libbeh*
If not is.JUS PCOP to-him hatred in-his-heart
qātoh, lē māḥī hwā lah.
for-her, not hit.JUS PCOP to-her.
'If he had had no hatred for her in his heart, he would not have hit her.'

b. *ʾIn layt hwā ʾīqārā gô libbeh,*
If there isn't PCOP respect in his-heart,
lē qābil hwā lah gô baytuhy.
not he-accept.JUS PCOP to-her in his-house.
'If there weren't respect in his heart for her, he wouldn't have welcomed her in his house.'

c. *Ḫa(d) pṣāḫā zʿorunā gô libbā d-ʿan*
One gladness tiny in heart of-those
mḫīlē. (*QC*: 30)
unfortunate-people.
'There is a tiny bit of gladness in the hearts of those unfortunate people.'

d. *B-yom trêyānā d-ʿēdā d-qyāmtā*
In-day second of-holiday of-resurrection
Kad libbī milyā pšāmtā
While my-heart fill.IN-PFT sadness.
'On the second day of Easter
My heart was filled with sadness.' (*Kul*: 9)

These four examples closely parallel the English usage. Not only is the HEART a CONTAINER, but it also stands for the PERSON. By attributing apparently emotional contents to the heart, speakers say something about the emotional state of the whole person. Emotions that typically can be contained in the heart include *hatred* and *respect*, as well as *love* and *anger*. Of these, only respect might seem odd in English. The last two examples, (1c) and (1d), in which the contained "emotions", *gladness* and *sadness*, respectively, are taken from literary works, come across as literary, as would also be the case with *happiness* (*ḫadutā*). While they resemble usages in English, they generally belong to a higher register.

Overall, the metaphor THE HEART IS A CONTAINER OF (PARTICULAR) EMOTIONS is less often used in NENA than in English. However, more commonly than English speakers, Assyrian speakers talk about a person being motivated in terms of his or her heart containing an unspecified "emotion":

(2) a. *ʾIn lē ylāh gô libbuḵ, lē māṣit*
If not it.COP in your heart, not you-can.JUS
d-yahḇitlāh qā (ʾ)ḫe(r)nē. (*API*: 55)
give.JUS-it for others.
'If it is not in your heart to do a good deed, you cannot do it (successfully) for others.'

b. *ʾIn lē ylāh b-libbeh, lā ʿḇodun ḫaylā ʿaleh.*
If not it.COP in-his-heart, not do.IMP force on-him. (*API*: 398)
'If it is not in his heart, do not push it on him.'

c. *Libbeh mlēleh.*
His-heart it-fill.PFT.
'His heart filled up (i.e. he could no longer contain his emotions).'

Since that which is contained in the heart is left unspecified, it is important to infer what is implied to be contained. Native speakers are likely to infer that (2b) pertains to a situation of say a man not feeling love (*ḫubbā*) for a woman, and that uttering it exhorts those who know the man to refrain from pressuring him to court her. This being the case, *love* is a prototypical member of the category of those "emotions" that are containable in the heart.

As for (2a), what "it" is that is in the heart is likewise left unstated, but the sense requires that "it" be something like a "character trait" or "temperamental disposition" – in other words, a state or potentiality of a person that is more lasting than the expression of an "emotion" or a "feeling", both of which individuals experience episodically. How episodes of emotion, emotions, feelings, dispositions, character traits, as well as affects, moods, desires, etc. are related to each other, including how they are to be defined, is far from a trivial matter. While cognitive linguists have not overlooked this knotty issue (see esp. Kövecses 1990 and 2000), philosophers and psychologists have expended much more effort discriminating among the various concepts (e.g. Roberts 2003: 112–115, 310–312, 318–323; various authors in Ekman and Davidson 1994: 49–96; and Goldie: esp. 11–49 and 2004: 11–13).

In the study of related expressions across languages, as the Introduction to this volume suggests, it would be wise to avoid being chained to language-specific concepts, such as those activated by the English words "emotion", "feeling", and the rest. Loosening the chains to language-specific concepts is, however, as students of linguistic relativity have learned, a deceptively daunting task (e.g. Lucy 2004). At this early stage of apprehending emotion and bodily-feeling concepts in cross-cultural perspective, I suggest turning back to one of the most thoughtful attempts to establish a framework of concepts – Aristotle's. Not only does Aristotle help us to avoid the pitfall of universalizing conceptual meanings as a result of specifically English frames, he brings us closer to the spirit of Assyrians' conceptualizations than investigators writing in English.[6]

Aristotle distinguished carefully among essences and potencies of things and persons, as well as the actualizations of these in concrete instances. More importantly for our purposes, he was careful to distinguish both potencies and actualizations according to their *duration*. Below are Aristotle's concepts that are of importance for us, moving from the most fleeting to the most perduring:

- *pathos* (pl. *pathē*): a particular type of quality (*poion*) as it is actualized in ways that are alterable in the short term, such as feeling shame, manifested in blushing, and fear, manifested in paleness. They may be glossed as 'episodic affective qualities' or 'transitory emotions'.
- *diathesis* (pl. *diatheseis*): an arrangement of parts, thus bearing comparison with a 'disposition', but as it applies to persons it designates their temperamental make-up that is neither fleeting nor enduring, such as warmth, coldness, healthiness, etc.
- *hexis* (pl. *hexeis*): if a *diathesis* "sticks", becoming hard to displace, it becomes a *hexis*. "*Hexeis*" designates the traits or qualities of a person that constitute the measure of moral goodness or badness in a person (cf. Hutchinson 1986: 37–38). Although "emotions" may often be judged to be "positive" or "negative" (Kövecses 2000: 44–45), *hexeis* are inherently so. Put technically, *hexeis* have "intrinsically evaluative predications" (e.g. Bierwiaczonek 1993). Usually glossed as 'habit', *hexis* has also been translated as 'state', 'active state', or 'disposition'. 'Character trait' may be more apposite in this paper's frame of discussion. In the *Nicomachean Ethics*, Aristotle argues that virtues and vices are above all *hexeis*, and this lends support to the choice of 'character traits'.

– *dunamis/dunameis*: as applied to persons, this is a virtually unchangeable potency that is the source of the particular actualizations or instantiations of *pathē*, *diatheseis*, and *hexeis*. It approximates what might be described in English as 'inborn natural capacities to develop qualities'.

What is designated in the cognitive linguistic literature as "EMOTION" leaves the issue of duration vague. While in their usage of "emotions", English speakers often approximate *pathē*, or emotions in the episodic sense, they can also use "emotions" to index longer-lasting psychological states, closer to *diatheseis*.[7]

The examples in (2) now deserve reexamination. While each sentence can be understood in different ways, the primary sense of what is being contained in the HEART is a HEXIS in (a), a DIATHESIS in (b), and a PATHOS in (c). Though it may be that in both English and in NENA such diversity exists in what can be contained in a body part, it is also likely that the emphasis on types of contents will vary from language to language. Indeed, this paper will suggest that NENA shows a propensity for containing *HEXEIS* in the heart, probably more so than in English.

The process of unfettering from the English concept of EMOTION now ensues by scrutinizing a series of expressions of *libbā* that build on the HEART AS A CONTAINER metaphor, but that also employ a CONTAINER FOR CONTAINED metonym. In these conventionalized usages, that which the *libbā* is both containing and standing for becomes a little less murky:

(3) a.

ʾIn	*rāʿyā*	*hāwēleh*	*libba,*	*gubtā*	*min*
If	a-shepherd	is.JUS-to-him	heart,	cheese	from

tiḇnā	*bit*	*dābiq.* (*API*: 58)
straw	FUT	seize.JUS.

'If a shepherd has heart, he will be able to extract cheese from straw' (i.e. 'If there is the will, there is a way').

b.

ʾIn	*rigālā*	*w-rakāḇā*	*hāwī*	*ḫa(d)*	*libbā,*
If	a-foot-soldier	and-cavalier	are.JUS	one	heart,

rigālā	*ḫa(d)čā*	*bit*	*rāyim*	*w-rakāḇā*	*ḫa(d)čā*
foot-soldier	a-little	FUT	rise.JUS	and-cavalier	a-little

bit	*kāʾip.* (*API*: 58)
FUT	bend-down.JUS.

'If a foot-soldier and a cavalier are of one heart (i.e. share the same will, and thus the same ultimate goal), the foot-soldier will rise and the cavalier will bend down.'

The latter expression is typically said to admonish a bickering couple to calm down. When these examples are looked at together with those in (2), it seems that that which is contained within the heart is something like PERDURING ORIENTATION TOWARD GOALS or WILL. In (3), one might be tempted to think of *libbā* as activating the idea of "will power", but it is probably better to think of it as the right or appropriate will or disposition. Here the overarching category of things being contained in the heart is better described as HEXEIS than "EMOTIONS".

The heart is conceived not merely as a passive CONTAINER, but as one *from* which those contents transfuse. This idea of TRANSFUSIBILITY is often communicated by *min gô*, 'from in' or simply 'from'. So the following expressions can be seen as instantiations of HEART AS A TRANSFUSIBLE CONTAINER:

(4) a. *Bāsmā* *gānuḵ,* *min* *gô* *libbī*
Be-pleased.JUS your-self, from in my-heart
humzimluḵ. (*API*: 418)
you-spoke.PFT.
'Congratulations, you spoke what was in my heart (i.e. I agree wholeheartedly with you; you took the words right out of my mouth).'

b. *Min* *kuleh* *libbī* *bā'in* *ḫāzinuḵ*
From all-of-it my-heart I-want.JUS I-see.JUS-you
ḫidyā *'am* *bnay* *bayta.* (*API*: 419)
happy with children-of house.
'With all my heart, I want to see you happy with all the family members.'

c. *'Ḵīlī* *libboḵ* *tā* *minnī* *'āzaḫ.*
I-eat.PFT your-heart come.IMP with-me we-go.JUS.
'May I eat your heart; come with me, let's go! (i.e. pretty please, come with me and we'll go).'

What these usages suggest is that what is in the heart should ultimately not be thought of as the private, exclusive property of the person whose heart it is. The last example, (4c) is particularly interesting. 'May I eat your heart' (*'Ḵīlī libboḵ*) is a very common expression meaning 'Please …', and it is used in an almost pleading tone, and especially toward a child, perhaps a touch jocularly.[8] I have glossed it with the modern colloquialism, 'pretty please'. *'Ḵīlī libboḵ* should be interpreted as a CONTAINER FOR CONTAINED metonymy, in which the contents are on the order of PERDURING GOALS, so the sense is of a parent assuring the child that if he were to come along as he should, his parent would be fully in tune with his perduring goals, that their basic orientations and wishes would be in unison, as if these contents of their hearts were totally transfusible. Do note that if the contents of the heart were presumed to be EMOTIONS, this interpretation would not be possible.

In this vein of the transfusibility of the heart, it is very common to talk about persons' hearts being *open* or *closed*, suggesting generosity or greed, respectively. But not only can the openness of heart be used to talk about such HEXEIS ('CHARACTER TRAITS') as these, they can also be used to talk about moods, as in the following example:

(5)	*Qā*	*(')ḫe(r)nē*	*libbeh*	*ptīḫā,*	*w-qā*	*d-'ān(y)*
	For	others	his-heart	open.IN-PFT,	and-for	of-that
	d-gāneh	*qurpiḫā.* (*API*: 285)				
	of-himself	dash.IN-PFT.				

'When he is off among others his heart is open (i.e. he is cheerful and outgoing), but when he is in his own surroundings, his heart is dashed (i.e. he is morose and his behavior has an inhibiting effect).'

It is curious that "open-hearted" (*libbā ptīḫā*) can give off the sense of both "generous" and "cheerful/outgoing", but it is less strange if once we are reminded that this description pertains to describing THE HEART AS A CONTAINER that holds those HEXEIS rather than "EMOTIONS". It can be understood as an entailment of the concept of a TRANSFUSIBLE CONTAINER, that if the container of the heart is left open, its contents can be properly circulated, and the situation will be conducive to the proper circulation of those contents.

Also in this vein of TRANSFUSIBILITY of the contents of the heart, there is the following interesting complex usage, which builds on top of the metaphor THE HEART IS A TRANSFUSIBLE CONTAINER OF HEXEIS a metonymy of the form of WORDS FOR THE CONCEPTS THEY EXPRESS (as in "I'll put in a word for you"). In it we have 'word' (*ḫabrā*) standing for that HEXIS, or perduring attitude, contained in the HEART:

(6)	*(')Nāšā*	*l-pumā*	*kē*	*gāšiq,*	*min*	*libbā*	*mud*
	A-person	to-the-mouth	PRST	look.JUS,	from	heart	what
	ḫabrā	*'ītleh?* (*API*: 66)					
	word	he-has.					

'If you look at his mouth, what is his heart's word? (i.e. if you just listen to what he is saying, what do his real intentions matter?)'

This saying typically is used as a critical response to someone who has justified, or made an excuse for, saying something bad or offensive, on the pretext that their intentions were blameless, or at any rate not so evil. In other words, it would be said in a situation in which an American might say, "I don't care what his intention was. What matters is that he said something bad."

We close this section by considering a final expression that (1) in an unusually clear way brings out the need to turn back to an archaic model; and (2) sheds further light on the knotty issue of whether talking about what is going on with the heart is a way of talking about a person's "EMOTIONS", or about something else such as HEXEIS. The expression does not involve a container metaphor, but it seems to be an example of a HEART FOR PERSON metonymy:

(7) *Ṭlībleh* *minnī* *ʿisrā* *dulārē, w-libbī*
He-request.PFT from-me ten dollars, and-my-heart
ḫupḫipleh. (cf. *API*: 379)
palpitate.PFT.
'He asked me for $10, and my heart palpitated (i.e. but I was reluctant to give it).'

The apparent metonymic relation between the heart and something like the PERSON stems from the circumstance that the ACTION OF THE HEART, specified by *palpitation*, is standing for an AFFECTIVELY-IMBUED DECISION PROCESS OF A WHOLE PERSON, which we might gloss as 'reluctance'. We sense a causal connection between the palpitation of the heart and the reluctance of the person, but we do not know whether the reluctance is a result of the palpitation, the palpitation is a physiological effect of the emotion of reluctance, or whether both the palpitation and reluctance are effects of a more fundamental physical-mental process (such as heating, drying, building up pressure, etc.). Native speakers of the language do not know either.

"He asked me for $10, and my *heart palpitated*" is a clear case of what Riemer has called a *post-metonymy* (2002: esp. 393–394). "Palpitation ⇒ reluctance" is an entailment of the mapping from one subdomain to another (ACTION OF THE HEART to another subdomain AFFECTIVELY-IMBUED DECISION PROCESS OF A WHOLE PERSON), which have some kind of causal link within a larger domain, but whatever this link is, *it belongs to an outdated model that no longer exists in the consciousness of current speakers* – this is what makes it a post-metonymy.

One more important point regarding this example: "palpitation of the heart" in NENA, while activating imagination of the same physiological process as in English, designates something rather different. In English, "palpitation of the heart" typically activates the idea of anxiety over love, and would certainly not be used to describe the reluctance to give money. The same embodied experience (heart palpitation) is giving rise to two very different construals of emotion in the two cultures. Universality is out of the question. It may indeed be that *libbī ḫupḫipleh* and "my heart palpitated" both activate the idea of the same physiological feeling (anxiety), but we have no way to know this.

The word "construal" (of emotion) was not chosen by accident. As I intimated above, cognitive linguists working on the emotions have rather blithely overlooked the extensive philosophical literature concerning what constitutes "emotion" and how it is related to physiological processes, temperaments, character traits, morality, and cognition. One of the most scintillating of the philosophers who have worked on this problem, Robert C. Roberts, has come to the conclusion that emotions are "serious concern-based *construals*." By this he means (in brief) that what underpins an emotion, such as guilt, is an intentional feeling, because it is the construal of a feeling or sensation (such as "a twinge, gnaw, curdling, or throb" – or palpitation!) *in terms of* a concern about being in a particular condition (the socially-defined condition of being guilty) (2003: esp. 68) Translated into cognitive linguistic terms, Roberts is saying that there is an imagined, or idealized, co-occurrence between a CONCERN REGARDING A SOCIALLY-DEFINED CONDITION and a PHYSIOLOGICAL FEELING/SENSATION, and that we use the word "emotion" when the two are simultaneously activated, regardless of whether we express the whole schema as (1) CONCERN REGARDING A SOCIALLY-DEFINED CONDITION FOR PHYSIOLOGICAL FEELING/SENSATION, or (2) the reverse (PHYSIOLOGICAL FEELING/SENSATION FOR CONCERN REGARDING A SOCIALLY-DEFINED CONDITION). "Emotion" (in English) includes both of these two concepts, rather than either one in isolation, and so it is intrinsically metonymic. As Shweder has argued, in comparative research, we should focus on the components of emotion, especially "affects" (cf. Roberts's "physiological feeling/sensation" and Aristotle's *pathē*), rather than on the whole story of the "emotions", because in different cultures, the exact terms of the metonymical model of something like "emotion" may not coincide (Shweder 1994).[9]

It has been shown that although the heart is seen in NENA as, among other things, a CONTAINER, it is inadequate to see that which is contained to be "EMOTIONS", as is usually described in the literature. Some increase in accuracy came with calling the category of those contents HEXEIS (enduring states of a person with moral significance), but it is hard to see from this any *motivation* for such a conception. Current NENA usage really gives no indication of any underlying causal relations. Aside from HEXEIS, PERDURING ORIENTATIONS TOWARD GOALS (or "guiding dispositions") were also salient, but still, the contemporary-language examples have not given us a model, or a set of metonymically related concepts that might help us to understand why the heart has such contents, and gives us a context for understanding other usages of *libbā*.

To anticipate what will be found: there is an old model that includes a principle, perceived as a substance, that simultaneously generates those guiding dispositions (PERDURING ORIENTATIONS TOWARD GOALS) and those apparently EMOTIONAL STATES such as anger, happiness, hatred, and respect. This generative principle is covered by the specific idea of the 'animal spirit' (*ruḥā ḥayuwātnāyā*), located in the heart, and more generally by the concept of 'mixtures' (*muzāḡē*) of the elemental substances or temperaments. Of course it is not so simple. There is a great deal to be said about the ways in which an archaic model serves as the conceptual *source* of many usages today, only a few of which can be contextualized in such a way in this paper.

2.2. The place of *libbā* in the system of the organs

Before turning to the ancient model, it is important to situate the *libbā* in the system of body organs, as they are talked about in NENA.

Any Assyrian, if asked, will say that the *libbā* refers to that body organ that pumps blood throughout the body and is located in the chest. However, in certain usages, it includes more than that:

(8) a.	*ʾḴīlī*	*laḥmā*	*w-libbī*	*dḇīqleh.*
	I-eat.PFT	bread	and-my-heart	grip.PFT.

'When I ate the bread, my *libbā* got gripped (i.e. I lost my appetite).'

b. *Libbī* *g̱bīleh* *w-mudirrī.*
My heart mold.PFT vomit.PFT.
'I got nauseated and vomited.'[10]

c. *'In* *(ḫ)šīlā* *lē* *ywit* *(')ḵīlā,* *libbuḵ* *qāmo*
If cereal not you.COP eat.IN-PFT, your heart why
mrā'ā *yleh?* (*API*: 53; cf. Maclean 1895: 350)
sick it.COP?
'If you have not eaten cereal, why is your heart/stomach sick?'

The last example is used to express the idea "If there is a matter with which you have nothing to do, how come you are sticking your nose in it?" Or, "If you don't have a hand in it, why are you all worked up about it?" In any case, the import of the three examples is that when a person has what we call a "stomach ache", it is the *libbā* that feels it.

As for the other internal organs, they are seldom used figuratively in the modern language. The 'stomach' (*ke(r)sā*) can be thought to contain wind, which, when it is let out, prevents the accumulation of anger, and so there seems to be a trace of the metaphor THE STOMACH IS A CONTAINER OF ANGER:

(9) *Har* *šḇoq* *hamzim,* *lā* *jawibitleh,* *pôḫā*
Just leave.IMP speak.JUS, not you-answer.JUS-him, wind
d-ke(r)sā *paluṭē* *yleh.* (*API*: 436)
of-the-stomach exits.CONT it.COP.
'Just let him speak, and don't answer him, so that he can vent his stomach (cf. spleen).'

The liver (*kaḇdā*), which is productive of figurative uses in Classical Syriac, is not much so in NENA, but there are a number of usages expressing the idea that a liver on fire is anger:

(10) *(')Nāšā* *ros* *ḫa(d)čā* *mīyā* *l-kaḇduḵ,* *mā*
Man sprinkle.IMP some water to-your-liver, what
luḵ *d-hādḵā* *jaldē* *bikrābā* *ywit?* (*API*: 454)
to-you that-thus soon be-angry.CONT you.COP?
'Come on! Sprinkle some water on your liver. What's with you, getting angry so quickly?'

The other body organs – kidneys, lungs, the spleen, etc. – are all plainly unproductive of figurative usages in the modern language. However, there is a body part that contrasts saliently with the heart, and that is even more productive of figurative uses than it: the eye (*ʿaynā*). In a great many expressions, it is expressed as a CONTAINER, of something like the subcategory of HEXIS that has an appetite for things or covets them, and, according to the underlying cultural model, it is very difficult for it to get its fill:

(11) a. *Kullāh* *dunyē* *dīyeh* *ylāh,* *ʾīnā* *har* *ʿayneh*
All-of-it world of-him it.COP, but still his-eye
kpīntā *ylāh.* (*API*: 432)
hungry.IN-PFT it.COP.
'He has everything in the world, but his eye is still hungry (i.e. he still wants more).'

b. *ʿAynē* *ʾaẖčī* *min* *ʿaprā* *kē* *mālī.* (*API*: 267)
Eye only from dust PRST be-filled.JUS
'The eye will only be fulfilled when it is filled with dust.'

There are a great many common expressions about the eye and envy, and these alone deserve a paper in itself. The eye is variously expressed as "not satisfied" (*lā sḇaʿtā*), "not loaded/forbearing" (*lā ṭaʿntā*), or "going out" (*plāṭā*) to express the idea of being maliciously envious. Here are examples of the last of these:

(12) a. *Plāṭā* *ynā* *ʿaynuḵ* *d-biẖzāyā*
Go-out.CONT they.COP your-eyes that-see.CONT
ywit *b-haw* *mantuyē* *yleh.* (*API*: 438)
you.COP in-he succeed.CONT he.COP.
'Your eyes are popping out (i.e. you are feeling malicious envy) when you see that he is succeeding.'

b. *(ʾ)Nāšā* *ʿaynuhy* *pālṭī* *w-šimmeh* *lā* *pāliṭ.*
A-person his-eyes go-out.JUS his-name not go-out.JUS.
(*API*: 67)
'If a person's eyes pop out (i.e. covet too much), his reputation will not circulate (he will be of no account). '

The idea in general is that the eye should be properly, moderately filled, carrying the right amount of those appetitive desires, and also that, as these expressions exemplify, they *be secured in their proper place*. Only if all three of these conditions hold in a person's eye, will he/she be able to avoid the vice of envy. Here *being in their proper place* is being included to emphasize the importance of the theme of MOVEMENT, which will prove to be of cardinal importance in the archaic model, and the relationship between movement from the heart to the eye is an important part of this idea.

Finally, there is another peculiar usage often applied to the eye, which has its sources in the archaic model, and which also has great significance for the HEART, namely DARKNESS/LIGHT. Here is an example:

(13)	*ʿAynā*	*ẖiškāntā*	*qa(d)mêtā*	*ẖiškoh*	*qā*	*māroh*	*yleh,*
	Eye	dark	first	darkness	for	its-owner	is
	w-ʿaynā	*bahrāntā*	*qa(d)mêtā*	*bahroh*	*qā*	*šḇāyē.*	
	and-eye	radiant	first	its-radiance	for	neighbors.	

(API: 268)
'The darkness of a darkened eye is primarily for the person who has it, whereas the radiance of a radiant eye is primarily for neighbors.'

In this example, what is being implied (in a way very obvious to native speakers) is that darkness associated with the eye stands for the vice of greed and egoism, and is accompanied by gloominess, whereas the association of the eye with light has the implication of the virtue of generosity, and perhaps altruism, and is accompanied by a contagious cheerfulness. This theme of DARKNESS/LIGHT is also a key one.

As can be seen from this discussion of the EYE, it and the HEART are to be seen in NENA as the two main containers of HEXEIS, but of different kinds. Whereas the HEART contains primarily those that are PERDURING ORIENTATIONS TOWARD GOALS and those apparently EMOTIONAL STATES, the EYE contains primarily the APPETITIVE DESIRES, such as coveting and being acquisitive.

3. A Classical Syriac model

As noted, Classical Syriac (hereafter "CS") is the language that until this day is used in the liturgy of Christian speakers of NENA and of various other churches, including some whose members continue to speak other forms of Aramaic. It had an active literature from perhaps around 200 A.D. through the 13th century, but continues to be written until this day.

Linguists see the language as a separate one from NENA, and not a direct ancestor (e.g. Mutzafi 2005). We find claims of ancestry hard to evaluate (if only because it is based on the conceptual metaphor LANGUAGE HISTORY IS BIOLOGICAL HISTORY, which is part of a largely unanalyzed cultural model). What is clear is that CS (1) is not understood by a large majority of NENA speakers, and so is certainly not the same language as NENA; but that (2) it has long borne influence on NENA, and because of efforts to "purify" the modern language (in both writing and speaking), this influence may be growing.

The medicinal theory of the humors, associated with the Greek Galen, exerted a profound impact in the West regarding physiology and personality. Geeraerts and Grondelaers (1995) have argued that the cognitive models associated with Galen have continued to leave their mark on English and related languages. Prior to their absorption into European cultural heritages, the works of Galen also made a profound impact in the Middle East. Into CS they were translated earlier than they were into Arabic and Persian, and often retranslated. Already by the early 6th century, many works of Galen were translated into CS by Sergios bar Rēsh'ayna (d. 536), and again in the 9th century, esp. by Job of Edessa (Ayyub al-Ruhāwī) and Ḥunayn ibn Isḥāk (Degen 1981). Most of these many translations (certain works many times over) are no longer extant, but they left their mark on many other works.

What is distinctive about the Syriac writers' reception of Galenic medicine was their integration of it into Early Christian mystical thought. It was in this integrated form, rather different from the form it was known in Europe, that it bore its influence on CS – influence that can also be detected in contemporary NENA, as this paper will demonstrate. There is, however, no modern study of this model, and the historical record of it is scattered and spotty. Nevertheless, a number of works have been consulted in the original for this study. One of the most interesting syntheses of Galenic medicine and Syriac mystical philosophy can be found in a work by the physician Simon of Ṭaibutheh (died c. 680 A.D.), and his work (abbreviated *ST*) will be the main source for the following account. Details have also been added from an anonymous work attributing its ideas to ʾAḫudimmeh, probably also from the 7th century, and known to us in two versions (Nau 1909; Chabot 1965); and a philosophical work by the 9th century physician and Galen-translator, Job of Edessa (*JE*).[11]

3.1. The core model: temperaments, moods, mental faculties, character traits, spirit

The general strategy in the works just mentioned is first of all to lay out the basic elements of their system, and how these are related to fundamental forces. Then, at increasing levels of complexity, it will be shown how these elements form compounds, how these compounds are reflected in temperaments, moods, mental faculties and spirit, how these compounds are distributed in the body to form "mixtures" and to be typically associated with particular organs, and finally how different distributions of mixtures in individual persons characterize their overall health and qualities of temperament and character. The following account will summarize what transpires at these levels, aside from the last level, which is only intimated in these works.

3.1.1. Basic elements and forces

There are four basic elements and each of them "serves" one of the "natural powers":

Table 1. The four simple elements and their powers

Basic Element	Natural Power
cold (*qarīruṯa*)	attractive (*nāṯopa*)
dryness (*yabīšuṯa*)	astringent (*ʾāḥoḏa*)
heat (*ḥamīmuṯa*)	digestive (*pāšora*)
wetness/humidity (*rṭīḇuṯa*)	repulsive (*dāḥoya*)

(*ST*: 316b, 63 and *JE*: 298a, 5)[12]

These four simple elements combine to form compound elements, and out of these compound elements the different humors, as well as components of the human being are composed:

Table 2. The compound elements and their associations

Compound Elements	Elements Derived From	Associated Humor	Component of Human Being
earth (*ʾarʿa*)	cold and dryness	black bile	body (*pağra*)
fire (*nura*)	heat and dryness	red bile	movement (*zawʿa*)

Compound Elements	Elements Derived From	Associated Humor	Component of Human Being
water (*mayē*)	cold and wetness/humidity	Phlegm	feeling (*rgešta*)
air (*ʾaʾir*)	heat and wetness/humidity	Blood	life/vitality (*ḫayuṯa*)

(*JE*: 298b, 317a, 56, 37; Chabot 1965: 61[13])

Because of the way each of these compound elements compose aspects of the human beings, they are the building blocks, literally "bridges and stairs" to human beings in their capacity as sentient, intelligent creatures (*mlīlē*) (Chabot 1965: 61).

The next step is not explained very well in the documents, namely how various faculties tend to concentrate in particular bodily organs. Furthermore, the different texts are not consistent how they assign such faculties to the organs. However, we can give an idea of what associations they make:

Table 3. The organs, their forces and functions

Organ	Force (Nau 1909: 110)	Psychical Functions (*maʿbdānuta*) Chabot 1965: 62)
brain (*muḫa*)	feeling (*rgešṯa*)	
heart (*lebba*)	discernment (*pārošuṯa*)	cognition (*reʿyāna*), reflection (*tarʿīta*); reasoning (*maḫšaḇṯa*)
stomach (*ʾesṭumka*)	passion[14] (*rigṯa*)	avarice/covetousness and passion for food
liver (*kaḇda*)	wrath (*rug̅za*)	
kidneys (*kulyāṯa*)	longing (*miṯyaʾḇānuṯa*)	fervor (*ḫemṯa*) and wrath; perspicuity
mind (*muḫa*)		imagination (*hagāguṯa*), fantasy
skull (*qarqapta*)		perspicuity (*buyāna*), intellect (*sukāla*)
Diaphragm (*ḫelba*)		Perceivance (*maʿda*), understanding, laughter
Skin? (*gešta*)[15]		feeling (*rgešṯa*)

Organ	Force (Nau 1909: 110)	Psychical Functions (*ma'bdānuta*) Chabot 1965: 62)
gall bladder (*mrārta*)		passion (*regta*) for good works, passion for flesh in fat that is on the kidneys (!)
Spleen (*ṯḥāla*)		Fear and treachery
Lungs (*ra'ta)*		love
Arm (*drā'a*)		Might (*ḥayltānuta*)
Hands (*'īdayya*)		Courage (*lbību̱ta*)

What role does specifically the heart play in this system? First of all, a crucial animating force, the 'animal spirit' (*ruḥa ḥayuṯānāya*) inhabits the heart (*JE*: 318a, 39). It is, according to Simon, the locus of desire and 'heat/fervor' (*ḥemṯa*) (*ST*: 308a, 49). Through the veins and arteries, this spirit is brought to all the body, and at the same time, it can be strengthened and invigorated in the heart *JE*: 314b, 33). In contrast, the head is the 'dwelling place' (*bayt ma'mra*) of the rational faculty, or the 'psychical spirit' (*ruḥa napšānāya*) (*ST*: 308a, 49; *JE*: 324a, 48). But the two are closely linked. Simon actually has the psychical spirit form in the left ventricle of the heart, which he also says is its dwelling place, but has the heart rarefy and send up this spirit to the lobes of the brain where it is reconstituted as the force of cognition. Here is how he puts it:

> The heart is composed of solid flesh and nervous matter. In it dwells the natural heat that is within us, and from it heat emanates as from a fountain; and the heat and its form resembles a fire that is broad below but narrowing (*qaṭīna*) above, and it mounts up like a flame. The heart has two ventricles ... The right ventricle receives the blood from the liver, strains it, and sends it out purified to the brain and to the rest of the body; while the left ventricle is the dwelling of the psychical spirit, and it rarefies that spirit and sends it to the lobes of the brain where are constituted rationality, memory, and understanding. (*ST*: 317b, 65)

What is important is that the action of the heart, working with the other internal organs, produces the right "mixture" of the elements (heat and cold, dryness and wetness) for the brain:

> The heart ... inclines toward the spleen, and next to it are the lungs which act like cooling bellows. As to the thin top of the heart, which resembles a flame of fire, it is not placed vertically, lest it should send out the power of its heat and thus upset its mixture (of humors), because this mixture of the brain by nature is predominated by coldness; but the thin top of the heart inclines to the left, toward the spleen, so that the heat of the heart may mix with the cold of the spleen, the ribs and the black bile. Perceivance and discernment which are in the heart, and rationality and understanding which are in the brain, fall back down when the psychical spirit which is in the heart thickens as a result of indigestion and bad chymification. (ST: 318a, 65)

Thus the heart is the source of heat for the body, and the natural movement of heat is upward and of cold downward (cf. *JE*: 28). It is this heat which gives the heart its perpetual movement, *zaw'a* (*JE*: 40–41), which can in many contexts be translated as 'feeling' or 'emotion', and it is through this movement that the heart has its effects.

3.1.2. Mixtures

What is associated with the heart and other organs is conceived of in this model not so much as emotions, but rather as (1) mixtures (*muzāḡē*; sing. *muzāḡa*[16]) of the humors, thought of as compounds of fire, water, air, etc.; and (2) faculties such as common sense, cognition, etc. Faculties associated with the heart and other body organs and body parts have been listed above. From the spotty record, a consistent picture of which faculties are associated with which organs does not emerge, and so these will not be of significance to the rest of the argument. But the mixtures are crucial.

Both the simplest elements (heat and cold, wetness and dryness) and compound ones (earth, air, water and fire) can be thought of as the constituents or the formative conditions of these "mixtures", which form the concrete SOURCE subdomain for understanding the abstract TARGET subdomain of "TEMPERAMENTAL DISPOSITIONS", which correspond closely with Aristotle's *diatheseis*, inclusive of *hexeis* (this DIATHESEIS ARE MIXTURES metaphor is foreign to Aristotle). We say *sub*domain because they clearly are part of the same domain or model, and their relationship is not one of similarity, but rather of CAUSE/EFFECT, or MEMBER-OF-A-CATEGORY/CATEGORY – in any case metonymical.

The compound elements actually come together in the form of the humors, as in other versions of the Galenic system. However, not much is made of them in the works under scrutiny. Perhaps not coincidentally, *the humors are virtually unknown in NENA*. In the CS texts, it is assumed that the way the humors circulate has an impact on the particular mixtures of particular persons. For example, the four most extreme (unbalanced) mixtures of the heart are given as:

Table 4. The extreme mixture of the heart

Mixture (*muzāḡa*)	Temperament (cf. *diathesis*)
hot	Marked by deep breathing, by light, sensitive touch of the arteries, by audacity, hot temper (*ḫemṯa*, 'hotness'), and courage (*lḇīḇuṯa* < *lebba*)
cold	Feel of the arteries is inhibited; those who have this mixture are naturally timid, enfeebled, and heedless
dry	Makes arteries feel hard; person not disposed (*mṭayḇa*) to hot temper, but when aroused is fierce, and subsides with difficulty
wet	Arteries feel soft to the touch; person quickly is moved to anger (*ḫemṯa*), but this anger also subsides quickly

(*SyrMed*: 259, 293)

As an entailment of this concept of "mixture", the Syriac writers recognized that a person's distribution of mixtures depends on his conduct, saliently including his diet regime. Hence, the temperamental dispositions, like Aristotle's *diatheseis*, are susceptible to significant change – probably more than they are in Western models. Thus, unlike some philosophers, the Syrians did not see these as fixed within the same person. In many contexts these mixtures existing in a person at a particular moment indicate what we might call "moods", or contingent dispositions that unlike emotions are not focused on particular objects.

Another entailment of moods/temperaments, substantialized as MIXTURES, together with the idea that the forces of cognition and discernment are in the heart, is that traits such as courage and generosity are produced within the heart. The texts do not draw this connection explicitly, but when there is the combination of temperament or felt mood and discernment (cf. construal), it does follow that something like PERDURING ORIENTATIONS TOWARD GOALS (as explicated earlier) will come into play.

3.1.3. The heart and other organs as CONTAINERS and SOURCES or HOUSES

Lost in translation is the idea that the heart and other organs are not simply containers, but rather both a house (*bet ma‘mra*) and a source. Simon elaborated on this point still further:

... *lebba*	*beyṯ*	*tēgurta*	*(h)o*	*d-nap̌ša:*	*wa-l-mēna*
... heart	house	trade	it.COP	of-psyche	and-the-port

d-reḡšē	*wa-mtarsyāna*	*d-regṯa.* (*ST*: 304b, 44)
of-senses	and-nourisher	of-desire.

'... the heart is the marketplace of the psyche, the haven of the senses and the nourisher of desire.'

We will shortly see how recovering one's spiritual health is likened to putting the house of one's heart back in order.

3.1.4. Heat, darkness and light

A corollary of movement originating from the heart is that it spreads light or radiance to the faculties. An important theme of the mystical tradition was to keep up this radiance. If a person does not conduct himself properly, including eating a proper diet, the stomach rather than the heart will become the locus of heat, and this will cause blockage and darkness. Note how Simon describes this process:

> The more the pursuit of our disciplined practices slackens, the more glowing hot will be the furnace of the stomach ... And when the channels of the stomach become filled up and the organs which provide the light from the brain to the heart become blocked, gloom will gather in on the heart, all the house will be filled with smoke, the limbs will suffer numbness, dejectedness will reign, the mind will be troubled, the soul will darken, the discernment will become blind, knowledge will be hampered, cognition will be muddled, (evil) thoughts will be unleashed, pleasant memories will be

> erased from the heart, and the griefs[17] – the children of demons – will materialize, dance with joy, and clap their hands. (*ST*: 299a, cf. 35–36)

The wrong conduct displaces the center of (dry) heat from the heart to the stomach, inhibiting movement (as noted above, hot and dry constitute fire, which is associated with movement). A series of insalubrious consequences for a person's temperament will ensue, as described in Simon's passage.

The whole point of spiritual exercises in the mystical tradition represented by Simon was to "drive out the darkness from the house of the heart" (*ST*: 34). Note Simon's remarkable description of how this is to happen:

> *Ṭuḇôh l-ʿayna d-ʾanhar l-gô mennēh šrāḡa da-qnomēh b-nuhra d-ṭaybuṯa: wa-ḫza miṯyaḏʿānāʾīṯ l-šôtaptēh d-ṣāmḫa ba-špulay baytēh d-lebba ʾayḵ gpitta hḏīrta: w-la-ḇnôh ḫušāḇē kaḏ krīḵīn l-pātora d-lebbēh: wa-ḇ-ḫaduṯa d-lebba pāqēḏ l-hôna peyrma d-niʿbar nbasēm ʾenon: w-našpaʿ ʿalayhon mešḫa mbasma ḫuba: šayna ḫaduṯa w-buyāʾa.*
>
> [Blessed is he who has kindled within himself the lamp of his qnoma, by the light of grace, and has seen spiritually his consort shining up from the lowermost parts of his house of the heart like a splendid vine, and the thoughts, his children, (sitting) round about the table of his heart, and, with the happiness of his heart, has commanded his mind, the censer, to go perfume them and pour upon them the perfumed oil of love, peace, happiness, and comfort.] (*ST*: 295a–b, 29–30)

Here Simon has fully elaborated the conceptual metaphor of the HEART IS A HOUSE, as well as the concept of RADIANCE FOR A HEALTHY CONDITION, the idea that a healthy condition is marked by light. LIGHT is conceived of as entering the body through the brain, and it is therefore a high priority of spiritual exercises to keep open the channels that transmit the light from the brain to the heart. That discernment of the truth, which dwells in the heart, becomes obscured whenever insufficient light reaches the house of the heart (*ST*: 34).

4. Vestiges of the Classical Model in the modern language

We have seen that the classical model of the workings of the heart and other organs makes much use of the concepts of HEAT/COLD, MOVE-

MENT/STASIS), DARKNESS/LIGHT, and to a certain extent MOISTURE/DRYNESS. There are many conventionalized expressions in NENA that make use of these concepts as well. However, ordinary contemporary speakers of NENA are not aware of the archaic model that has been presented here. The idea of a person's *diathesis* (temperamental disposition) being the distribution of his "mixtures" of the basic and compound elements, and of the humors, is all unknown. Very few speakers have the slightest inkling of the humors – the word for them, *mirta* (now used for 'gall' and 'gall bladder'), is no longer used or understood as "humors" or "mixtures". Finally, the idea that light must come down through the brain to illuminate the heart, and that heat must rise to preserve the healthy movement of the spirits, is also unfamiliar.

Before turning to vestiges of the various causally-related components of the archaic model, in other words metonymic relations, it is worth saying a few words regarding how ideas concerning the function of the heart in the archaic CS model have been carried over into the contemporary language. One can indeed find what appear to be analogs of the old model in the modern language, but it is hard to assess their significance. For example, in the former, certain forces or psychical functions of discernment and cognition are associated with the heart. Modern speakers would not put the matter anything like this, but they can say "I am thinking with my heart" (*b-libbī biḥšāḇā*) (*Kul*: 12). This THE HEART IS AN INSTRUMENT OF THINKING metaphor may easily have other sources.

As for one of the pivotal elements of the archaic model, namely that the heart is the source and dwelling of the animal or the psychical spirit, this has a curious echo in NENA:

(14)	*Kē*	*mālyā*	*leh*	*libbuh*	*ḥdā*	*ruḥā*	*ḥêtā*. (*Kul*: 37)
	PRST	fill.JUS	to-him	his-heart	one	spirit	living.

'A living spirit fills his heart.'

The context of this usage is telling. It comes from a sardonic literary work by a priest, Q Šmuʾēyl b-bēyt Kulyā, who had knowledge of CS, though it is by no means certain that he consciously drew upon such knowledge. The line follows passages in which members of a troop of an imposter priest pretend to implore God to come to the rescue of a poor Russian muzhik's household. In the subsequent line, the muzhik kisses the hand of the supposed priest in gratitude. Saying that "a living spirit fills [the muzhik's] heart" thus suggests that the man's hopes are raised, his mood enlivened, his desires whetted, all in the expectation of God's bounty. In the light of the earlier discussion distinguishing the APPETITIVE DESIRES contained in, or carried by, the eye, from the PERDURING ORIENTATIONS TOWARD GOALS contained in the heart, this usage is untypical because it treats what is contained in the muzhik's heart as more like what is more commonly contained in the eye. However, it recalls Simon of Ṭaibutheh's characterizations of the "animal spirit", as the locus of desire, residing in the heart, and may accordingly be a vestige of the archaic CS model.

Now to the main thrust of the current section: the expressions making use of the old metonymic links to the heart are numerous but in general not very productive (at least not semantically; syntactically they are more pliant). This lack of productivity is indicative of them being "post-metonyms" – expressions that make use of the old metonymic relations without speakers being aware of these relations (Riemer 2002). As a result, they tend to come across as metaphorical usages. A nice example of such a post-metonym is the metaphor (in English) SADNESS IS DOWN (Barcelona 2000a: 9–10). Being drooped over has been imagined as a causal consequence of being depressed, but insofar as most users do not consciously apprehend this once-believed connection, a usage such as "Pete is *down* in the dumps" is to be characterized as metaphoric. The original foundation in the source domain (BEING DOWN *FOR* SADNESS) has receded from memory, and so whatever semantic extensions get produced, they are the result of metaphoric elaborations.

4.1. HEART AS A HOUSE

We start out with a carryover of a metaphor-like expression in the old model, namely THE HEART IS A HOUSE FOR HEXEIS, that exemplifies occasionally productive uses of the old concepts. Both examples are from upper-register 19th century literary texts, by writers who were fluent in CS:

(15) a.

Pṯoḇon	*dargāgē*	*d-libbayḵon*
Open(pl.).IMP	gates	of-your-hearts
Mbasmon	*l-ʿālmā*	*b-ṣāwātayḵon.* (*RabYon*: 124)
Please.IMP	the people	with-your-conversations.

'Open up the gates of your hearts
Please the people with your words.'

b. *B-heḇī* *ywin* *d-ʾānē* *nāṣīhatē* *bit* *dāḇqī*
In-my-hope I.COP these advice (pl.) FUT grab.JUS
binyat *gô* *libôḵon.* (*ZB* 51(10), October 1900:76)
foundation in your-hearts.
'I hope that these pieces of advice will secure a foundation in your heart.'

These are not ordinary everyday usages, but are easily understood. Note that it makes more sense to construe what the speakers are wishing to dwell in the house to be the perduring qualities of *hexeis*, with positive ethical implications, rather than emotions, which come and go.

4.2. Temperature

We said in the discussion of the Syriac version of the Galenic model that 'heat/fervor' (*ḥemṯā*) has its locus in the heart, and that this heat, when it combines with dryness, is associated with fire and brings about movement, and that indeed, it is the heart that is the source of the movement that is in the body. As if presupposing these concepts, speakers of NENA commonly use this idea of the heart being the locus of fire:

(16) a. *Libbī* *(y)qidleh* *qāteh,* *w-qam* *yahḇinneh*
My-heart burn.PFT-it for-him, and-PST I-give.JUS-him
zuzē.
money.
'My heart burned for him (i.e. I took pity on him), and gave him money.'

b. *ʾIn* *libbā* *lā* *yāqid* *dimʿē* *lē* *ṣālī.* (*API*: 55)
If heart not burn.JUS, tears not descend.JUS
'If your heart does not burn (i.e. if you are not caring), tears will not be shed (for you).'

c. *Libbī* *smiṭleh* *qāteh* *ʾīman* *d-ḥzēli*
My heart it-burn.PFT for-him when that-I-see.PFT-him
b-hay *ʾahwāl.* (*API*: 403)
in-this condition.
'My heart welled up with pity when I saw him in that condition.'

In the CS model, the sense of there being fire in the heart, because it was responsible for movement, had beneficial implications for bodily health. In general, this is also true in NENA, that having the heart burn has positive overtones inasmuch as it indicates that the person whose heart burns is a *caring* person (nb. the HEART FOR PERSON/MORAL SELF metonymy). In Persian, also, the heart burning for a person captures this sense of taking pity on or feeling sympathy for a person (Sharifian this volume: 253). While the heart burning in these examples can imply an unpleasant feeling, it does so more weakly than the English equivalent.

Underlining the idea that heat in the heart stands for caring, the opposite, namely the *cooling of the heart*, to mean a loss of interest, also applies:

(17) *Qam masqidīlī,* *'ḫe(r)nā* *libbī* *qišleh (qirreh)*
PST tire-out.JUS-me, again my-heart it-be-cold.PFT
w-lē *bā'yin* *d-ṭā'lin* *minnayh.*
and-not I-want.JUS that-I-play.JUS from-them.
'After they wore me down, my heart cooled and I no longer want to deal with them.'

This sense of heat as indicative of caring, or being sensitive or passionate, and the loss of heat, cooling, as calling attention to no longer caring, or losing interest, is productive inasmuch as it licenses creative new usages, as reflected in this poetic lines expressed by Rabbān Yônān:

(18) *'Aḫir* *kulayhy* *ṣāwāteḵy*
Finally all-of-them your-conversations
Nurā *(y)layh* *b-madḵiryāteḵy*
Fire they.COP in-your-remindings
L-kaḇdī *w-libbī* *maḫyimī*
To-my-liver and-my-heart they-heat.JUS
D-lā *qāṭi'* *biḵyā* *m-pumī* (*RabYon*: 46)
That-not it-be-interrupted.JUS cry.IN-PFT from-my-mouth
'All your words ultimately
Are reminders of fire
They heat up my liver and heart
So that my mouth not cease crying'

The context of the poem – the poet expressing grief for his mother – makes it clear that however unpleasant this feeling of grief is, it is a good thing to be so caring. The feeling he is expressing in the poem has praiseworthy implications for his *character* (part of his MORAL SELF, not just his person). As previously argued, far more than "EMOTIONS" are at stake here.

However, fire in the heart is not always a good thing. In other expressions, too much fire in the heart has the effect of disturbing a person's powers of reasoning, which are in the head. But at the same time that these modern expressions disregard those positive ramifications of fire being produced in the heart, they preserve the connection between heart and brain as spelled out in the CS model:

(19) a. *Nurā d-gô libbā qā rešā tinnā kē šādir.*
Fire which-in heart for head smoke PRST sends.
(*API*: 252)
'Fire in the heart sends up smoke to the head.'

b. *Karbā d-gô libbeh musmēleh l-hôneh.*
Anger which-in his heart it-blind.PFT-it his mind.
'Anger in the heart extinguishes one's reason.'

c. *Tinā d-gô rešeh musmīyêleh ʿaynuhy.*
Smoke which-in his head it-blind.IN-PFT-them his eyes.
(*API*: 321)[18]
'Smoke in the head blinds one's eyes.'

There is a good reason for this shift of meaning from CS to NENA. The word *ḫemṯā* means 'heat', 'fervor', and 'anger' (in both languages). The modern has preserved the underlying belief in the heart as the seat of *ḫemṯā*, emphasized one of its meanings as anger, and drawn out the entailments of that association.

In these examples, the relationship between the heart and the brain, as laid out in the CS model, is curiously being preserved. Recall that in Simon's work, the psychical spirit originated in the heart and then was pumped up to the brain by the movement of the heart, and then in the brain it reconstituted the powers of reason and cognition. Accordingly, though conscious belief in this psychical spirit scarcely survives, this principle of the replenishment of cognition is preserved. If too much anger characterizes the heart, this has ill consequences for what is sent up to the brain, and adversely affects the cognitive processes of the brain (it "blinds" them).

With the importance of the eye as a locus of desires in the modern language, a further entailment has been drawn from this principle of the heart sending out its heat or fire to the body:

(20) *Mabyunē ʾīt ẖa(d) mindī qā(d)m libbeh,*
It-appears.CONT there-is one thing in-front-of his heart,
nurā yleh plāṭā min ʿaynuhy. (*API*: 424)
fire it.COP go-out.CONT from his-eyes.
'It appears that he has something in front of his heart (i.e. he is about to carry something out impetuously), for fire is flaming out of his eyes (i.e. he is livid; cf. "he is breathing fire").'

Here, a judgment is being made about how a person is looking with his eyes, and how this is an indicator of something that is happening with his heart. To "have something in front of one's heart" means, as noted, to have in mind to do something without thinking it through with one's powers of reason (how this expression fits in within a larger system will have to be left uninvestigated here). But it is clear that the signs of ire that can be seen in a person's eyes indicate a change that has taken place in the heart, and this involves the sending up of heat to the brain, which is a key feature of the CS model.

4.3. Wetness and dryness

While not particularly elaborated in the CS writings consulted, the heart could be described as being wet or dry (esp. *SyrMed*). As reported above, the dry mixture of the heart means that arteries feel hard. A person with a dry mixture in his heart is not disposed to hot temper, but when aroused this temper is fierce, and does not subside easily. In contrast, a wet mixture means quick to anger, but quick also to recover from being angry. It may be that this idea has been carried over into NENA, but the evidence is soft:

(21) a. *Mud(y) libbā barozā ʾīt leh, d-qam māẖē*
What heart dry there-is to-him, that-PST hit.JUS
leh?
him?
'How dry is his heart to hit him (like that)?'

b. *ʾĀhā qiṣṣat mapšimāntā libbī mupširrah.*
That story sad my-heart it-melts.PFT.
'That sad story melted my heart (i.e. made me feel sad, sympathetic).' (often said sarcastically)

As (21a) shows, the heart in NENA can be talked about as "dry" to mean "unfeeling" or "merciless", approximating "hard-hearted" in English. This construction could have developed from the idea of a slow-to-subside temper: the sentence may presume that being angry too long is the cause of unreasonably hitting someone like that. Speakers of the contemporary language do not talk about the heart being wet (in CS it can be so described), but, as (21b) shows, they do talk about it as melting, and give it a different sense from that in English. Having a melted heart does seem close to being the opposite of a dry heart: the person with a melted heart feels compassion and mercy, which is consistent with having a quickly subsiding bad temper. In NENA, a dry heart is unfeeling, and a wet (melted) one feeling. More historical evidence would have to be marshalled in to show whether these modern usages really are ultimately entailments of the CS model, or whether they emerge from other sources.

Other modern instantiations of WETNESS/DRYNESS are not common. One speaker of the Jīlu dialect did report that she had heard expressions such as *libbā brīzā* ('dry heart') and *libbeh ṭrīnneh* ('his heart hardened' [and thereby dried out]), to mean something like 'to lose desire'. However, these usages are mostly unfamiliar and are (even by her) too sketchily recalled to say anything more about.

4.4. Darkness and light

In NENA it is possible to talk about the heart darkening. As we saw (by way of extensively quoting Simon), a central theme of maintaining physical and spiritual health is to make sure the body – including above all the heart – remain illuminated. The idea of light not reaching the heart as an indicator of an unhealthy condition is indeed preserved in NENA:

(22) *Ḫḇēleh* *libbī* *w-npīlī.* (cf. *API*: 374)
It-darken.PFT my-heart and-I-fall.PFT.
'My heart darkened (i.e. I got faint) and I keeled over.'

As in CS, there is also a link between the darkening of the heart and of the eyes. *Ḫḇēlon ʿaynē*, 'my eyes darkened', is a way of talking about not only getting dizzy, but also becoming angry (*API*: 374). Such anger is an indicator of that unhealthy spiritual condition brought about by darkness, as laid out in the CS model.

4.5. Movement

The richest quarry of remains of the CS model can be found in expressions concerning movement of the heart. As noted, both in CS and in NENA, the word for "movement" (*zawʿā*) is the word that most closely corresponds to what is meant in English by "emotion". To recapitulate one of the main points of the CS model about the heart: the heart is the source of movement, and movement is essential to the healthy physical and psychical condition of the body and person. While contemporary speakers cannot articulate this aspect of the obsolete CS model, the striking productivity of the metaphor BEING HEALTHY IS INTERNAL MOVEMENT along the lines of the old model indicates that at some less than fully conscious level, this facet of the model is alive and well.

In NENA there is first of all a very common usage expressing the idea that the heart has reached stasis, through the employment of the verbal root Pʾ Š, 'to remain/stay put':

(23) a. *Pišleh* *libbī* *bīyeh/l-kisleh* *sābab*
It-stay-put.PFT my heart in/by-him because
muqrimleh *kulleh* *zuzē*. (cf. *API*: 440)
he-lose.PFT-it all-of-it his-money.
'My heart stayed by him (i.e. I took pity on him) because he lost all his money.'

b. *Du(y)waḫ* *ʿḇīrē* *b-šinnē* *w-(ʾ)ḫe(r)nā*
So-as-we it-pass.PFT in-years and-again
laytlan *pyāšā* *haw* *libbā*
there-is-not-to-us stay-put.CONT it heart
d-qābil. (*API*: 404)
that-accepts.JUS.
'Now that we have become older, our hearts are no longer accepting (i.e. we cannot take it anymore).'[19]

c. *Mupišloḵ* *libbī* *ʾīman* *d-ʾmīrroḵ*
You-make-stay-put.PFT my-heart when you-tell.PFT
mirʿyā *yleh*. (cf. *API*: 410)
that-he-be-sick.IN-PFT COP.
'You made my heart stay put (i.e. you made me upset and uneasy) when you told me that he is sick.'

d. *Tānit* *ʿal* *libbī* *yleh* *(y)tīḇā*. (*API*: 467)
You say.JUS on my heart he.COP sit.IN-PFT.

'It is like he is sitting on my-heart (i.e. I find him burdensome; he is putting me in a bad mood).'

e. *Ḫzīlī* *ʿayḇē,* *libbī* *ḫlīṣlēh.*
I-see.PFT clouds, my-heart tighten-PFT.
'When I saw clouds, my heart tightened (became melancholy).'

Native speakers are at a loss to explain these very idiomatic usages. Cognizance of the CS model, however, makes such usages perfectly understandable. In all these cases, the idea of the heart staying put is a way of describing the transitory temperamental state (*diathesis*) signaled by the movement generated by the heart coming to stasis – whether a person has come to this state on his own [examples (23a) and (23b)], or through the action of others [(23c) and (23d)]. It describes that condition in which the mixture of the heart loses its hot and dry quality, not much differently from the idea of the heart cooling. This explains why a number of native speakers have described these expressions concerning the heart staying put as being rather close to the idea of the heart cooling. The resulting condition, one in which the heart fails to generate movement, corresponds again to the unhealthful blocked condition that Simon described as so unpleasant.

It is also common to express in NENA the opposite phenomenon, namely the heart moving. Here are three examples:

(24) a. *Libbī* *bīrḫāšā* *yleh* *d-ʾāḵlin* *duḫwā.*
My-heart go.CONT it.COP that-I-eat.JUS dukhwa.
'My heart is going (i.e. I have a craving) to eat dukhwa (a barley, yogurt, and meat stew).'

b. *Libbī* *bēʾzālā* *yleh* *l-d-hay* *brātā.* (a vulgar usage)
My-heart go.CONT it.COP to-that girl.
'I'm horny for that girl.'

c. *Priḫleh* *libbah* *gô* *hāwā* *min* *ḫādutā.*
It-fly.PFT her-heart in air from happiness.
'Her heart was flying in the air out of happiness (cf. she was in seventh heaven).'

What is not self-evident in (24a) and (24b) is that the moods they describe are *pleasant* ones, that the heart being in a state of movement is healthful. Yet that is the sense of them to native speakers. Again, although the connection to the CS model is lost, the concept expressed by these usages depends on that model for its motivation to be apparent. Looking at these expressions only in a contemporary context, it might appear that what generates them is the metonym a MOVING HEART FOR THE ACTIVATION OF DESIRES, which in turn is produced by the combination of the metaphor THE HEART IS A CONTAINER OF DESIRES/MOODS, the metonym CONTAINER FOR CONTAINED, and the metaphor ACTIVATING IS MOVING. However, juxtaposing such usages with (24c), in which there is a more extreme form of motion, flying, that clearly activates a sense of more intense pleasure, makes obvious the connection with the archaic model, which linked the movement of the heart to psychical health. *Flying* not only is an intensified form of motion, but it also suggests lightness and upward movement, which, according to the old model, were other concomitants of the fire (in the heart) that generated such movement. Thus contextualization of these expressions in an outdated model of how the body and mind work ascribes an understandable motivation to them, even if such a motivation is no longer conscious or unconsciously effective. Clearly the model has not evaporated completely: splotches of it remain.

4.6. Courage

In a less subtle way, we also have the very common metaphor of COURAGE IS HEART. Earlier we saw that the ideal condition of the heart is to be hot, and that when a person is characterized by the hot mixture of the temperaments, he has courage, *lḇīḇuṯā* (an abstract noun from the word for heart, *libbā*). The identification of the heart with courage is still very much in effect:

(25) a. *Bāʿē libbā d-māḥe l-dibbā.* (*API*: 96)
It-wants.JUS heart that-beats.JUS a-bear.
'One needs heart (courage) to slay a bear.'

b. *Gô kul libbā ḥa(d) ʾaryā dmīḵā yleh.*
In every heart one lion sleep.IN-PFT it.COP.
(*API*:112)
'In every heart there is a sleeping lion.' (said of a person who thinks highly of himself)

c. *Laytleh haw libbā.* (*API*: 404–405)
There-is-not-to-him he heart.
'He does not have any heart (courage).'

d. *Qam šāmiṭleh libbī b-ẖabrānuhy sikrānē.*
PST break.JUS-it my-heart with-his-news harsh.
(*API*: 463)
'The harsh news caused my heart to break (i.e. was disheartening/ discouraging).'

In keeping with this theme of *libbā* cueing notions of a courageous disposition, we also have the common usages of *yahbil leh libbā* ('He gives him heart'), meaning to 'encourage', *paliṭ lah libbā* ('He extricates her heart'), meaning to 'dishearten'. These usages activate the metaphor THE HEART IS A TRANSFUSIBLE CONTAINER OF HEXEIS as well.

It is also worth noting that the adjective corresponding to "courageous" is *libbāna*, simply the adjectival form of *libbā*. All these data makes it look like we simply have the metaphor COURAGE IS HEART. Were the old model to still be alive, these usages of *libbā* would be activating a model of the temperaments in which the ideal state of the heart is characterized by courage. Just as in Lakoff's classical example of an ICM, in which mention of an unqualified, unmarked "mother" activates a cluster of schemas concerning the idealized mother (Lakoff 1987: 74ff.), mention of the heart would also be activating the various heart-implicating schemas that entail courage, including the ideal hot mixture of the heart. In the consciousness of modern speakers, it is hard to imagine that such activation is taking place, but the usage survives as a relic, a vestige of the old model – a post-metonymy.

5. Heart and face: intimations of another model

To allay any suspicion that this paper is claiming that all the figurative uses of *libbā* in NENA are either vestiges of an archaic model of the temperaments, or instantiations of near-universal metaphors, a small sample of other usages will now be offered that point to the existence of another general, underlying metaphor that is by no means universal but it is widespread.

Assyrians often talk about respectable people as being *clean*, *pure*, and *white*, and disgraceful people as *dirty*, *blemished*, and *black*. Such discourse corresponds with the general conceptual metaphor BEING MORAL IS BEING CLEAN AND WHITE that Yu has identified in Chinese (this volume: 143). Also as in Chinese and many other languages, these qualities are often expressed by invoking the face. Assyrians have a proverb that in effect asserts that the face is a metonym for the heart:

(26) *Gāšiq ʿal pāteh w-(y)diʿ libbeh.* (*API*: 123)
Look.IMP at his-face and-know.IMP his-heart.
'If you look at a person's face, you will know their heart.'

Just why or how the face might be a reflection of the heart, we do not know. Nevertheless it is clear that the proverb presumes a conceptual key HEART IS/FOR MORAL CHARACTER. Because we do not know the schematic relation between HEART and MORAL CHARACTER, I use "conceptual key", following Charteris-Black (2003: 293–94), to call attention to a superordinate category that includes both conceptual metaphor and conceptual metonym. The following proverb reinforces the idea that the HEART stands for the perduring character, not only *hexis*, but also *dunamis*, 'potency', of the person:

(27) *Kul ḫa(d) ʾayḵ libbeh ʾAlāhā kē yāḇileh.*
Every one like his-heart God PRST give.JUS-it.
(*API*: 205)
'Everyone has the heart with which God has provided them.'

Just like the face, the heart can be described as black:

(28) *Kul mindī d-ʿāḇdī qāteh ʾīnā libbeh har*
All thing that-they-do.JUS for-him but his-heart still
kumā yleh. (*API*: 403)
black it.COP.
'In spite of all the things they have done for him, his heart is still black.'

Having a *black* heart reflects poorly on a person's moral character. If one has a black *face*, this implies that the person's reputation is bad; if they have a black heart this means that they are immoral, specifically that they are distrustful and have malicious intentions. Just as commonly, when a man is referred to as having a *dirty* heart (*šiḫtānā*), this indicates that he has only licentious desires when he looks at women. Although the heart can also be described as "dirty" in Persian, in that language it means something rather different, namely being upset or dissatisfied with a person (Sharifian this volume: 252). It thus pertains to a feeling rather than a *hexis* or character trait as it does in NENA.

By contrast with having a dirty heart, a person with a 'clean' or 'pure heart' (*libbā tammiz* or *ṣipyā*, respectively) is above suspicion. However,

BEING MORAL IS BEING CLEAN AND WHITE is not fully productive inasmuch as a "white heart" is seldom, if ever used.[20] Having a 'white face' (*pāʾtā ḫwārtā*), however, is extremely common to describe a person who has managed to dodge being disgraced.

It can readily be seen that these usages do not have an obvious connection either with that archaic Syriac model of the temperaments, or with any metaphors that are candidates for universal status: how can there be a universal embodied experience of the heart as black and this having any connection with malevolence? This brief sampling of another figurative use of the heart makes it clear that no one model can underpin a vast array of figurative usages in a particular domain of a language. There is indeed a vast array of figurative usages applied to the heart in Northeastern Neo-Aramaic, and only a small segment of them have been touched upon in this paper.

6. Conclusion

As it has developed through the last 20 years or so, the cross-cultural study of metaphors and metonyms from a cognitive-linguistic point of view has spawned two opposed strategies. On the one side are those who have been so impressed by the apparent universality of certain "general" or "primary" metaphors such as THE BODY IS A CONTAINER OF THE EMOTIONS, that they think such metaphors stem from the correlated embodied experience of all human beings. From this insight, they draw the methodological lesson that the best way to pinpoint cross-cultural differences is to see the expressions used in particular languages as specific *elaborations* of these general, near-universal metaphors.

On the other side, there are those who are not ready to bank on the universality of basic conceptual metaphors. Their methodological priority is the identification and elaboration of culturally-specific models that exert much, perhaps decisive, influence on the specific contents of metaphoric, metonymic, and other forms of figurative expression in particular languages and language varieties.[21]

Sharifian, Dirven, Yu, and Niemeier (this volume) support this strategy, and most of the papers in this volume follow it. At the heart of this approach is the insight that a great many, possibly all, specifically elaborated metaphors are actually generated out of metonyms, which is to say out of such believed relations as cause and effect, part and whole, category and member of category (e.g. Barcelona 2000c, Radden 2002, and Riemer

2002). So it is incumbent to focus on sets of interrelated metonyms that are pivotally generative of specific linguistic expression and discourse. These sets are none other than what in cognitive linguistics have been called "domains" (Langacker 1987: 147–182), "idealized cognitive models (ICMs)" (e.g. Lakoff 1987), or "schemas" (D'Andrade 1995: 122–147). Since in real life they are never the complete and idiosyncratic invention of individuals, these domains of meaning or models are also always cultural (cf. D'Andrade 1995: 130–132, on "cultural schemas"). Hence sets of interrelated metonyms also deserve to be equated with and called simply *cultural models*.[22]

In a series of recent works, Kövecses has tried to reconcile these opposed approaches (e.g. 1995, 2000, and 2005). While aiming to make room for vast cultural differences, he does not abandon the conviction that a certain degree of universal embodied experience of human beings constrains the existential conceptualizations that underpin particular culturally-specific linguistic expressions.

Here is not the place to assess how successful Kövecses has been in his attempt, but a recapping of some of the components of his synthesis helps to contextualize this paper within a larger theoretical frame.[23] First, he sees universal human embodied experience not as determinative of near-universal metaphors, but rather as *constraining* what metaphors might arise, and, more strikingly, what ones are unlikely to. Thus he interprets the curiously widespread and uniform, though not universal, compound metaphor THE ANGRY PERSON IS A PRESSURIZED CONTAINER as a natural, though not inevitable, outgrowth of several "universal correlations in bodily experience" (2000: 155–160; cf. 2005: 38). For him, it is a high priority to find such "generic-level" metaphors that have appeared in unrelated cultures.

Second, he has identified a number of dimensions of difference of "congruent metaphors", i.e. the instantiations of the same generic-level metaphors. These variable aspects include the degree of elaboration, the range of the target domain (the set of source domains that map onto the target domain), the scope of the source domain (the set of target domains it gets mapped onto), and the preferences the speakers of a particular language, or in a particular culture, have for selecting metaphors.

And third, he has interpreted these variable culturally-specific instantiations in the light of such variable factors as differential personal and historical experience, the cultural context (comprising the salient concepts and values in that culture), and different concerns and interests (often worked out as ideologies).

What is the place of this paper in Kövecses's synthesis? The identification of an archaic cultural model, and the detection of its traces in contemporary speech (and writing) falls squarely under the rubric of "cultural context" (cf. Kövecses 2000: 167–169 and 2005: 234–236). In other words, this paper has followed the strategy of cobbling together the components of an old cultural model that still exerts its influence on contemporary figurative usages regarding the HEART, *libbā*. Concordant with Kövecses's insistence on allowing for the possibility of universal correlations in embodied experience constraining linguistic expression, it kicked off its analysis by taking as its cue what may appear as a near-universal metaphor, THE BODY IS A CONTAINER OF THE EMOTIONS.[24]

Nobody knows what is in store for us once we come to grips with the broader historico-cultural contexts that underpin figurative usages, through the accumulation of detailed case studies such as this one. It will not be surprising if the results of such studies reshape, and perhaps whittle away at, what we posit as those universal constraints. Aside from uncovering an archaic model of the temperaments which has exerted lasting effects on figurative uses of language in a form of Aramaic, this paper has advocated two (relatively) new tactics to assess the influence of cultural models: (1) the exploration of subtle differences among emotions, temperaments, moods, dispositions, character traits, etc., with the help of a philosophy (Aristotle's) that lies intermediate between our cultural concepts and those of NENA speakers; and (2) special attention to metonymical relations in an archaic cultural model, with an assessment of how living and productive they are among contemporary speakers. A brief retracing of this paper's journey will shed light on how these tactics were pursued and what they yielded.

The starting point of analysis was the supposed elaboration of THE CONTAINER OF THE EMOTIONS source domain in NENA, as it may be mapped onto the target domain of the HEART. Such an investigation quickly encountered problems: the meanings of concepts that might initially be activated by "EMOTION", not to mention "HEART", were by no means self-evident (the practice of putting them in small caps, to emphasize that concepts rather than lexemes are being alluded to, has not saved them). In other words, it was seen that it is premature to impute any "universal correlations in bodily experience" to such a generic metaphor as THE BODY IS A CONTAINER OF THE EMOTIONS. The reason is that it is premature to imagine that there is anything universal by what is meant by "EMOTION".

Furthermore, as is already well-recognized, the heart, in the sense of the organ that pumps blood to the body, is conceived in widely diverse ways in different societies in its relation to anything resembling EMOTIONS. For this reason, the issue of how the "heart" (*libbā*) related to the other body organs, especially the "liver", the "stomach" and above all the "eye", was taken up. A number of expressions concerning what is contained in the *libbā* as opposed to the "eye" (*ʿaynā*) made it clear that in NENA these two body organs are somewhat different loci or agents of desires, moods, emotion, or volition, both with respect to the types of these they contain, and the degree of emphasis on being filled in proper measure (for the eye this proper measure was emphasized and for the heart it was not). In many contexts, it was found helpful to shift the focus from transitory or episodic emotions (cf. *pathē*) to more durable dispositions, such as temperamental dispositions (*diatheseis*) and, above all, moral character traits (more precisely: *hexeis*).

After these initial orientations, the paper shifted back to ancient Classical Syriac sources in order to reconstruct, preliminarily, an old cultural model concerning the workings of the heart and other body organs, and their implications for emotions, character, and temperament. Classical Syriac, like languages in the West, was deeply influenced by the appropriation of the old Greek humors theory associated with the medical writings of Galen, but there were specifically Syriac syntheses and elaborations of such a theory.

The subsequent section of the paper explicated a number of diverse figurative expressions in NENA that hark back to certain features of the old cultural model, even though the metonymic connections of this obsolete model (causal, part/whole, category/member) are no longer active in contemporary consciousness. These vestiges included (1) the concept that THE HEART IS A HOUSE-LIKE CONTAINER of not so much emotions but certain kinds of guiding principles; (2) TEMPERATURE to describe the (a) PERDURING STATE OF THE HEART as a way of talking about a person's temperament or moral character, and (b) CHANGES IN THE CONDITION OF THE HEART as a way of talking about TRANSITORY EMOTIONS; (3) DARKNESS as a way of talking about an unhealthy condition of the HEART (this is a minimal vestige of a once much more elaborated part of the model); and (4) the remarkably productive domain of MOVEMENT (staying-put/moving) for expressing those TRANSITORY EMOTIONS.

While the old Classical Syriac model of the organs informs a great many figurative usages in the contemporary language, it certainly does not account for all of them. Mention was made of another model that informs some other expressions regarding the heart. This other model, only hinted at, concerns usages involving the heart that impute moral character to a person in ways that are related to talking about a person's face. It has been included to drive home the point that the sources of contemporary conceptual metaphors concerning the heart are diverse, and should not be sought in a single cultural model.

The study of figurative expression in the context of a culture's historical legacy of models of how the world works is still in its infancy. Since such models may leave their mark on a language long after they have disappeared from the consciousness of its speakers, the need to mine the written record in order to stitch together those models is an urgent priority.

But there is another, no less important priority. This paper has sought to address the issue of decomposing certain key concepts that cognitive linguists have not been careful enough about. Specifically, the category of "emotions" has been too vague in the literature. Attention has to be paid to such issues as the difference between ephemeral feelings and perdurable dispositions and everything in between. Aside from the issue of the transitoriness vs. durability of such states included under or related to "emotions", the issue of the degree to, and manner in which, such states are construed within moral frameworks needs also to be taken up. If this paper has helped to get the ball rolling on these everyday philosophical matters, it has achieved its main goal.

Acknowledgments

My primary debt is to Oraham Yalda Oraham for his collaboration, who has been unstinting in sharing his vast knowledge of the Assyrians' language with me. He has provided many of the examples in the paper, explained and clarified a great many matters in both Northeastern Neo-Aramaic and Classical Syriac, and carefully checked over the examples. He also brought the main source of examples (Dinḫā and Beytšamʿon 2004) to my attention. Eve Pinsker has also been of enormous assistance from nearly the inception of the project. She has been a sounding board and critical commentator for a great many of the ideas expressed in the paper. Not being able to attend the conference myself, she graciously volunteered to read the paper in my place. She also exerted most of the effort to transform the written version of this paper into an oral presentation. Both she and Oraham have been, in a sense, co-authors. I also would like to acknowledge that I got valuable help from Sam Fox, Chip Coakley, and Hezy Mutzafi. Finally, it was one of the editors of this volume, Farzad Sharifian, who originally gave me the idea for the paper, provided my initial orientations, and has been patient and supportive throughout.

Notes

* The following abbreviations for the sources of expressions in Northeastern Neo-Aramaic (NENA), and of evidence for the Classical Syriac (CS) sources will be used: *API* for Dinḫā and Beytšamʿon (2004); *JE* for Job of Job of Edessa (1935), *Kul* for Kulyā (1967), *MalkDict* for Malko (Gewargis Ashitha) (1997), *QC* for Dānīʾeyl (1961), *RabYon* for Malko [Gīwargīs] ʾĀšīṯā (1997), *ST* for Simon of Ṭaibūtheh (1934), *SyrMed* for Budge (1913), and ZB for *Zahrīrē d-Bahrā*.

1. It remains a tricky issue what to call this language. Speakers rather consistently call it "Modern Assyrian" or simply "Assyrian". Oraham prefers that this label be used in scholarly work as well. My own preference would be to call the language "Modern Eastern Syriac", as a translation of *Lešānā Surāyā Madnḫāyā ḫa(d)tā*, primarily because this would be in keeping with the anthropological principle of calling a language what its speakers prefer it to be called, but also to stress its *cultural* connections with Classical Syriac, to be drawn on below, which has endured as a written language for nearly 2000 years.
2. The difference between the CS *lebba* and the NENA *libbā* is not in the original languages, but only in the conventions of transliteration.
3. To form the continuous present in the koinē and many dialects, this stem is preceded by the prefix b- and followed by a clitic copula form to indicate the agent (cf. Fox 1997: 23). In some dialects, however, the b-prefix is omitted. In our interlinear translation, we will simply overlook this prefix, specify the agent, and indicate that the word is a continuous stem.
4. In the various copular forms, this paper will not distinguish between independent and clitic forms.
5. A note on the convention to be followed in citing texts of NENA expressions: a great many examples are taken from Dinḫā and Beytšamʿon 2004 (*API*). This work is a recent, rather remarkable collection of proverbs and idioms in NENA. If "cf." is noted in the citation, this means that Oraham has generated a sentence with the proverb or idiom provided. Otherwise, the example is lifted verbatim from this *Treasury*. Uncited sources are expressions provided by Oraham or by other speakers and confirmed by him. Oraham's own testimony is marked by "OYO".
6. It would be presumptuous to pretend to represent Aristotle's framework accurately in a short summary, especially since he there are subtle differences between his accounts in his *Categories*, *Metaphysics*, *On the Soul*, and *Nicomachean Ethics*. In what follows, I offer a non-expert account, based mainly on Tredennick (1933) and Cooke and Tredennick (1938), with further help from Sacks (1999), Hutchinson (1986), and Savarino (1993). My purpose is to loosen the grip of English concepts on us as we try to grasp NENA concepts, not to explicate Aristotle per se.

7. In an interesting discussion, Roberts takes Goldie (2000) to task for taking this latter sense of "emotion" as a multifaceted, lasting state to be the primary sense of the term (Roberts 2001: 544–548). Such disagreement among sophisticated philosophers indicates just how underspecified "emotions" are in English, as well as terms for specific emotions, with respect to their duration (and degree of inclusiveness in a narrative context).
8. Even more jocularly, some people play with the expression, saying to a little baby boy, *'k̲īlī buṭuk̲* ('May I eat your penis') or (using a jussive stem), *'Āk̲linhon anē 'iškātē* ('I would eat those testicles'). Both would be typically used, say, to express hope that the baby boy would cooperate as his diaper is being changed.
9. In different terms, Shweder makes an argument similar to Roberts's:

 > The "emotion" is the whole story, the whole package deal – a kind of somatic event (fatigue, chest pain, goose flesh) and/or affective event (panic, expansiveness, emptiness) experienced as a kind of perception (of loss, gain, threat, possibility) linked to a kind of plan (attack, withdraw, confess, hide, explore). (1994: 38)

 That is, "emotion" can be broken down into the components "somatic" and/or "affective event" (cf. "sensations"), perception (cf. "concern-based construal"), and a "plan" (that instigates a course of action). Kövecses suggests that depending on the emotion, "desire" can result from, that is, be "spurred" on by the emotion, or it can consist in "having" the emotion, as in *hungering for love* (2000: 45), which I would interpret as prompting or triggering the emotion. In this context, Kövecses is using "desire" as an equivalent of Shweder's "plan", and he grants that it can be "viewed as a special kind of emotion". In consonance with both Roberts and Shweder, it makes sense, at least in regard to English, to conceptualize this "desire" as part of the whole frame or model invoked by "emotion". In NENA, there is a "whole package deal" as well, though it should be assumed that it differs somewhat from the one in English. This paper does suggest that NENA may emphasize the link to moralized character traits, or *hexeis*, more than English does, but this has to remain only a conjecture until more research can be undertaken.
10. While this expression is common, there is confusion in meaning on account of homonymy, and, possibly, dialect difference. There are two distinct verbs, with the roots *GB̲L* and *GWL*, that sound precisely the same (both are to be found in CS, but the second one is rare in NENA). That is, what NENA linguists phonemically represent as *ləbbi gwəlle* can be spelled *libbī gb̲īleh* or *libbī gwīleh*. The former verb means 'to shape, mold, work into', as in shaping kofta or meatballs, whereas the latter can mean 'to stir, move, repel, cast out'. *MalkDict*

includes both possibilities, with *libbī gḇīleh* meaning 'to be faint' and the *libbī gwīleh* 'to vomit'. Most native speakers questioned on the meaning of *ləbbi gwəlle* find it to mean 'I was nauseated'. This includes OYO as well as other speakers of the Lower Ṭyārē group, (including some reported by Hezy Mutzafi, personal communication), but speakers have been found who report it to mean 'he vomited'. Whether they interpret the expression to mean 'he was nauseated' or 'he vomited', native speakers as a rule know neither which root they are using (*GḆL* or *GWL*), nor the decomposed meaning of the expression. Whichever meaning is invoked, however, the point being made in the present context still stands – *libbā* may encompass the stomach.

11. Degen (1972) has been very helpful in directing us to these published texts. The best known source on medicine in CS can be found in a collection of several anonymous works now known collectively as the *Book of Medicines*, dating probably from the 7th to 9th centuries and including many passages translated from Galen (*SyrMed*). Only one small section of this collection was found to be very helpful for our purposes here. Other important extant sources are mostly unpublished.
12. In these citations, we include the original CS page location first, followed by the translation.
13. Only the original CS is cited, for the text is translated into Latin only.
14. Mingana calls this 'desire', which is nourished by the heart.
15. Chabot translates this as *tactu*, 'sense of touch' – the usual translation – but it can also mean the 'pulse', and, possibly, the skin, by virtue of the correlation between feeling one's pulse and touching one's skin.
16. I follow the vocalization followed in *SyrMed*, rather than as in the standard dictionaries.
17. *ḥašē*. In CS this means 'disease', the pain and suffering which are the product of disease, and the passions, such as anger, fear, sadness and gladness. As the English "grief", the meaning of *ḥašē* has become more restricted in the modern language (NENA), and now applies primarily to the feeling of grief, especially for a person who has died.
18. Examples (19b) and (19c) go together. The authors present (c) as the proverb (attributed to Q Dinkha) and add (b) in their commentary as the sense of the proverb. Both lines, however, can be interpreted as richly figurative. Presenting them together, in this order, makes the cultural model slightly easier to grasp.
19. Notice that *laytlan pyāšā* has the literal sense of 'we do not have staying-put', but this is the ordinary, idiomatic way of saying, "We no longer ..."
20. *MalkDict* does include 'white heart' (*libbā ḥwārā*) to mean the same as a 'pure heart', but this usage could not be confirmed.
21. Consistently with this strategy (and the argument followed in this paper), Traugott concluded that the "schema" of an apparently dead metaphor "may

still provide the template for further innovations" (1985: 50). What she calls a "schema" is the equivalent of a "conceptual key", which could be a conceptual metaphor or metonymy, and in any case is part of an old cultural model. Geeraerts and Grondelaers (1995) elaborated this strategy of anchoring analysis in the retrieval of archaic cultural models, and helped to inspire the current paper.

22. See Sharifian, Dirven, Yu, and Niemeier (this volume: 9–13) and the references therein, for more background on cultural models. For a sophisticated discussion of the relation between the universal and the cultural in the study of conceptual metaphor, see also Zinken, Hellsten, and Nerlich (2003).
23. The following examples of Kövecses's moves are culled from Kövecses (2005).
24. I am not aware that anyone has actually asserted this metaphor to be universal, but it has been claimed to be widespread (e.g. Taylor and Mbense 1998: 194 and Kövecses 2002: 184), and much of the cognitive-linguistic literature on emotion does address this metaphor.

References

Barcelona, Antonio

2000a Introduction: The cognitive theory of metaphor and metonymy. In Barcelona (2000b), 1–28.

2000b *Metaphor and Metonymy at the Crossroads: A Cognitive Approach.* Berlin and New York: Mouton de Gruyter.

2000c On the plausibility of claiming a metonymic motivation for conceptual metaphor. In Barcelona (2000b), 31–58.

Bierwiaczonek, Bogusław

1993 The axiological domain in the description of evaluative predications – an exercise in cognitive semantics. In *Images from the Cognitive Scene*, Elżbieta Górska, (ed.), 27–42. Kraków: Universitas.

Budge, E. A. Wallis (ed. and trans.)

1913 *Syrian Anatomy, Pathology and Therapeutics: Or, "The Book of Medicines"*. London: Oxford University Press. Two volumes.

Chabot, J.-B. (ed. and trans.)

1965 Notice sur deux manuscrits contenant les œuvres du moine isaac de rabban isho et du métropolitain aḥoudemmeh. In *Notices et extraits des manuscrits de la Bibliothèque nationale et autres bibliothèques*, volume 43, 43–76. Paris: Imprimerie Nationale and Librairie C. Klincksieck.

Charteris-Black, Jonathan
2003 Speaking with forked tongue: A comparative study of metaphor and metonymy in English and Malay phraseology. *Metaphor and Symbol*, 18(4): 289–310.
Chyet, Michael L.
1995 Neo-Aramaic and Kurdish: An interdisciplinary consideration of their influence on each other. In *Language and Culture in the Near East*. (Israel Oriental Studies 15.) Shlomo Izre'el and Rina Drory, (eds.), 219–252. Leiden: E.J. Brill.
Cooke, Harold P. and Hugh Tredennick, (eds.)
1938 *Aristotle: The Categories of Interpretation; Prior Analytics*. (Loeb Classical Library 325). Cambridge, MA: Harvard University Press.
D'Andrade, Roy
1995 *The Development of Cognitive Anthropology*. Cambridge: Cambridge University Press.
Dānī'eyl, Wilyam D. Ṣ.
1961 *Qāṭīnē ga(n)bārā: Ḫdā mšuḫat-ga(n)bārē* [Qatinee Gabara: An Epic Poem]. Tehran: Maṭba'tā d-Sī'tā Siprêtā d-'Eleymē 'Ātorāyē (Assyrian Youth Cultural Society Press).
Degen, Rainer
1972 Ein corpus medicorum syriacorum. *Medizin historisches Journal* 7: 114–122.
1981 Galen im Syrischen: Eine Übersicht über die syrische Überlieferung der Werke Galens. In *Galen: Problems and Prospects*, Vivian Nutton, (ed.), 131–166. [London]: Wellcome Institute for the History of Medicine.
Dinḫā, Q Šmo'eyl [Rev. Samuel Dinkha] and Rāman Beytšam'on [Raman Bitshamoon]
2004 *Sīmtā d-matlē w-ḫāšḫātē 'ammāyē d-'Atorāyē = A Treasury of Popular Assyrian Proverbs and Idioms*. San Jose: Adiabene Publications.
Dirven, René and Ralf Pörings, (eds.)
2002 *Metaphor and Metonymy in Comparison and Contrast*. Berlin and New York: Mouton de Gruyter.
Ekman, Paul and Richard J. Davidson, (eds.)
1994 *The Nature of Emotion: Fundamental Questions*. New York: Oxford University Press.
Fox, Samuel Ethan
1997 *The Neo-Aramaic Dialect of Jilu*. Wiesbaden: Harrassowitz Verlag.
Geeraerts, Dirk and Stefan Grondelaers
1995 Looking back at anger: Cultural traditions and metaphorical patterns. In *Language and the Cognitive Construal of the World*, John Taylor and Robert E. MacLaury, (eds.), 153–180. Berlin: Mouton de Gruyter.

Goldie, Peter
2000 The Emotions: A Philosophical Exploration. Oxford: Clarendon Press.
2004 *On Personality*. London and New York: Routledge.

Hutchinson, D. S.
1986 *The Virtues of Aristotle*. London and New York: Routledge and Kegan Paul.

Job of Edessa
1935 *Encyclopedia of Philosophical and Natural Sciences as Taught in Baghdad about A.D. 817: Or, Book of Treasures*. Cambridge: W. Heffer & Sons.

Kövecses, Zoltán
1990 *Emotion Concepts*. New York: Springer-Verlag.
1995 Anger: Its language, conceptualization, and physiology in the light of cross-cultural evidence. In *Language and the Cognitive Construal of the World*, John Taylor and Robert E. MacLaury (eds.), 181–195. Berlin: Mouton de Gruyter.
2000 *Metaphor and Emotion: Language, Culture, and Body in Human Feeling*. Cambridge: Cambridge University Press.
2002 *Metaphor: A Practical Introduction*. Oxford and New York: Oxford University Press.
2005 *Metaphor in Culture: Universality and Variation*. Cambridge: Cambridge University Press.

Kulyā, Q Šmuʾeyl Yôsip d-beyt, d-Gogtāpāh
1967 *Ḫāĉāqôgē, yan "plaḫtā d-ḫiltunyutā" b-hamāntā l-duglē* [Thieves of the cross, or "the strong delusion" of believing in lies]. Tehran: Maṭbaʿtā d-Sīʿtā Siprêtā d-ʿEleymē ʾĀtorāyē (Assyrian Youth Cultural Society Press). Original edition, 1963.

Lakoff, George
1987 *Women, Fire, and Dangerous Things: What Categories Reveal about the Mind*. Chicago: University of Chicago Press.

Lakoff, George, and Mark Johnson
1999 *Philosophy in the Flesh: The Embodied Mind and Its Challenge to Western Thought*. New York: Basic Books.

Langacker, Ronald
1987 *Foundations of Cognitive Grammar*, volume 1: *Theoretical Perspectives*. Stanford: Stanford University Press.

Lucy, John A.
2004 Language, culture, and mind in comparative perspective. In *Language, Culture and Mind*, Michel Achard and Suzanne Kemmer, (eds.), 1–22. Stanford: CSLI Publications.

Maclean, Arthur John
1895 *Grammar of the Dialects of Vernacular Syriac as Spoken by the Eastern Syrians of Kurdistan, North-west Persia and the Plain of Mosul*. Cambridge: Cambridge University Press.

Malko (Gewargis Ashitha), Odisho
1997 *Hilqa de leshana: Assyrian-Arabic Dictionary*. Baghdad: Al-Maghreb Printing (Offset Colour Press)

Malko [Gīwargīs] ʾĀšīṯā, ʿAḇdīšoʿ [Odisho M. Gewargis Ashitha], (ed.)
1997 *Gujī, ʿam syāmē (ʾ)ẖrenē d-Rabbān Šm. Yônān Tẖumnāyā = Guggi, with other poems, Rabban Yonan Tkhomnaia*. Baghdad: Maṭbaʿat Ūfsīt al-Maghrib.

Mutzafi, Hezy
2005 Etymological notes on North-Eastern Neo-Aramaic. *Aramaic Studies* 3 (1): 83–107.

Nau, F. (ed. and trans.)
1909 Traité d-aḥoudemmeh sur l'homme. *Patrologia Orientalis* 3: 97–119.

Niemeier, Susanne
2000 Straight from the heart – metonymic and metaphorical explorations. In Barcelona (2000b), 195–213.
this vol. To be in control: Kind-hearted and cool-headed: The head-heart dichotomy in English.

Peña Cervel, M. Sandra
2003 *Topology and Cognition: What Image-schemas Reveal about the Metaphorical Language of Emotions*. Muenchen: Lincom Europa.

Radden, Günther
2002 How metonymic are metaphors? In *Metaphor and Metonymy in Comparison and Contrast*, René Dirven and Ralf Pörings (eds), 407–434. Berlin and New York: Mouton de Gruyter.

Riemer, Nick
2002 When is a metonymy no longer a metonymy? In *Metaphor and Metonymy in Comparison and Contrast*, René Dirven and Ralf Pörings (eds), 379–406. Berlin and New York: Mouton de Gruyter.

Roberts, Robert C.
2001 An essay review of Peter Goldie's *The Emotions: A Philosophical Exploration*. *Philosophia Christi* [Series 2] 3 (2): 543–552.
2003 *Emotions: An Essay in Aid of Moral Psychology*. Cambridge: Cambridge University Press.

Sacks, Joe, (ed.)
1999 *Aristotle's Metaphysics*. Santa Fe, NM: Green Lion Press.

Savarino, Mary Ella
1993 Toward an ontology of virtue ethics. *Journal of Philosophical Research* 18: 243–259.

Sharifian, Farzad
this vol. Conceptualisations of Del 'heart-stomach' in Persian.

Sharifian, Farzad, René Dirven, Ning Yu, and Susanne Niemeier
this vol. Culture and language: Looking for the mind inside the body.

Shweder, Richard A.
1994 'You're not sick, you're just in love': Emotion as an interpretive system. In *The Nature of Emotion: Fundamental Questions*, Paul Ekman and Richard J. Davidson (eds.), 32–44. New York: Oxford University Press.

Simon of Ṭaibūtheh
1934 Medico-mystical work. In *Early Christian Mystics*, A. Mingana, (ed.), 1–69, 281–320. (Woodbrooke Studies: Christian Documents in Syriac, Arabic, and Garshūni 7.) Cambridge: W. Heffer & Sons.

Taylor, John R. and Thandi G. Mbense
1998 Red dogs and rotten mealies: How Zulus talk about anger. In *Speaking of Emotions: Conceptualisation and Expression*, Angeliki Athanasiadou and Elżbieta Tabakowska, (eds.), 190–226. Berlin and New York: Mouton de Gruyter.

Traugott, Elizabeth Closs
1985 'Conventional' and 'dead' metaphors revisited. In *The Ubiquity of Metaphor*, Wolf Paprotté and René Dirven, (eds.), 17–57. Amsterdam and Philadelphia: John Benjamins.

Tredennick, Hugh, (ed.)
1933 *Aristotle: The Metaphysics, Books I–IX.* (Loeb Classical Library 271.) Cambridge, MA: Harvard University Press.

Yu, Ning
2002 Body and emotion: Body parts in Chinese expression of emotion. *Pragmatics and Cognition* 10 (1/2): 341–367.
this vol. The Chinese heart as the central faculty of cognition.

Zahrīrē d-Bahrā
1900 Mamllā qā millat suryêtā [Speech for the Syrian millet]. 51 (10): 76. By Mar(y) Yonan.

Zinken, Jörg, Iina Hellsten, and Brigitte Nerlich
2003 What is cultural about conceptual metaphors? *International Journal of Communication* 13 (1–2): 5–29.

Hearts and (angry) minds in Old English

Dirk Geeraerts and Caroline Gevaert

Abstract

On the basis of *heart* expressions in Old English, we argue that an onomasiological analysis is indispensable to get a good insight into the (relative) importance of embodied imagery of the mind and the soul. As a first step, an analysis of *heart* and *mood* compounds in Old English reveals three things. First, *mood* represents an integrated conception of mental life, in which cognition, emotion, and volition occur side by side. Second, the heart is seen as an embodied seat of the mind and the soul. Third, however, the importance of this second recognition is severely mitigated by the fact that *heart* imagery constitutes a minority in comparison with literal *mood* denominations. In a second step, the dominance of literal concepts is further corroborated by an analysis of the concept 'anger'. We will identify ten different motifs that shape the Old English expressions for anger; applying two indirect criteria for distinguishing between literal and figurative denominations, we will then be able to show that figurative imagery occupies a minor role. On a methodological level, this leads to the conclusion that an onomasiological analysis is indispensable to get a good insight into the importance of embodied imagery. On the theoretical level, we will have a look at recent discussions in conceptual metaphor theory about the universality of metaphors. In recent work, Kövecses (2005) has radically changed his earlier position and accepted the point voiced by Geeraerts and Grondelaers (1995): metaphors are not necessarily universal, and variation in metaphorical conceptualization may be sensitive to cultural influences. Our results further support this view.

Keywords: heart, mood, anger, cognition, emotion, volition, onomasiology, literal meaning.

1. Introducing the questions

To what extent could we say that the cultural model of the mind that may be detected in Old English sources is in some sense 'cardiocentric'? The question involves a confrontation with two possible kinds of dualism. First, if the heart is seen as the bodily locus of the mind, a cardiocentric conception assumes an embodied concept of the mind, in contrast with so-called Cartesian conceptions that assume a dualist separation between body and mind. Second, if the heart is seen as the bodily locus of the mind, what kind of mind are we talking about? Is it the rational mind, that some contemporary cultural models now tend to dualistically dissociate from the emotions? Or, given that the same contemporary cultural models associate reason with the brain and feelings with the heart, does the cardiocentrically embodied mind of the Old English period perhaps involve the emotions only?

We will deal with both questions, with an emphasis on the first one. With regard to the second point, we will analyze the Old English word *mood*, and show that it exhibits an encompassing view of the life of the mind, i.e. a view in which emotions and reasons are not relegated to clearly separate spheres. With regard to the first point, we will show that indeed there are indications in the extant sources for a cardiocentric conception, but those indications should be treated with extreme care, because it can not be established that the cardiocentric view is the dominant one in Old English – at least not in comparison with basic literal expressions. Because the argumentation for the latter point involves a methodological issue that we consider to be the central aspect of our paper, let us now first introduce the methodological point.

Cognitive linguists are well aware that actives and passives do not mean the same thing: if each construction in the language has its own meaning, active and passive construals would be semi-synonymous at most. Still, many cognitive linguists seem to adhere to the view that (1) equals (2).

(1) Source concept X conceptualizes target concept Y
(2) Target concept Y is conceptualized by source concept X

If (2) is interpreted as an existential quantification, implying that there are at least some instances in which Y is seen from the perspective of X, nothing is wrong with the transition from (1) to (2), but if (2) is assigned a universal quantification, it clearly does not follow from (1). A lot of the research inspired by Conceptual Metaphor Theory, however, seems to rely precisely on such an unwarranted claim. Studies on 'the conceptualization of Y in language Z' more often than not content themselves with listing the various X's that appear in the expression of Y in Z, without asking the critical question what weight these metaphorical expressions carry with regard to each other, and with regard to alternative expressions of Y. Target concept Y may very well be conceptualized by source concept X, but if it is *not only* conceptualized by X, you will want to know exactly what the impact of X is in comparison with the other types of expression.

To illustrate the danger of ignoring the relative salience of alternative expressions, we present a case study that involves *heart* as the seat of feeling and thought in Old English. We show that there is clear evidence that feelings and thoughts are indeed conceptualized as being metaphorically situated in the heart, but at the same time, we point out that this perspective is very much secondary with regard to an alternative, more literal one that involves *mood* rather than *heart*. The apparent necessity to take into account the relative position of competing expressions is further backed up by an analysis of the expressions for anger in English. As in the *heart* case, literal expressions for anger appear to be dominant throughout the early history of English: an exclusive focus on metaphorical conceptualizations would blatantly distort the picture.

The approach advocated and illustrated here, involving a comparison of the relative weight of the alternative expressions for a given concept, is obviously an *onomasiological* one. The importance for Cognitive Linguistics of a switch from a predominantly semasiological to an onomasiological perspective has been repeatedly argued by the first author of the present paper (see the papers in Geeraerts 2006), but the illustrations provided so far have not been specifically concerned with metaphor analysis. The materials used for illustration and argumentation are treated in more detail in the doctoral dissertation of the second author of the paper.

The methodological point we intend to make has a general validity, but we will illustrate it with diachronic data from Old English. Focusing on such a remote period has the advantage of enhancing the importance of methodological cautiousness: processes of interpretation that would seem to be self-evident from a synchronic point of view (even when they are not) are more easily recognized as problematic when dealing with unfamiliar data.

2. Heart and mood

An efficient way of exploring the semantic range of the word *heart* in the earlier stages of English is to have a look at the different compounds in which the term appears: the presence of concepts like 'happy-hearted' or 'mad-hearted' gives us an indication of which aspects of mental life could be associated with the heart. In the first subsection, we will have a look at the Old English compounds with *heart*. We will derive the conclusion that the heart is indeed seen as a possible seat of the life of the mind and the soul, but a subsequent comparison with the compounds of *mood*, in subsection 2.2, will lead to an important mitigation of that conclusion. (As a terminological remark, note that we deliberately refer to soul and mind here.

Mind conventionally refers to the mental faculty, to the seat of thought, feeling, and will: that is also the meaning that is relevant for the following discussion of Old English. However, a glance at the dictionary suffices to see that in the contemporary semantic profile of *mind*, memory and cognition play a predominant role, rather than the other, non-intellectual mental faculties. To avoid a restrictive interpretation of *mind*, then, we add a reference to *soul*. But similarly, the soul we invoke is not specifically the religious or moral aspect of our mental life, but the spiritual part of human life at large – intellectual, emotional, and volitional.)

2.1. Heart

Table 1 alphabetically lists the compounds with *heart* that may be found in Old English. The first column presents the compound, the second column contains a gloss that clarifies the construction of the compound, and the third column provides an explanatory definition or synonym. The data are based on the Thesaurus of Old English and the Anglo-Saxon Dictionary. Question marks indicate that these sources express uncertainty about the exact origin or reading of a word. The definitional third column is not 'normalized', that is to say, we have adopted the descriptions that we found in our sources (sometimes in a simplified form), rather than trying to bring the various definitions together into a classificatory system of our own. For instance, while definitional glosses like 'gentleness' and 'kindness' are very close to each other, we have not brought them together under a single label.

Table 1. Old English compounds with *heart*

bliðheort	happy-hearted	happiness; kindness, merciful
cealdheort	cold-hearted	hardheartedness, cruelty, severity
clænheort	clean-hearted	purity, moral cleanness
earmheort(nes)	poor-hearted	cowardice, pusillanimity; pity (cp. miseri-cors)
efenheort	even/equal-hearted	peace, tranquillity; concord, harmony
gramheort	fierce/hostile-hearted	hostility
hatheort	hot-hearted	ardour, fervour, strong feeling; anger
heahheort	high-hearted	pride

heardheort	hard-hearted	obstinacy; wrath, sternness, displeasure
hearmheort	harm-hearted	malice
hellheort	hell = hell or cp. M.H.Germ. hell=weak	cowardice, pusillanimity; faint-hearted?
mildheort	mild-hearted	gentleness; compassion
oferheortnes	over-heartedness	ardour, fervour, strong feeling
rihtheort	upright in heart	righteousness, rectitude
rumheort(lice/nes)	liberal-hearted	freedom from trouble, comfort, security; favour, kindness, grace; bountifulness, munificence
samheort	of the same heart	peace, tranquility; of one heart
stearcheort	strong-hearted	boldness
ungeheort	not hearted	cowardice, pusillanimity
unmildheort	not mild-hearted	pitilessness
wedenheort(nes)	mad-hearted	madness, frenzy, folly; insanity
wodheortnes	mad-heartedness	madness, frenzy, folly
wulfheort	wolf-hearted	cruel

On the basis of Table 1, it would be deceptively easy to conclude that 'the heart is the seat of feeling and thought', i.e. that the mind is conceived of in terms of the heart (and more generally, that the mind is embodied). Apart from the paramount recognition that all of the expressions refer to the life of the mind, either in the form of temporary feelings or in the form of more lasting emotional and temperamental characteristics, the conclusion may be supported by a comparison with compounds formed on the basis of the word *mood*, a literal expression for the overall mental (cognitive and/or emotional) state of a person. The best contemporary rendering of *mood* would probably be *mind,* or maybe *soul*, but for formal identification, we will use *mood.*

2.2. Mood

Before comparing the compounds of *mood* with those of *heart*, we should first have a look at the semantic range of *mood.* The full set of *mood*-compounds, following the same conventions as used for Table 1, is given in the Appendix. What aspects of mental life are represented by *mood*? Does *mood* cover feelings only, or does it constitute a broader conception of mental life, bridging the dichotomy between cognition and emotion? Such a dichotomy does not, by the way, do full justice to the width of phenomena that belong to the realm of the mind and the soul: next to thoughts and feelings, the *will* (the domain of *motivation* or *volition* next to that of cognition and emotion) would traditionally be seen as belonging to the human mind and soul. If, then, *mood* represents an integrated, non-dichotomous or non-trichotomous conception of the mind and the soul, we expect to find compounds on all three subdomains.

Table 2 presents a subset of the items included in the Appendix, showing that *mood* does indeed cover the three domains of cogniton, emotion, and volition. With regard to each of the three domains, we find a variety of specific phenomena, either permanent features and dispositions, or temporary characteristics and states. It could perhaps be remarked that in the entirety of the Appendix, the cognitive domain is less well represented than the other two, and that some pertinent cognitive traits, like cleverness and stupidity, are not represented. However, there are some obvious feelings missing as well, like love or mirth. We may say, in other words, that *mood* covers at least partially the three domains of cognition, emotion, and volition, and as such, exemplifies an integrated conception of mental life.

Table 2. The semantic range of Old English compounds with *mood*

COGNITION	
madness, insanity	*æmod* 'without mood';
sagacity	*gleawmod* 'wise/skilful-mood/mind'; *þancolmod* 'wise-mood/mind'
sincerity	*meagolmod(nes)* 'earnest-mood/mind'
insincerity, wantonness	*leasmodnes* 'not truthful-mood/mind'; *leohtmod(nes)* 'light-mood/mind'
EMOTION	
anger	*bolgenmod* 'swollen-mood/mind'; *irremod* 'angry-mood/mind'; *tornmod* 'anger-mood/mind'; *wraþmod* 'anger'
fear	*acolmod* 'frightened-mood/mind'; *forhtmod* 'fearful-mood/mind'; *sceohmod* 'fearful-mood/mind'
sadness	*angmod(nes)* 'narrow-mood/mind'; *freorigmod* 'chilled/sad-mood/mind'; *gealgmod* 'sad-mood/mind'; *geomormod* 'sad-mood/mind';

sadness	*hefigmod* 'heavy-mood/mind'; *hreowigmod* 'sad-mood/mind'; *reom/nigmod* 'sad-mood/mind'; *sarigmod* 'sad-mood/mind'; *unrotmod* 'sad-mood/mind'
VOLITION	
obstinacy	*heardmod* 'hard-mood/mind'; *stearcmod* 'strong-mood/mind'
determination, resolution	*stiþmod* 'strong/fierce-mood/mind'; *styrnmod* 'stern-mood/mind'; *swiþmod(nes)* 'very much-mood/mind'
disinclination to act, listlessness	*sleacmodnes* 'slack-mood/mind'; *swærmod* 'heavy-mood/mind'

Now, if we compare the compounds with *heart* with the compounds with *mood*, can we confirm our initial conclusion that the heart is indeed conceived of as a possible seat for the full gamut of mental experience? Two observations support that conclusion.

First, the same feelings may be expressed by *heart* compounds and *mood* compounds. Anger, for instance, is expressed by the words *gealgmod, irremod, reþemod, tornmod, wraþmod* and by *gramheort*. In all of these cases, the first element of the compound has a literal meaning, to the extent, for instance, that *gram* as such means 'angry, hostile, fierce'. If we look for expressions with a more figurative first element, we may align *weamod* and *bolgenmod* with *hatheort* and *heardheort*. Similar cases of semantic overlap involve happiness (*bliðheort* next to *bliðemod, glaedmod, heahmod*) and fortitude (*heardheort* and *stearcheort* next to *deormod, hwætmod, styrnmod, unforhtmod, heardmod, stearcmod, strangmod*).

Second, the identical semantic ranges of the *heart* and *mood* compounds receives a particularly striking expression when both words are combined with the same first element. Table 3 enumerates the cases in which the morphological structure of the *heart* and *mood* compounds is completely identical.

Table 3. Similarities between *heart* and *mood* compounds

bliðemod	happy mood/mind	happiness
bliðheort	happy-hearted	happiness, kindness, merciful
efenheort	even/equal-hearted	peace, tranquillity; concord, harmony
efenmodlice	even-mood/mind	gentleness, meekness, composure

gramheort	fierce/hostile-hearted	hostility
grammod	fierce-mood/mind	hardheartedness, cruelty, severity
heahheort	high-hearted	pride
heahmod	high-mood/mind	pride, exultation, joy; magnanimity
heardheort	hard-hearted	obstinacy; wrath, sternness, displeasure
heardmod	hard-mood/mind	strength, fortitude; obstinacy

There is sufficient evidence, in short, that the heart may be seen in Old English as the embodied locus of the mind. However, we cannot simply conclude from that observation that the mind is generally and predominantly perceived as being embodied – a conclusion that cognitive linguists might be all too willing to arrive at, given their interest in the concrete experiential grounding of abstract concepts. If we compare the Appendix with Table 1, the simple truth of the matter is that there is an overall asymmetry between the *heart* and the *mood* compounds: we find 78 compounds with *mood*, versus only 23 with *heart.*

The *heart* imagery, in other words, is secondary with regard to the literal denomination *mood*, and an exclusive focus on imagery would simply distort the facts. Furthermore, we cannot even claim that our own investigation covers all the relevant observations because we have not included other sources of imagery into the study. A necessary next step, then, will be to focus on just one aspect of the *heart* imagery, and compile a systematic and exhaustive overview of the other types of expression (figurative or otherwise) that express the concept concerned.

3. Anger

From the point of view of Cognitive Linguistics, the expression *hatheort* that we came across in Table 1 is specifically interesting because it seems to illustrate a metaphorical pattern that is well entrenched in the Cognitive Linguistics literature: ANGER IS HEAT (a.o. see Kövecses 1995, and compare Geeraerts and Grondelaers 1995). But before we can draw any conclusions from the presence of *hatheort*, we will want to know how strong its presence actually is: what other themes conceptualize anger, and which are the salient ones? Restricting ourselves to Old English materials from before 850, we can group the expressions into ten etymological themes: anger as a strong emotion, a wrong emotion, fierceness, unmildness, affliction, sadness, insanity, swelling, heat, or a synaesthetic experience.

We call these themes 'etymological' because they are initially based on the etymological sources of the expressions. The moot question, of course, is whether these etymological source domains correspond with actual imagery for the speakers of Old English. For lack of more direct methods, we will apply two indirect criteria to classify a specific expression as figurative. First, if the 'anger' reading is the dominant sense of the word, it is considered to be literal. Conversely, figurative expressions are non-prototypical ones. Second (and regardless of the first criterion), compound expressions that readily receive a figurative interpretation, like *hatheort*, are taken to be non-literal.

In both cases, our estimate of literalness may be rather conservative, i.e. we may be attributing more figurativeness to Old English than there actually was. The compositional criterion, in fact, overlooks the possibility of dead metaphors, and the prototypical criterion simply assumes that non-central meanings simply evoke the central ones. However, even given this possibly minimalistic attitude with regard to literalness, the results of the analysis will show that literal expressions are dominant.

Unless stated otherwise, all definitions in the following pages are taken from Bosworth and Toller (1898) and/or Toller and Campbell (1921). All bibliographical references to primary texts are those of the Toronto Corpus (DiPaolo Healey and Venezky 1980). The etymological themes are identified in small caps.

3.1. Anger as a strong emotion

The hyperonym *anda* 'emotion of mind,- malice, envy, hatred, anger, zeal, annoyance, vexation' finds its origin in the Indo-European stem *ant* 'breath'. Out of the notion of 'breath', the notion of 'soul, mind' developed in several languages, the best known example of which is Latin *anima.* As a hyperonym expressing STRONG EMOTION, the reference of *anda* is rather vague and only contextual information reveals whether it refers to anger or not. The noun *anda* occurs twice with reference to anger, but given the hyperonymic status of the word, it seems safer not to regard the anger reading as a separate sense of *anda*, but rather as a contextual specification. In no case, however, can it be considered a figurative reading.

(3) ða wearð bliðemod burga aldor, gealp gramlice gode on andan, cwæð þæt his hergas hyrran wæron and mihtigran mannum to friðe þonne Israela ece drihten. (Dan 712)
'So the lord over cities grew merry and boasted immoderately, to the anger of God. He declared that his armies were greater and more effectual for people's protection than the everlasting Lord of the Israelites.' (Bradley 1982: 85)

(4) Se ðone lig tosceaf, halig and heofonbeorht, hatan fyres, tosweop hine and toswende þurh þa swiðan miht, ligges leoman, þæt hyra lice ne wæs owiht geegled, ac he on andan sloh fyr on feondas for fyrendædum. (Dan 339)
'Holy and heavenly bright, he thrust aside the hot flame of the fire; the glare of the flame he swept back and brushed aside by his very great might so that their body was not a whit harmed – rather, he flung the fire in anger upon their adversaries, for their wicked actions.' (Bradley 1982: 76)

3.2. Anger as a wrong emotion

The social connotation of anger may be a negative one: it is a WRONG EMOTION. Such a negative connotation can be found in the etymological basis of the expression *irre*. It is derived from an Indo-European *er(a)s*-root, which is also present in Latin *errare*, and means 'to flow, to move'and hence 'to stray, to wander'. According to the *OED,* 'the transition to the sense 'angry' […] arose from the consideration of anger as a wandering or aberration of the mind'. Significantly, *irre* is still sometimes used in the sense 'gone astray' in Old English. The *ASD* cites a.o.

(5) Oþ ðæt his eage biþ æfþancum full yrre geworden (Sol I 994)
'until his eye is filled with evil thoughts and gone astray' (ASD p. 600)

(6) Ealle synt yrre ða ðe unwise heora heortan hige healdaþ mid dysige (Ps 75.4)
'All are gone astray who unwisely hold their hearts high with folly.'

Irre is the most neutral and most frequently used word to indicate anger, and 'anger' is definitely the prototypical reading of the word, i.e. we may consider *irre* to be the literal expression for anger par excellence in Old

English. There are forty-five attestations of *irre*, spread over fourteen texts of different genres. It is used as a noun (*yrre*, *eorre*, *ierre*, fourteen attestations), an adjective (*yrre*, *eorre*, eighteen attestations) and an adverb (*yrre*, *yrringa*, *yrremod*, thirteen attestations).

(7) Hy þa þurh yrre Affricanus, fæder fæmnan ageaf on feonda geweald Heliseo. (Jul 158)
'Then in his fury, the virgin's father Africanus handed her over to the disposal of her enemies, to Eleusius.' (Bradley 1982: 306)

(8) þa wearð yrre anmod cyning, het he ofn onhætan to cwale cnihta feorum forðam þe hie his cræftas onsocon. (Dan 224)
'Then the obstinate king grew furious. He ordered a furnace to be heated for the destruction of the young men's lives because they challenged his powers. (Bradley 1982: 73)

(9) Ða to Euan god yrringa spræc: 'Wend þe from wynne.' (Gen A,B 918)
'Then to Eve God angrily spoke: 'Set yourself aside from happiness'.' (Bradley 1982: 37)

Occasionally, *irre* is combined with the metaphor *-þweorh* 'crooked, cross'. *þweorh* is derived from the Indo-European *twerek*-root, meaning 'twist, turn around', which also gave rise to Latin *torquere* 'to twist' and Dutch *dwars* 'transverse'.

(10) Eode þa fromlice fæmnan to spræce, anræd ond yreþweorg, yrre gebolgen, þær he glædmode geonge wiste wic weardian. (Jul 89)
'He promptly went to speak to the virgin, one-minded and perverted with anger, puffed up with anger, to where he knew the cheerful young girl kept residence.'

3.3. Anger as fierceness

A concept that typically seems to be associated with anger is FIERCENESS, as persons who are angry tend to be violent. In this group, *reðe*, *wrað* and *gram* are to be found, which all mean 'fierce' besides 'angry', as shown in examples (11) to (13).

(11) Wiht unhælo, grim ond grædig, gearo sona wæs, reoc ond reþe, ond on ræste genam þritig þegna, þanon eft gewat huþe hremig to ham faran, mid þære wælfylle wica neosan. (Beo 120)
'This creature beyond redemption, savage and voracious, fierce and violent, was instantly prepared, and from their resting-place he snatched thirty thanes. From there, exulting in his plunder, he went journeying back homewards, seeking his lairs with that feast of carion.' (Bradley 1982: 414)

(12) Feala ic on þam beorge gebiden hæbbe wraðra wyrda. (Dream 50)
(the holy cross bearing witness to the crucifixion) 'Many cruel happenings I have experienced on that hill' (Bradley 1982: 161)

(13) Hie ða fromlice leton forð fleogan flana scuras, hildenædran, of hornbogan, strælas stedehearde; styrmdon hlude grame guðfrecan, garas sendon in heardra gemang. (Jud 220)
'Firmly entrenched, they vigorously let fly from the curved bow showers of darts, arrows, the serpents of battle. Loudly the fierce fighting-men roared and sent spears into their cruel enemies' midst.' (Bradley 1982: 501)

Reðe is derived from the Indo-European root *ret* 'to roar'. Unlike *wrað* and *gram*, 'fierce' is clearly its basic meaning. It is only sporadically used with clear reference to anger. It appears twice as a simple adjective, and twice in the compound adjective *reðemod*.

(14) ða wæs ellenwod, yrre ond reþe, frecne ond ferðgrim, fæder wið dehter. (Jul 140)
'Then the father was mad with rage, furious and incensed, menacing and savage-minded towards his daughter.' (Bradley 1982: 305)

(15) Cyning biþ reþe, sigora syllend, þonne synnum fah E W ond U acle bidað hwæt him æfter dædum deman wille lifes to leane. (Jul 704)
'The King, Giver of victories, will be wrathful when, stained with sins, terrified, E, W and U [...] awaits what he wills to decree them according to their deeds as reward for their life.' (Bradley 1982: 319)

(16) þa wearð reðemod rices ðeoden, unhold þeodum þam þe æhte geaf. (Dan 33)
'Then the Prince of power grew angry and disaffected towards the people to whom he had granted possession.' (Bradley 1982: 68)

(17) þa com halig god wera cneorissa weorc sceawigan, beorna burhfæsten, and þæt beacen somed, þe to roderum up ræran ongunnon Adames eaforan, and þæs unrædes stiðferhð cyning steore gefremede, þa he reðemod reorde gesette eorðbuendum ungelice, þæt hie þære spæce sped ne ahton. (Gen A,B 1678)
'Then came the holy god to look at the work of the generation of men, the fortress of the chiefs, as well as the beacon that the heirs of Adam had begun to raise up towards the heavens and for this ill-advised course the stern-minded king set a punishment: he angrily made the languages of the earth-dwellers different, so that they had no advantage from speech.'

For both *gram* and *wrað*, the meanings 'fierce' and 'angry' occur side by side, neither of them clearly being the basic meaning. From the point of view of literalness, then, these would be undecided cases. *Gram* is derived from the Indo-European root *ghrem* 'to thunder, growl' and hence metonymically 'to be angry' (growling being an external sign of anger). With reference to anger, the adjective *gram* occurs twice. *Gremian*, derived from as *gram*, occurs once as a past participle with the meaning 'angered', as becomes clear through the use of the synonym *irre*.

(18) Hete hæfde he æt his hearran gewunnen, hyldo hæfde his ferlorene, gram wearð him se goda on his mode. (Gen A,B 301)
'He had won his lord's hate and he had lost his affection. God became angry with him in his mind.'

(19) Ealles þæs forgeton siððan grame wurdon Egypta cyn ymb an twig; ða heo his mægwinum morðor fremedon, wroht berenedon, wære fræton. (Ex 144)
'All this they forgot when the Egyptians became angry because of a disagreement; then they murdered his relatives, accused them falsely and broke the convenant.'

(20) Hæfde styrne mod, gegremed grymme, grap on wraðe faum folmum, and him on fæðm gebræc yrre on mode; æðele bescyrede his wiðerbrecan wuldorgestealdum. (Gen A,B 60)
'He had a stern heart, unrelentingly provoked; with hostile hands he snatched up his enemies and crushed them in his grasp, irate in his heart, and cut off his adversaries from their native home, from the heavenly mansions.' (Bradley 1982: 14)

Wrað is derived from the Indo-European root *wreit* 'to twist' as in present-day English *writhe*. In Old Norse and Old English, the meaning developed into 'fierce, angry', which is also present in present-day Dutch *wreed* 'cruel'. *Wrað* is basically used as an adjective, either in simple form or in the compound *wraðmod*. Of the twelve attestations, nine are adjectives, two are adverbs and one is a noun.

(21) Him wæs hælend god wrað geworden for womcwidum. (Christ A,B,C 280)
'God the Saviour had become enraged against them for their blasphemies.' (Bradley 1982: 94)

(22) Nys unc wuht beforan to scursceade, ne sceattes wiht to mete gemearcod, ac unc is mihtig god, waldend wraðmod. (Gen A,B 812)
'There is nothing in front of us as a defence against the storm nor any provision made for food, but mighty God the Ruler is in angry mood towards us.' (Bradley 1982: 34)

3.4. Anger as unmildness

Perhaps as a parallel to the conceptualization in terms of fierceness, anger is also conceptualized as UNMILDNESS. This conceptualization is to be found in the one occurrence of the expression *unmiltse*, a noun derived from the Indo-European roots *ante* 'opposite, not' and *meldh* 'soft, mild', like present-day English and Dutch *mild*. Similar to the interpretation we suggested for *anda*, we will consider the 'anger' reading to be a contextual specification of *unmiltse*, with attenuating and euphemistic overtones, not a separate reading.

(23) hæbbe he Godes unmiltse & sancte Martines se þe ðis awende & þere stowe oþbrede. (Ch 287 4.1)
'may he have God's and Saint Martin's wrath who changes this and takes away this place.'

3.5. Anger as affliction

Another important domain that is associated with anger is AFFLICTION. Anger is conceptualized as something that hurts and making someone angry as hurting someone. The image is best interpreted as metonymical: if a person is psychologically hurt, he may react with anger. This conceptualization is present in the expressions *torn* and *sare*. They are used to refer to affliction in a physical sense, as illustrated by the following examples.

(24) Wæs seo hwil micel; XII wintra tid torn geþolode wine Scyldinga, weana gehwelcne, sidra sorga. (Beo 146)
(referring to Grendal's slaughters) 'It was a great while: for a space of twelve years the friend and lord of the Scyldings suffered pain, every kind of misery and profound grief.' (Bradley 1982: 415)

(25) Hwæt we wieton ðæt sio diegle wund bið sarre ðonne sio opene, forðam ðæt worsm ðæt ðærinne gehweled bið, gif hit bið utforlæten, ðonne bið sio wund geopenod to hælo ðæs sares. (CP 38.273.21)
'We know that a hidden wound is sorer than an open wound because of the pus that is inflamed therein, and if it is let out, then the wound is opened for the healing of the pain.'

These words can also be used to refer to psychological rather than physical affliction. There is but a thin line between anger and other kinds of psychological affliction, so it is often difficult to assign the meaning 'anger' to these expressions with a considerable degree of certainty.

Torn is derived from the Indo-European root *der* 'to hurt, to skin', which also gave rise to present-day English *to tear*. There are ten attestations, eight of which are nouns, one an adjective and one an adverb. As with all the expressions involving the affliction theme it is contextual information only that enables us to assign the meaning 'anger' to *torn*. In (26), it is the co-occurrence of the synonym *gebolgen*. In (27) it is the wider encyclopaedic knowledge that the god of the Old Testament is not a human being whose feelings can be hurt, but a powerful ruler who demands obedience from his servants or else is angry and takes revenge.

(26) Beoð þa gebolgne, þa þec breodwiað, tredað þec ond tergað, ond hyra torn wrecað, toberað þec blodgum lastum; gif þu ure bidan þencest, we þec niþa genægað. (Guth A,B 287)
'Then they will be enraged; then they will knock you down and tread on you and harass you and wreak their anger upon you and scatter you in bloody remnants. If you presume to face us out we shall attack you with afflictions.' (Bradley 1982: 257)

(27) Nu me Sethes bearn torn niwiað and him to nimað mægeð to gemæccum minra feonda; þær wifa wlite onwod grome, idesa ansien, and ece feond folcdriht wera, þa ær on friðe wæron. (Gen A,B 1257)
'Now the children of Seth renew the anger in me and are taking to themselves women as their mates from among my foes; there the beauty of females, the appearance of the women, and the eternal fiend have malevolently insinuated themselves into the nation of men who were previously in concord.' (Bradley 1982: 41)

Another representative of this conceptualization is the word *sare*, derived from the Indo-European root *sai* 'pain'. This word is used as an intensifier with words in the conceptual field of AFFLICTION (cp. Molina 2005). It is the Old English counterpart of present-day *sorely*, which can be considered a synonym of *very*, but is clearly associated with the domain of AFFLICTION because of the *sore*-root. The same combination is to be found in its Dutch counterpart *zeer*, which means 'very' as well as 'pain'.

(28) þa reordade rodora waldend wrað moncynne and þa worde cwæð: 'Ne syndon me on ferhðe freo from gewitene cneorisn Caines, ac me þæt cynn hafað sare abolgen. (Gen A,B 1253)
'Then the Ruler of the skies spoke, angry with humankind, and said these words: 'The tribe of Cain have not gone from my mind free men, for that family has sorely enraged me. ' (Bradley 1982: 41)

3.6. Anger as sadness

Also expressions referring to a more specific form of psychological affliction, namely SADNESS, are metonymically used to refer to anger. This is the case with *unbliðe* 'not glad' and *gealgmod* 'sad-minded, sad-spirited'. *Unbliðe* is derived from the Indo-European roots *ante* 'opposite, not' and *bhlei* 'shine' and hence 'glad, friendly' (cp. Dutch *blij*) and occurs once in the sense 'angry' in the corpus. *Gealgmod* occurs twice and is derived from the Indo-European roots *ghalgh* 'seriously thinking' and *mod* 'mind'.

(29) ða wearð unbliðe Abrahames cwen, hire worcþeowe wrað on mode, heard and hreðe, higeteonan spræc fræcne on fæmnan. (Gen A,B 2261)
'Then Abraham's wife became angry, harsh and cruel at heart with her slave girl, and spoke injuriously against the woman.'

(30) ða se gerefa het, gealgmod guma, Iulianan of þam engan hofe ut gelædan on hyge halge hæþnum to spræce to his domsetle. (Jul 530)
'Then the prefect, a furious man, ordered Juliana to be led out of the narrow cell, in a holy state of mind, to his judgement seat for a pagan judgement.'

(31) þa se dema wearð hreoh ond hygegrim, ongon his hrægl teran, swylce he grennade ond gristbitade, wedde on gewitte swa wilde deor, grymetade gealgmod ond his godu tælde, þæs þe hy ne meahtun mægne wiþstondan wifes willan. (Jul 594)
'Then the judge grew wild and savage-minded and began to tear his robe. He bared his teeth too and ground them. He grew crazed in his wits as a wild beast. Rabid-hearted, he stormed and abused the gods because they with their power could not withstand the will of a woman.' (Bradley 1982: 316)

3.7. Anger as insanity

INSANITY is also closely related to anger, as is shown in the one occurrence of *ellenwod*, which is a combination of *ellen* 'strength' and *wod* 'mad', from Indo-European *uat* 'excited, inspired'. (A contemporary example along the same line would be *rage*.) As it is difficult to determine whether anger is literally taken as a form of madness, we will consider this case as undecided.

(32) ða wæs ellenwod, yrre ond reþe, frecne ond ferðgrim, fæder wið dehter. (Jul 140)
'Then the father was mad with rage, furious and incensed, menacing and savage-minded towards his daughter.' (Bradley 1982: 305)

3.8. Anger as a synaesthetic experience

The anger-domain also finds a source domain in the domain of bodily sensations: anger is experienced as something dark, bitter and heavy. These SYNAESTHETIC conceptualizations are expressed in the words *sweorcan* 'to become dark' (one attestation), *biter* 'bitter' (two attestations), and *hefig* 'heavy' (one attestation) denoting anger. These words can be traced back to the Indo-European roots *suordos* 'black', *bheid* 'split' (hence 'bite, biting') and *kap* 'take' (hence 'contain' and hence 'heavy') respectively. According to Ingersoll (1978: 164), *biter* is 'generally used in OE to intensify verbs of strong emotion, as *bitterly* still is in ModE'.

(33) Geswearc þa swiðferð sweor æfter worde, þære fæmnan fæder, ferðlocan onspeon (Jul 78)
'Then the headstrong man, the virgin's father, grew furious, and after this speech swore and disclosed his heart' (Bradley 1982: 304)

(34) Gesawon ða æfter wætere wyrmcynnes fela, sellice sædracan, sund cunnian, swylce on næshleoðum nicras licgean, ða on undernmæl oft bewitigað sorhfulne sið on seglrade, wyrmas ond wildeor; hie on weg hruron, bitere ond gebolgne, bearhtm ongeaton, guðhorn galan. (Beo 1425)
'Then they observed the multitude of a species of serpent, strange sea-dragons taking to the water, as well as amphibian monsters lying on the slopes of the cliff – reptiles and savage beasts which often at their morning mealtime make their woe-fraught way into the path of sail-driven ships. Aroused and puffed up with anger they plunged away: they had registered the clarion sound of the battle-horn crying out.' (Bradley 1982: 449)

(35) þæt ðam godan wæs hreow on hreðre, hygesorga mæst; wende se wisa ðæt he wealdende ofer ealde riht, ecean dryhtne, bitre gebulge. (Beo 2327)
'To that virtuous man this was a heartfelt grief, the greatest of anxieties upon his mind. The wise man imagined that in breach of ancient law he had severely provoked God, the ruler, the everlasting Lord;' (Bradley 1982: 472)

(36) þæt wæs torn were hefig æt heortan. (Gen A,B 979)
'This was (a cause of) anger that was oppressive in the man's heart.'

3.9. Anger as heat

Anger is also conceptualized as HEAT, as we have already seen. This conceptualization can be found in the expressions *hatheortnysse* and *hygewælm*, which both occur once in the corpus. The compound noun *hatheortnysse* consists of *hat* 'hot', from Indo-European *kai* 'heat', and *heorte* 'heart', from Indo-European *kerd* 'heart'. It appears in a letter from Saint Boniface to abbess Eadburga, in a passage translating a verse from a Latin psalm, in which it translates Latin *furor*.

(37) And him þuhte þæt þa englas wæron swilce hig byrnende wæron, þa þe hine læddon ut of þam lichaman, and he ne mihte nænig þinga locian on hig for þære micclan beorhtnesse þe hig mid ymbseted wæron, and hig sungon swiðe wynsumum stefnum and swiðe geswegum, and hig cwædon, Domine, ne in ira tua arguas me, neque in furore tuo corripias me, þæt is, Drihten, ne þrea þu us in þinum yrre, ne þu us ne steor in þinre hatheortnysse. (Let 1 (Sisam) 17)
'And he thought that it was angels – so much did they burn – who led him out of his body, and he could in no way look at them because of the great brightness with which they were surrounded, and they sung in very pleasant voices and very harmoniously, saying: 'Domine, ne in ira tua arguas me, neque in furore tuo corripias me', that is, 'Lord, do not rebuke us in your anger and do not reprove us in your wrath.''

In *hygewælm*, the conceptualization of HEAT is more specific. Its first component is *hyge* 'mind, heart, soul'. *Wælm* does not simply mean 'heat', but 'that which wells, boils'. It is derived from Old Germanic *walmi-z* 'boiling', which is itself derived from Indo-European *wel* 'to roll'. Thus, this conceptualization combines the ideas of motion and of heat in the idea of a boiling fluid. Which of these two ideas is more basic synchronically speaking is hard to say.

(38) Hygewælm asteah beorne on breostum, blatende nið, yrre for æfstum. (Gen A,B 980)
'Indignation surged up in the man's bosom, a livid fury, anger out of envy.' (Bradley 1982: 39)

3.10. Anger as swelling

The conceptualization ANGER IS SWELLING is expressed by forms of the verbs *belgan*, *abelgan*, *gebelgan* '(to cause one) to swell with anger', by the adjective *bolgenmod* 'angry' and by the nouns *abylg* and *abylgþ* 'anger', all derived from the Indo-European root *bhelgh* 'swell'. These expressions occur four times as nouns, eight times as verbs, twenty times as adjectives (most of which are past participles), and once as an adverb. With thirty-three attestations, spread over eleven texts, the SWELLING-conceptualization takes up a very considerable part of the conceptual field of anger.

(39) Let þa up faran eadige sawle, Adames cyn, ac ne moste Efe þa gyt wlitan in wuldre ær heo wordum cwæð: Ic þe æne abealh, ece drihten, þa wit Adam twa eaples þigdon þurh næddran nið, swa wit na ne sceoldon. (Christ A,B,C 405)
'Then he let the blessed souls, Adam's kin, ascend – but Eve could not yet look upon heaven before she declared aloud: 'Once I provoked you, everlasting Lord, when we two, Adam and I, through the serpent's malice ate of an apple as we should never have done.' ' (Bradley 1982: 96)

(40) þa wearð se mihtiga gebolgen, hehsta heofones waldend, wearp hine of þan hean stole. (Gen A,B 299)
'Then the mighty and supreme Ruler of heaven grew angered and threw him down from the lofty throne.' (Bradley 1982: 20)

(41) Ða him bolgenmod Babilone weard yrre andswarode, eorlum onmælde grimme þam gingum, and geocre oncwæð, þæt hie gegnunga gyldan sceolde oððe þrowigean þreanied micel, frecne fyres wylm, nymðe hie friðes wolde wilnian to þam wyrrestan, weras Ebrea, guman to þam golde, þe he him to gode teode. (Dan 209)
'Then, enraged, the ruler of Babylon furiously answered them and fiercely informed the young men and harshly told them that they must forthwith do homage or else suffer great torture, the terrible swirling of fire, unless they, the Hebrew men, would pray for indemnity to that most evil thing, to the gold which he had ordained for himself as a god.' (Bradley 1982: 72–73)

(42) Huru, ic nolde sylf þurh gielpcwide gæstes mines frofre gelettan, ne fæder mines æfre geæfnan, æbylg godes. (Guth A,B 1234)
'Truly I myself have not wished to jeopardize the comfort of my soul through bragging, nor ever to cause the anger of God my Father.' (Bradley 1982: 280)

3.11. The resulting picture

The overall results of the analysis are summarized in Table 4. The table mentions the etymological themes that we used as a starting point, the actual expressions (in a reduced form, with reference to the stem of the words only), the semantic mechanism they exhibit, and the number of attestations. A final column categorizes the whole set into two fundamental groups: a literal one (top half of the table) and a figurative one (lower half of the table).

Table 4. Literal and figurative expressions for anger in Old English

theme	expressions	semantics	n	
WRONG EMOTION	*irre*	literal	46	
FIERCE	*gram, wrað*	literal or hyperonymy	15	
INSANE	*ellenwod*	literal or hyperonymy	1	
STRONG EMOTION	*anda*	hyperonymy	2	
UNMILD	*unmiltse*	hyperonymy	1	65
AFFLICTION	*torn, sare*	metonymy	11	
SADNESS	*unblide, gealgmode*	metonymy	3	
SWELLING	*belgan*	metaphor	33	
SYNAESTHESIA	*sweorcan, biter, hefig*	metaphor	3	
FIERCE	*reðe*	metaphor	4	
HEAT	*hatheort, hygewaelm*	metaphor	2	56

Three conclusions may be retained from the table and the foregoing discussion. First, as in the previous section, the literal denominations dominate, given that the hyperonymic uses should be considered literal. In fact, if we are interested in comparing literal names with ones based on imaginery in general and bodily imagery in specific, the hyperonymic names (and the ones that are uncertain between hyperonymy and literalness) should be counted with the literal ones, since they do not involve figurative imagery. Second, the figurative expressions are not all metaphorical: about a quarter of them are best interpreted as metonymical. And third, the heat conceptualization is definitely a minor one. Again, as in the previous section, the importance of an onomasiological approach is established beyond doubt: jumping to conclusions on the basis of preselected examples that fit an interesting model is simply bad linguistics.

4. Coming to conclusions

In this paper, we have argued that an onomasiological analysis is indispensable to get a good insight into the (relative) importance of embodied imagery of the mind and the soul. We took two steps in the argumentation.

As a first step, an analysis of *heart* and *mood* compounds in Old English revealed three things. First, *mood* represents an integrated, non-dualistic conception of mental life, in which cognition, emotion, and volition occur side by side. In that sense, *mood* represents the full spectrum of mental life – the rational mind just as well as the feelings and the will. Second, the heart is seen as an embodied seat of the mind and the soul: *heart* compounds and *mood* compounds cover the same semantic domain. Third, however, the importance of this second recognition is severely mitigated by the fact that *heart* imagery constitutes a minority in comparison with literal *mood* denominations. The embodied model, in other words, seems to be less entrenched than a mere literal reference to *mood* as encompassing cognitive, emotive, volitional phenomena.

As a second step, the dominance of literal naming was further corroborated by an analysis of the concept 'anger'. We identified ten different motifs that shape the Old English expressions for anger, and applied two indirect criteria for distinguishing between literal and figurative denominations. As in the *heart/mood* study, it became clear that figurative imagery occupies a minor role. In particular, the ANGER IS HEAT imagery that is represented by a compound like *hatheort*, has a peripheral position within the entire range of expressions.

To place these results in a wider context, we should have a look at recent discussions in conceptual metaphor theory about the universality of metaphors. In recent work, Kövecses (2005) has radically changed his earlier position and accepted the point voiced by Geeraerts and Grondelaers (1995): metaphors are not necessarily universal, and variation in metaphorical conceptualization may be sensitive to cultural influences. In Kövecses's words: 'Universal experiences do not necessarily lead to universal metaphors; bodily experiences may be selectively used in the creation of metaphors; bodily experience may be overridden by both culture and cognitive processes' (2005: 4).

At the same time, the assumption of a universal basis for anger conceptualizations has not been given up. Referring to evidence from English, Hungarian, Japanese, Zulu, Polish, Wolof, and Chinese (see Kövecses 2000), Kövecses postulates a universal metaphorical model THE ANGRY PERSON IS A PRESSURIZED CONTAINER. The model has a physiological basis (and hence, is metonymically motivated rather than purely metaphorically): 'My proposal here was that conceptualized physiological characteristics (i.e., the conceptual metonymies) provide the cognitive motivation for people to conceptualize the angry person metaphorically as a PRESSURIZED CONTAINER. … If conceptualized physiological responses include an increase in internal pressure as a major response in a given culture, people in this culture find the use of the PRESSURIZED CONTAINER metaphor natural' (Kövecses 2005: 41).

Given this background, one way in which variation occurs is through a different embedding of the metaphor in culture-specific conceptual frameworks. Referring to Geeraerts and Grondelaers (1995), Matsuki (1995), and Yu (1998) on culture-specific elements in English, Japanese, and Chinese respectively, Kövecses (2005: 234–235) argues that the more schematic PRESSURIZED CONTAINER model is realized at a lower, more specific level in terms of culture-specific theories about the four humors, Japanese *hara*, and Chinese *qi*. Roughly speaking, the substance that is pressurized is conceived of differently in the various cultures.

Apart from these cultural differences, historical differences may also lie at the basis of metaphor variation. Kövecses (2005: 247–250) at length cites Gevaert (2001) where she shows the historical shifts in the conceptualization of anger in Old and Middle English. Kövecses suggests that the difference between the early, Old English presence of SWELL metaphors (exemplified by words like *belgan*) and the HOT FLUID IN A CONTAINER metaphor that surfaces centuries later with the rise of the humoral theory are both historically specific ways of highlighting distinct components of the underlying model. The specific-level conceptual metaphors that surface at one point in time may be based on one feature of the universal embodiment associated with the target concept, while those that arise at another point may focus on another aspect or dimension.

Needless to say, we welcome Kövecses's recognition of the importance of culture and variation in metaphor studies, but at the same time, against the background of the data discussed in the present paper, we are not entirely convinced by the way in which he tries to reconciliate the alleged presence of a universalist embodiment model with actual cultural and historical variation. To begin with, if the emergence of specific-level conceptual metaphors that allegedly link up with the PRESSURIZED CONTAINER metaphor has a physiological basis, one would expect that physiological association to show up in the actual contexts of use of the conceptual metaphors. As we argued, the presence of a compound like *hatheort* as such is not sufficient to talk about a physiological model: to the extent that a physiological interpretation plays a role at all, the rest of the *heart* and *mood* compounds in English demonstrates that it plays a secondary role at most. In Gevaert (2007: 232–249), this contextual analysis is repeated for the entire set of anger expressions in Old and Middle English. Her conclusion for motifs like ANGER AS HEAT and ANGER AS FIRE is that contextual references that indicate that the conceptualization may have been embodied, in the sense of evoking a physiological framework, are extremely scarce in the Old English and the early Middle English period. It is only with the rise of the humoral theory that unmistakable physiological references show up. In other words: rather than the emergence of conceptual frameworks being triggered by physiological associations, it seems that the physiological associations are triggered by the emergence of the conceptual frameworks.

Further, on a more general level, it is not clear how Kövecses's position can be saved from vacuousness if he accepts that 'it also seems possible that universal physical of biological embodiment is entirely ignored in conceptualization' (2005: 251). He cites Lutz (1988) on anger expressions in Ifaluk as an illustration that 'we know of at least one culture in which the angry person is not, or is only to an insignificant degree, viewed as a pressurized container' (2005: 251). Against the background of this remark, what exactly is the universal nature of the embodiment hypothesis? The interpretation definitely cannot be that all languages activate the model: Kövecses himself provides us with a counterexample. Although Kövecses does not specify an alternative interpretation, we can think of two ways of nuancing the universalist hypothesis: a quantitative one and a qualitative one. In both cases, however, the type of historical evidence that we present in this paper sheds doubts on the validity of the universalist hypothesis.

The qualitative reinterpretation could take the following form: 'If a language conceptualizes anger in an embodied way, the embodiment model takes the form of the PRESSURIZED CONTAINER metaphor'. That is to say, languages do not necessarily have to opt for an embodied understanding of anger, but if they do, the underlying model they use will be that of a container under pressure. However, already in the materials presented here, it is not easy to find a counterexample: the AFFLICTION motif is definitely an embodied one (pain is a physiological phenomenon), but it is totally independent of the PRESSURIZED CONTAINER metaphor.

The quantitative reinterpretation would see the universalist embodiment model as a statistical universal: there would be an overriding tendency in the languages of the world to activate the model. However, how strong should the tendency actually be? As long as we do not agree on the expected distribution or the required strength of a universal tendency, how can we test it? Moreover, testing would not only have to take into account the mere presence of a certain pattern in a language, but would also have to allow for the salience of the pattern in each of the languages. If the pattern is only weakly present in a given language, then surely that language counts less in the overall calculation than languages in which embodiment is the prime model. The data shown in the present paper make clear that English, which is usually presented as an example of the latter is actually an example of the former. So how much would be left of the quantitative reinterpretation if we apply this type of analysis more systematically to the languages involved?

The overall methodological conclusion, then, that we would like to derive from the two case studies presented above may be formulated as follows: an exclusive focus on metaphorical conceptualizations (*a fortiori*, on embodied metaphorical conceptualizations) is likely to distort the picture of the actual cultural models at work. It is only by systematically comparing the weight of the embodied and/or metaphorical conceptualizations with the alternative ones that their true salience can be assessed.

Appendix

Mood compounds in Old English

æmod	without mood	madness; cowardice
æwiscmod	ashamed-mood/mind	shame
acolmod	frightened-mood/mind	fear
ætrenmod	poisonous-mood/mind	malevolence, malice
angmod(nes)	narrow-mood/mind	sadness
anmod(nes/lice)	one-mood/mind	strength, unanimity, constancy
bliðemod	happy-mood/mind	happiness
bolgenmod	swollen-mood/mind	anger
clænmod	clean-mood/mind	purity, moral cleanness
deormod	bold-mood/mind	strength, fortitude
dreorigmod	sad-mood/mind	bad feeling, sadness
eaþmod(heort)	easy-mood/mind	humble, obedient
efenmodlice	even-mood/mind	gentleness, meekness, composure
fæstmod	firm-mood/mind	constancy
felamodig	very-bold	boldness
forhtmod	fearful-mood/mind	fear
freorigmod	chilled/sad-mood/mind	bad feelings, sadness
gealgmod	sad-mood/mind	bad feelings, sadness, anger
geomormod	sad-mod/mind	bad feeling, sadness, dejection

geþyldmod(nes)	patience-mood/mind	patience
gewealdenmod	controlled-mood/mind	restraint, temperance
glædmod	glad-mood/mind	gladness, cheerfulness; favour, kindness, grace; promptitude, readiness
gleawmod	wise/skilful-mood/mind	sagacity
grammod	fierce-mood/mind	hardheartedness, cruelty, severity
guþmod	war-mood/mind	courage, boldness, valour
heahmod	high-mood/mind	pride; exultation , joy; magnanimity, greatness of soul
heanmod	low/poor-mood/mind	dejection
heardmod	hard-mood/mind	strength, fortitude; obstinacy
hefigmod	heavy-mood/mind	bad feeling, sadness; oppressive disposition
hohmod	careful/thoughtful-mood/mind	bad feeling, sadness
hreohmod	hreow=sad; hreoh=fierce (mixed use)	bad feeling, sadness; fierceness
hreowigmod	sad-mood/mind	dejection
hwætmod	swift/bold-mood/mind	boldness
irremod	angry-mood/mind	anger
langmod	long-mood/mind	patience
laþwendemod	evilly disposed mind	hatred
leasmodnes	not truthful-mood/mind	wantonness, levity, frivolity
leohtmod(nes)	light-mood/mind	wantonness, levity, frivolity
lytelmod	small-mood/mind	cowardice, pusillanimity
madmod	senseless-arrogance	pride
meagolmod(nes)	earnest-mood/mind	earnestness
micelmod	great/much-mood/mind	magnanimity, greatness of soul
mihtmod	powerful-mood/mind	anger; strong feeling, passion
ofermod	excessive pride	pride, arrogance
ormod(nes)	without mood	despair
reom/nigmod	sad-mood/mind	sadness
reþemod	fierce-mood/mind	fierceness; wrath, sternness, displeasure
rummod(lice/nes)	liberal-mood/mind	abundance; bountifulness, munificence; overliberality, waste of money
sarigmod	sad-mood/mind	bad feeling, sadness
sceohmod	fearful-mood/mind	fear

seocmod	sick-mood/mind	mental weakness; not strong-minded
sleacmodnes	slack-mood/mind	disinclination to act, listlessness
stearcmod	strong-mood/mind	obstinacy
stiþmod	strong/fierce-mood/mind	will, determination, resolution; obstinacy; hardheartedness, cruelty, severity
strangmod	strong-mood/mind	will, determination, resolution
streccanmod	extended/spread-mood/mind	perseverance
styrnmod	stern-mood/mind	will, determination, resolution
swærmod	heavy-mood/mind	disinclination to act, listlessness
swearcmodnes	dark?-mood/mind	cowardice, pusillanimity
swiþmod(nes)	very much-mood/mind	will, determination, resolution; proud, arrogant; magnanimity, greatness of soul
þancolmod	wise-mood/mind	sagacity
þearlmod	severe-mood/mind	hardheartedness, cruelty, severity; strictness, austerity, severity
þol(e)mod	suffer-mood/mind	patience
tilmodig	good-mood/minded	favour, kindness, grace; noble-minded
torhtmod	splendid-mood/mind	glory, splendour, magnificence
tornmod	anger-mood/mind	anger
unbliþemede	not glad-mood/mind	bad feeling, sadness
unforhtmod	not frightened-mood	fearlessness
ungemod(nes)	not of one mind	quarrel, contentiousness, strife
unmod	without mood	dejection; without courage, not proud
unrotmod	sad-mood/mind	bad feeling, sadness
unþolemodnes	not suffer-mood/mind	impatience
wacmod(nes)	weak-mood/mind	moral weakness; cowardice, pusillanimity
weamod(nes)	weamod=anger; wea=affliction/malice	anger

welrummod		favour, kindness, grace
werigmod	tired/sad-mood/mind	dejection
wiþermod(nes)	against-mood/mind	hostility, contention, opposition

References

Bosworth, Joseph & Thomas N. Toller
1898 *An Anglo-Saxon Dictionary*. Oxford: Oxford University Press.

DiPaolo Healey, Antonette & Richard L. Venezky
1980 *A Microfiche Concordance to Old English: The List of Texts and Index of Editions*. Toronto: University of Toronto.

Geeraerts, Dirk
2006 *Words and Other Wonders. Papers on Lexical and Semantic Topics*. Berlin: Mouton de Gruyter.

Geeraerts, Dirk & Stefan Grondelaers
1995 Looking back at anger. Cultural traditions and metaphorical patterns. In John R. Taylor & Robert E. MacLaury (eds.), *Language and the Construal of the World* 153–180. Berlin: Mouton de Gruyter.

Gevaert, Caroline
2001 Anger in Old and Middle English. A 'hot' topic? *Belgian Essays on Language and Literature* 89–101.
2007 *The History of Anger. The Lexical Field of Anger from Old to Early Modern English*. PhD Thesis, Department of Linguistics, University of Leuven.

Kövecses, Zoltan
1995 Metaphor and the folk understanding of the anger. In James A. Russell & José-Miguel Fernandez-Dols (eds.), *Everyday Conceptions of Emotions: An Introduction to the Psychology, Anthropology, and Linguistics of Emotion* 49–71. Dordrecht: Kluwer.
2000 *Metaphor and Emotion*. Cambridge: Cambridge University Press.
2005 *Metaphor in Culture*. Cambridge: Cambridge University Press.

Lutz, Catherine A.
1988 *Unnatural Emotions. Everyday Sentiments on a Micronesian Atoll and their Challenge to Western Theory*. Chicago: University of Chicago Press.

Matsuki Keiko
1995 Metaphors of anger in Japanese. In John R. Taylor & Robert E. MacLaury (eds.), *Language and the Construal of the World* 137–151. Berlin: Mouton de Gruyter.

Molina, Clara
2005 On the role of onomasiological profiles in merger discontinuations. In Nicole Delbecque, Johan Van der Auwera & Dirk Geeraerts (eds.), *Perspectives on Variation* 177–194. Berlin: Mouton de Gruyter.

Toller, Thomas N. & Alistair Campbell
1921 *An Anglo-Saxon Dictionary: Supplement.* Oxford: Oxford University Press.

Yu, Ning
1998 *The Contemporary Theory of Metaphor.* A Perspective from Chinese. Amsterdam: Benjamins.

To be in control: kind-hearted and cool-headed. The head-heart dichotomy in English

Susanne Niemeier

Abstract

This contribution aims at exploring two body part conceptualizations as they show up in English. It will explore linguistic expressions and image schemata concerning the body parts *head* and *heart* and relate them to their underlying cultural models (cf. Quinn 1987; Quinn and Holland 1987; and Quinn 1991) and ethnomedical traditions and motivations. The aim of this discussion is to show that these conceptualizations are far from arbitrary, and that they in fact reflect the concepts that are or were at a certain historical stage present in the culture in question. Even if today's advanced medical knowledge no longer agrees with these traditional concepts, linguistic expressions that are so deeply entrenched that they have come to belong to unconscious knowledge far outlive such a change in conscious knowledge and still function as a basic demarcation line in allocating the locus of emotions and intellect. The most salient cultural model in this respect is the dualism between the body and the mind/reason propagated by Descartes and his successors. As the analysis shows, this distinction is still to be found in present-day English and comes to the fore in non-literal, mainly metaphorical language usage. However, once we trace the metaphors back to their origins, we will find that they are indeed metonymically motivated and are firmly linked to (equally largely unconscious) historical beliefs and traditions. In present-day understanding what seems like a metaphor may have developed over time out of a metonymy.

Keywords: cultural model, Descartes, embodiment, emotion, image schema, metaphor, metonymy, reason

1. Introduction

Since its very beginnings, cognitive linguistics has stressed the notion of embodiment, that is, the role of bodily experiences in conceptualization. On the one hand, these kinds of conceptualization have their origin in the fact that our very first experiences of image schemata in this world are body-related. This is evident, for instance, in the basic role played by spatial prepositions for contact, support, containment, verticality, proximity etc. On the other hand, these kinds of conceptualizations are to a large extent rooted in our cultural traditions, and this is evident in the large number of cultural models and in culture-bound body-related metaphors and metonymies.

Given the embodied view of our experience (Johnson 1987), underlying metonymizations are especially obvious in expressions referring to emotions and to mental activity. I will therefore focus on these two domains, namely the cultural model of the heart as the locus of emotions and the cultural model of the head as the locus of intellect as they exist in present-day English. A cultural model, as a representation of a community's wisdom in a given domain, can be seen as both the synthesis and simultaneously as the source of many clusters of metaphors and metonymies and also of many single individual metaphors and metonymies. In this respect, it differs fundamentally from Lakoff's notion of conceptual metaphor which sees one global conceptual domain mapped onto some other domain. A cultural model, in contrast, combines a variety of single and often contradictory metonymic and metaphorical views, crystallizing in a more dominant conceptual frame. This ties in with the contention of Quinn (1987), Quinn and Holland (1987) and Quinn (1991) that cultural models are at the basis of metaphor, and not the other way round as Lakoff claims. As a bricolage[1] of diverse experiences in a domain, a cultural model usually also contains various sub-models that complement each other as different facets of the more overarching cultural model. It should also be borne in mind that such a model, because it is by necessity culturally based, tends to contradict and even be in conflict with scientific knowledge. The present neurological and medical view of *head*, for instance, is one where feelings influence the brain and are at the same time claimed to be controlled by the brain. The cultural model view sees this aspect quite differently and locates emotions in the heart, as the following compounds clearly suggest: *cool-headed* is positive, but *cool-hearted* is negative, whereas *warm-hearted* and *hot-headed* invoke the opposite axiology.

The present contribution will first summarize an analysis of a sample of "heart"-expressions (Section 2) (for a more extensive analysis see Niemeier 2000) and then present an analysis of a sample of "head"-expressions (Section 3). Finally, an attempt will be made to relate the head-heart dichotomy to traditional cultural and ethnomedical beliefs, arguing that this dichotomy in English relates to a specific period in language and thought development, starting from the 17th century, when Descartes and his successors propagated the latent dualism between emotions and reason (Section 4).

2. Western viz. Anglo conceptualizations of *heart*

In my previous analysis (Niemeier 2000), four categorically different kinds of *heart* expressions were distinguished.[2] In each of these categories, the relation between the metaphorical expression and the underlying metonymy becomes weaker by degree.

2.1. *Heart* as a metonymy for A PERSON'S FEELINGS

As the concept "feelings" comprises emotions, courage and attitudes, amongst other things, we are dealing here with the most specific level of meaning: the heart is the most salient body part in the traditional Western cultural model of emotions. The prototypical emotion associated with the heart is romantic love, a uniquely human sensation. If asked to symbolize romantic love, most Western people would invoke a prototype figure by drawing the stereotypical picture of a red heart. This close association of the feeling of love with the heart can be seen as *prima facie* evidence for the cultural model of the heart as the seat of emotions. Since at least the Middle Ages, the heart has been, and still is, taken as a symbol for romantic love.[3]

Still, we do not find one single, clearly delimited cultural model, but rather a conglomerate of different sub-models, which are clearly reflected in specific linguistic expressions, each one focusing on a different aspect, involving different shades of meaning. For example, the heart may be conceptualized as A MOVEABLE OBJECT (*set one's heart on sth*), or as CHANGEABLE IN SIZE (*have a big heart*). Furthermore, not only positive emotions such as love are conceptualized in *heart* metaphors. We also find negative emotions associated with the absence of heart or its diminished size, referring for example to a lack of hope and enthusiasm (*to lose heart*).

A common conceptualization of the heart is the cultural sub-model of HEART AS A SOLID, which is also present in other conceptualizations, for example in the case of negative emotions, where the heart is conceptualized as consisting of certain materials with which different qualities are associated. The *heart* can either be seen as being made from soft materials and thus as housing soft, altruistic feelings (e.g., *soft heart, tender heart*), or it can be seen as made from heavy and hard materials and thus as housing hard feelings (*hardness of heart, heart of stone*).[4] In the latter case, this hardness of material is reinterpreted as an emotionless attitude and thus metaphorically mapped onto the domain of the heart. In a further step, these metaphors are then used metonymically insofar as they pick out one salient detail of a person's disposition which then stands for the whole moral outfit of that person. The expression *heart of stone*, for example, suggests the disappearance of the heart as the centre of emotions and its substitution by a stone.

The heart's temperature is also used in describing a person's attitude towards others: we find expressions such as *warm-heartedness* or *cold-heartedness*, respectively used of a person who emits warmth and sympathy towards others and a person who lacks such behaviour. These expressions refer to heated versus non-heated body parts, a transfer which cannot be accounted for in a Lakovian view of metaphor, since in a mere mapping from source to target domain the physical level is non-existent. Therefore only a multi-domain model (cf. Fauconnier and Turner 2002), integrating the physical domain (cold/warm), the physiological domain (heart beat, rhythm) and the psychological domain (emotion itself) makes sense here.

In the sub-category of the heart as AN ANTHROPOMORPHIZED ENTITY[5] (*being one's heart's desire*), the heart itself is seen as desiring something and metonymically stands for the whole person. A further cultural sub-model visualizes the FULL or WHOLE ENTITY OF THE HEART, which may be only half present, or even totally absent (*half-hearted*, *heartless*). *Half-hearted* stands for a lack of enthusiasm and/or courage[6], i.e. it refers to a dysfunction, as half a heart is supposed to be valueless. *Heartless* again invokes the *heart of stone* concept, where emotions have been replaced by hardness.

2.2. *Heart* as a metonymy for a PERSON AS A WHOLE

Metonymic conceptualizations of the heart as a PERSON AS A WHOLE entail that the heart is not only seen as a part of a person but in a more generalized way as referring to the person as a whole. This comes close to what Ibarretxe-Antuñano (This volume) calls "external reference", that is reference to a physical area (the person), where the organ (the heart) acts as a focalized centre.

An obvious example of such metonymic focalization can be observed in the description of a love partner, where the reference to the actual person may dwindle down to seeing him/her and even oneself as just a heart (*all hearts throb for her; two hearts that beat as one*). However, we know that these *hearts* are not independent or autonomous in nature as their beating may still be conceptualized as being dependent on or regulated by other physiological processes, such as blood pressure or the rhythm of the heart beat.

The conceptualization of the heart as a metonymy for a PERSON AS A WHOLE comes most often to the fore when negative emotions felt within oneself are described as causing damage to the heart. Here again, we find different cultural sub-models focusing on different aspects of the overarching cultural model: the damage to the heart can be described in various ways, either generally (*aching heart, broken heart*), or more explicitly describing the wounds that the heart suffered (*my heart is bleeding*), or the instrument that did the damage (*heart-burning*).[7] Finally, the heart can be also be destroyed (*heart-rending*). In all the examples, the heart is metonymically conceptualized as acting or being acted upon on its own, without naming its "container", that is the person.

This second category (*heart* for the PERSON AS A WHOLE) is closely related to the first category (*heart* for a PERSON'S FEELINGS). However, the perspective has widened and the heart is no longer considered as referring only to the emotional part of a person but to the complete person. Consequently, the metonymies involved are less basic ones and often rely on double metonymizations or even appear as metaphors. Nevertheless, the underlying metonymic basis is kept intact.

2.3. *Heart* as an OBJECT OF VALUE

At an even more general level of conceptualization, the heart is seen as a kind of treasure chest. Implicitly, emotions are involved as well but they are not named or singled out. Here again, diverse cultural sub-models are involved which help to structure this domain of meaning. The single emotions or the connection between the heart and the person who possesses it are not foregrounded, but in an ever-narrowing perspective the heart is seen as an entity and as an object of value. This level of meaning is even more culture-specific – in the sense of Anglo versus other Western cultures – than the two previous ones because the notion of value is highly culture-determined, and is conceived of in terms of the source domain of economics, that is of "prize", "win", "lose" etc.

Each of the cultural sub-models involved takes a different perspective on the *heart* as an object of value: the heart may be seen as a COVETED

ENTITY in the sub-models HEART AS A PRIZE (*win someone's heart*) and HEART AS BOOTY (*to steal every heart*). These expressions imply that one no longer wants to or is no longer able to be in control of one's emotions but lets another person take control. Similarly, the heart can be conceptualized as a MANIPULABLE OR FRAGILE OBJECT (*broken-hearted*). Manipulation is also involved in expressions referring to courage, that is active human behaviour (*to put heart into*); the same is true for the opposite notion, that is passive human behaviour and thus the lack of manipulation (*to lose heart*).[8]

It is interesting to note that *to lose heart* only refers to (a lack of) courage, whereas *to lose one's heart* refers to love, and not only to romantic love, one can also lose one's heart to objects, for example. When the complete heart is lost – when another person, or something else one is taking a vital interest in, has "won" it – it is transferred to this other person or to the object. In the case of *to lose heart*, only that part of the heart is "lost" that is not connected to love. The notion of the heart as the seat of courage can also be transferred to other persons. For example causing somebody else to feel anger or discontentment can be conceptualized as *to dishearten* this person, that is to discourage him/her from doing something which s/he intended to do or from continuing along a particular path of action.

Finally, other negative feelings are also conceptualized as being related to the heart. Thus, resenting something or being hurt by somebody may be expressed as *to take it to heart*. If one takes something to heart, one thinks about it very seriously. This expression may well be reminiscent of the Old or Middle English concept of the heart as the seat of knowledge (comparable to *by heart*).

In this third category we still find a conceptual closeness to the non-figurative understanding of "heart". Even if the metonymic undertones are not always immediately noticeable and are not as obvious as in the first two categories, they are still present in the general understanding of the expressions belonging to this category.

2.4. *Heart* as a CONTAINER

The human body in general as well as many of its major parts and internal organs, including the head, the heart, the chest, and the womb, are frequently conceptualized as containers.[9] As physiological evidence was not available for a long time, all kinds of cultural models could flourish, seeing the heart as different types of containers, depending on the aspects focused upon in conceptualization. These views will be traced in a narrowing perspective from the outer qualities of the container, including its handlability, via its inner qualities to its internal structure.

In a static view, the container is seen as covered by a lid which may be opened or closed (*heart* as a LID CONTAINER). When opened, there is free access to a speaker's emotions (*to open one's heart to sb*). When the heart is conceptualized as closed (*sth locked up in one's heart*), there is no access to that person's emotional world. There may also be instances when the heart is full of emotions but one does not wish it to flow over on its own. In these cases, the heart is seen as a MANIPULABLE CONTAINER that can be dealt with at one's own will (*to pour one's heart out to sb*).

If there are too many feelings involved, the container might be too small to contain them all, in which case it might overflow. In this case it has to be open, that is, unhindered by a lid or, like any cooking pot does when the fluid is too hot, lifting the lid by itself. This may be illustrated by the dynamic view of an OPEN CONTAINER FOR LIQUIDS (= positive emotions), as in *a heart overflowing (with gratitude),* where the path of the liquid's movement from the container is visualized. The opposite kind of movement, namely towards the container, is highlighted in *this filled my heart (with joy).* Thus, the way from feelings to heart is not a one-way street but works both ways, as feelings may enter the heart but may also leave it.

Another cultural sub-model sees the heart as a CONTAINER WITH GREAT DEPTH, filled with positive emotions. This type of container is conceptualized as having a bottom representing the location where one's most intense feelings are stored (*from (the bottom of) one's heart*). The container may be conceptualized as integrating a further INTERNAL CONTAINER, hosting feelings that are even more intense and thus further accentuated (*in one's heart of hearts,* a Shakespearian coinage for the "deepest and most hidden thoughts and feelings" [Cassell 1997: 132]).

The heart may furthermore be conceptualized as a sort of storehouse where feelings of different kinds are grouped, labelled and stored (*I could not find it in my heart*). In this storehouse, not only feelings may be stored, but also opinions and knowledge. Thus, if one memorizes texts or poems, one puts them into this STOREHOUSE CONTAINER (i.e. learning them *by heart)* in order to be able to take them out again when necessary. This is a very intriguing idea as memory is usually seen in connection with the head/the brain – which also acts as a storehouse for information (see Section 3.4) – and not so much with the heart. This may be a remnant of earlier cultural models of the *heart* from Old or Middle English, where the heart was still conceptualized as not only being the locus of emotions but also of rational thought.

We find an even greater degree of specification of the container in the expression *have a place in every heart*, referring to a person who seems attractive and desirable to many others who reserve some space for her or him in their inner emotional sanctums. Here, the heart is clearly seen as a COMPARTMENTALIZED CONTAINER, able to integrate different types of objects into different slots.

All these different-faceted conceptualizations of the heart as a CONTAINER show that we are not dealing with a one-track conceptualization, but instead with a conglomerate of different cultural sub-models. Nevertheless, compared to the more culture-specific nature of the other categories discussed so far, the container schema seems to be relatively more universal. Its higher level of generality is further underlined by the fact that it is not unique to the cultural model of the heart but works in lots of different contexts as well (words as containers for meaning, the head as container for rational thought, etc.). This comes to the fore also in the fact that the image of the heart as a container seems to be much more "metaphtonymically" structured (see Goossens 1995) than, for example, that of the heart as an object of value. As the underlying metonymies are not as obvious as they are in the other categories, this fourth category seems to be dependent on the existence of the other categories which provide its metonymic basis.

We know more about the way the heart functions than we know about the brain or the mind. The pulse is one of the symptoms of life which can be measured for and by every individual. Here again, we find a metonymic perspectivization in our cultural model of the heart. Nobody has so far been able to precisely locate the site of feelings. An important turning point in the conceptualization of feelings was the work of William James (1842–1910), who saw feelings as the mental aspect of emotion. He argued that an emotion is based on a stimulus followed by physiological changes in the body, which are then perceived by the mind as a specific feeling. In other words, our feelings rely purely on our mind's perception of a certain physiological condition and not vice versa, as had previously been assumed.[10] Today's ever-increasing neurophysiological knowledge still basically agrees with James. It sees emotions as connections between the body and the cortex, via the sympathetic and/or the parasympathetic nervous systems (depending on the kind of emotion). Feelings are understood to be the mental concepts that we come up with when we experience specific bodily sensations. Although the actual processes in the brain are far better known today, we still face problems when we have to explain precisely how and where feelings are generated – in the brain or in the blood circulation system or in other bodily systems.

Thus, starting with their most basic bodily experiences, people can feel their heart acting differently according to the emotions they experience and the adrenaline that is set free in this process by neurochemical impulses: the heartbeat accelerates, slows down, becomes irregular, etc. Already in ancient times emotions were assumed to reside in the most conspicuous body organ (the only one everybody is able to draw – although anatomically and physiologically incorrectly) – at least in Western cultures since the Middle Ages (see Fn 3) – and the heart was taken as a metonymy for a person's feelings and could stand for the whole person experiencing a specific emotion. It is presumably on the basis of this archetypical metonymy that the other understandings could arise and flourish.

We are indeed able to experience certain physiological sensations connected to our emotions, for example to that of the prototypical emotion of romantic love. It is not actually the heart that loves, but the whole person – the brain is involved to a large extent, as endorphins and further hormones are set free in order to create the physiological sensation of "being in love", which is basically brain-controlled. As such, the metonymic character of the heart expressions mentioned in this sub-section is obvious. The bodily sensations are used metonymically to refer to the feelings and these metonymies may then be used as source domains for either new metonymies or for metaphors.

Apart from love, less prototypical emotions, such as compassion, pity, envy, or inhumanity, and types of behaviour like courage or cowardice – where one should rather speak of feelings than of emotions – are conceptualized on a more general level as residing in the heart, and can also be claimed to be based on metonymic mappings.

3. Conceptualizations of *head*

The corpus serving as the basis for the following analysis stems mainly from Roget's Thesaurus. Further material comes from the dictionaries listed at the end of the paper. The expressions for *head* found in the corpus may tentatively be grouped under three different headings for the sake of clarity. I will discuss the examples and their underlying conceptualizations and cultural sub-models separately for each category, although it will become obvious that the categories overlap.

At first sight, conceptualizations concerning the head stand in contrast to those concerning the heart, as the head in Anglo culture is primarily seen as the locus of rational thought, standing in opposition to emotions. Although there are slight overlaps and contact zones between the conceptualizations of these two body parts so that the polarity between them is not complete, this polarity still seems to be quite entrenched, and for this there is a historical explanation.

The notion of reason or ratio goes back very far in philosophical thinking. It can be traced to Aristotle and Plato's ideas, for example. The Greek word "logos" (meaning "word, speech, discourse", but also "reason") was translated into Latin "ratio"[11] (meaning "reckoning, understanding, motive, cause") and then French "raison", from which the English word "reason" was derived (cf. en.wikipedia.org/wiki/Reason). The first time that the English word "reason" was employed in order to mean "to think in a logical manner" was in 1593 (cf. www.etymonline.com). A quasi-contemporary of Descartes (1596–1650), the French philosopher Blaise Pascal (1623–1662), formulated his well-known saying according to the beliefs of those days: *Le coeur a ses raisons que la raison ne connaît point*,[12] and this dualism is still noticeable in Western language usage today.

The basic meaning of *head*, similar to that of *heart*, is the reference to the body part, but frequently *head* expressions are used to refer to the head's presumed content, that is the brain, the mind, human ratio, intelligence.[13] Concerning the brain, there is neurophysiological evidence that it occupies 90% of the upper head and that it is the location of abilities such as perception, spatial imagination, creativity, and logic. Therefore, *brain* expressions will not enter the analysis because they tend to be less figurative – although of course research on the brain's functioning is ongoing and thus current theories may be inadequate. Also the complete meaning cluster of HEAD AS THE TOPMOST OR FRONT POSITION[14] will be left out as it does not deal with the conceptualization of mental life. What is more interesting for the current analysis are expressions referring to the mind, to rationality and to intelligence. In Dirven's (2002) metonymic chain, these expressions are situated hierarchically lower than *brain* expressions, which in turn are situated hierarchically lower than *head* expressions and are thus closer to the metaphoric pole of the metaphor-metonymy continuum. The location of the mind, rationality and intelligence has been presumed to be the head since the mid-1600s, when brain anatomy started. Only recently have these views been challenged, for example by Damasio (1994) who argues that mere intellectual reasoning is an impossibility and that human thinking is always supported and strengthened by emotions. He further argues that emotions are created in the brain, which controls the accompanying bodily reactions and thus links physiological body changes with emotions. In his view, reasoning and emotions cooperate and should not be seen as separate entities. Even the brain regions for emotions and for reasoning overlap and, according to Damasio, they cannot therefore be unrelated. He called his influential publication "Descartes' Error", hinting at the fact that Descartes changed philosophical thought by claiming that mind and body are basically unrelated and that the mind/head has the topmost priority – "*cogito ergo sum*" ("I think therefore I am"). According to Damasio, the two verbs in this phrase should be turned around: *I am therefore I think*. In other words, only the fact that we have a body allows us to adequately use our minds. Dualism is of course older than Descartes (it goes back to Plato and Aristotle), but was taken up as a general trend in Western thought in the 17th and 18th centuries, and Descartes is the prototypical, if not the only, exponent of this tradition.[15]

Nonetheless, such recent scientific views have not yet influenced the use of entrenched linguistic expressions, as language change is not usually a quick process. Therefore, the analysis will refer to traditional cultural models and views within the Western, here English-speaking, world.

Rationality being an abstract concept, we can find a multitude of metaphorical and metonymic expressions. One might hypothesize that it is easier to find metonymic bases for *head* expressions than for *heart* expressions. The ascribed functions of the head seem to be within the body part or domain matrix – even if metaphorically reified because, according to Dirven (2002), the "culturally conditioned" language user construes the neurological domain and the mental domain of thinking as conceptually contiguous, which means that they share one domain matrix. In contrast, concerning the heart, physiologically there is no visible or tangible, but only laboratory-accessible, proof for the neurological processes underlying feelings or emotions, which means that we are – from today's perspective at least – dealing with two different domains.

We can differentiate three different categories of *head*-expressions, each displaying a different degree of connection between the metaphors and the metonymies, although all categories are interlinked. The first category (3.1. *Head* as LOCUS OF REASONING) refers to the most specific level of meaning where there is a clear metonymic basis to be found insofar as the head is considered to be an autonomous entity in a whole-for-part metonymy, irrespective of the human being it belongs to. This is not surprising, considering the relative closeness of the hierarchical positions of *head* and *thinking* in Dirven's metonymic chain (2002). In the second category (3.2. *Head* as CONTROL CENTRE), the metonymic basis is less obvious, although it is still possible to establish the connection between the metaphors and metonymies. Here, the head is seen as the main locus for decisions and for the exertion of power – these aspects are related to the first category and to reasoning but exceed them, because they are more generalized. Thus, the distance between the *head* expressions as indicated in Dirven's (2002) idea of a metonymic chain has become larger as the idea of "control" approaches the area of *mind* expressions. In the third category (3.3. The *head* as a CONTAINER), the metaphors seem at first sight to be unrelated to any prior metonymization, but once the complete cultural model is focused upon, the relationship becomes evident.

3.1. *Head* as LOCUS OF REASONING

The basic meaning of *head* is the reference to the body part. This body part contains the brain, the five senses (vision, hearing, smell, taste, touch) and is the locus of reasoning. Thus, the head is named in order to refer to the brain (whole-for-part) and, in a further whole-for-part metonymization, to the process of rational thinking that is taking place in the brain. Dirven (2002) calls this a metonymic chain which goes from *head* via *brain* via *grey cells* via *thinking or thought processes* via *mind* via *thought* to *intelligence* (cf. also Fn 13).

In the corpus, we find plenty of *head* expressions referring to rational thinking and to intelligence, such as

(1) *We need some good heads to solve the problem*
(2) *Use your head!*
(3) *Do some headwork!*
(4) *head-hunter* (relating to business)

These examples allude to the major function of the brain that Anglo cultures attribute to it. If we need "some good heads" (1), we are thinking of intelligent people who are able to solve the problem in question. On a "good-bad" scale, we differentiate between people who are more gifted for reasoning and intelligence and people who are less gifted in these respects. The appeals in (2) and (3) work in the same direction as they order the addressees to do some thinking. Example (4) is a newer usage, referring to somebody who tracks down intelligent people wanted by potential employers, that is, someone who is on the look-out for "good heads".

While the above examples refer to the general human gift of reasoning, we also find more specialized expressions referring to a particular kind of reasoning (or its absence). An example of this kind of metonymy is

(5) *She has no head for figures*

This expression clearly relates to the brain as part of the head, which – in this case – is unable to deal with arithmetic. On the one hand, this underlines the folk view that mathematical problems are to be dealt with in our heads/brains, but on the other hand, it goes beyond that, as it also underlines the somewhat newer perspective that our intelligence is not a uniform whole but that there are various kinds of intelligence involved, among them mathematical, spatial, emotional, social or linguistic intelligences.[16] The person in question is thus seen as lacking a specific kind of intelligence.

3.2. *Head* as CONTROL CENTRE

In this category the head is seen as the main site for decision making. This aspect is related to the previous category but, being more generalized, exceeds it. Here, we do not look at rational or irrational behaviours of an individual but are taking some kind of outward perspective on the general processes which we assume to be taking place in our heads. Thus, the head is not primarily seen as the locus of reasoning but as some kind of general control centre (cf. also Geeraerts 2002, who discusses the metaphor THE HEAD IS THE LOCUS OF ONE'S SELF-CONTROL). Such a control centre is certainly related to our idea of reasoning, but if we see reasoning as carrying out mental processes and using one's intelligence as a tool, it is evident that such notions are absent from the examples in this sub-category. Instead, the examples in the present sub-category focus on the importance of some kind of balance in our heads/minds as a precondition for reasoning, and as such they are less specific.

It is difficult to trace back the origin of the metaphor of the balanced mind, but generally it refers to stability of mind and body. Already the ancient Greeks believed in the balance of mind and body, in some kind of harmony in dualism.[17] Interestingly enough, nearly all examples provided below are negative, that is, they refer to a lack of balance rather than to an actual balance. The metonymic character of these expressions is not immediately visible, as it can only be seen after tracing back various metonymic steps. Thus, the brain is located in the head (metonymy), reasoning is one of the brain's functions (metonymy), and a balanced mind is a precondition for reasoning (metonymy again). There are several causes that can upset such a balance (mostly expressed by metaphor), so that at first sight we seem to be dealing more with metaphor than with the underlying metonymies.

This is, for example, the case when we talk about outward influences on the head, such as in

(6) *to have one's head in the clouds/in the sand*

Here, the bodily ability of vision – which, as an important sensory function, is part of our cultural model of senses and mind[18] – is addressed. When one's head is in the clouds, one can't perceive clearly anymore and is thus lost in dreams, and the same is true for a head stuck in the sand, where one can't see anything anymore and is in a state of denial, not wanting to face the facts. If we can't really see – due to outward influences such as clouds or sand – or understand reality (KNOWING IS SEEING), we are also unable to judge and our rational control centre can't work properly.

Other examples refer to further causes for the imbalance in our heads/minds:

(7) *my head is spinning*
(8) *the success went right to her head*
(9) *to keep a clear head*
(10) *to keep a cool head*

If the head is spinning (7), it is quite literally impossible to keep one's balance and one is unable to think straight. Also, (8) is hinting at an imbalance created by the person's success, which prevents her from thinking in a normal way because conceit might have put her off-balance rationally. Similarly, (9) and (10) stress the necessity of avoiding such an imbalance. These examples show rather nicely that the "emotional" can take hold of the "rational" in our dualistic cultural model of the head.

We also find examples which go beyond a potential imbalance in our minds and instead suggest that the head is already absent, and therefore that this person has no ability to judge rationally anymore, in other words, that this person has lost control:

(11) *to be off one's head*
(12) *to lose one's head*
(13) *headless*

Especially the last example shows very clearly that without a balanced control centre, rational control is completely absent.

As mentioned above, the underlying metonymies do not show up at first sight, but a headless person behaves as if s/he has lost the capacity for rational thought which is supposed to reside in the head – thus, although this expression seems to be completely metaphorical, there is indeed a metonymic motivation.

3.3. *Head* as CONTAINER

As categories have fuzzy boundaries, this last category shows some overlap with the previous categories, since all of them allude to something "inside" the head. However, in the other categories the aspect of containment is not explicitly expressed but just tacitly assumed. Still, there are a number of expressions in English in which the head is explicitly conceptualized as a container and which focus either on its fullness[19] vs. emptiness or on the dynamic process of filling or emptying it.

Examples for the first variety are expressions such as

(14) *my head is exploding*
(15) *empty-headed*

In the first case, the head is perceived as being so full of content (information, problems) that nothing more can enter, and the person feels the pressure inside the head so much that s/he has the impression that the head is exploding due to overload. Example (15) refers to a stupid person, whose head is conceived of as empty, without any content, and thus unable to perform rational thought processes – brainless persons can't think.

The other type of expression, relating to the process of filling or emptying the head, yields more examples. Although still to be researched in corpora, it seems that this dynamic action of filling and/or emptying is more salient and more entrenched in the Anglo cultural model of the head than a more durable full/empty state.

(16) *It never entered my head that he might be lying.*
(17) *When will you get it into your head that you must not lie?*
(18) *Put these weird thoughts out of your head.*
(19) *Who's put such bizarre ideas into your head?*
(20) *He has a head like a sieve.*

As can be gathered from the examples, filling and emptying the head can happen in different ways. In (16) no agent is named but information enters (or not) on its own. In contrast, (17) and (18) presuppose the ego to be the active party in getting or putting something into one's own head, whereas in (19) a third party is putting ideas into someone's head while the ego remains passive. Example (20) is somewhat different, since it does not focus on the dynamic transfer but on the state of the container, which is apparently full of holes, making containment difficult and allowing for a slow emptying.

Once again, when speaking of containers and containment, the metonymic motivation of the expressions is kept intact. The cultural model assumes that the head has certain content on different levels – the brain, the mind, ideas – and the expressions in this category focus on its acting as a container. This is a metaphor, but it is based on the underlying metonymy of how we conceptualize the head. Today we do not have a completely clear idea of how the head, or more precisely, the brain works and even if we had such an idea, entrenched expressions would stay in common usage for a long time to come, as language change is a very slow process.[20]

As for the analysis of *heart* expressions above (cf. Niemeier 2000), we can conclude here that the broader and the more general the perspective, the fewer metonymies and the more metaphors are involved (and vice versa), although the metaphors retain a metonymic basis.

4. Conclusion

It goes without saying that the main meaning clusters of *head* expressions discussed above not only overlap but that they are also compatible with one another. They are all facets of a culturally shared constructed model of the head as the locus of rational thought. Rational thought is seen as being located somewhere inside the head and as being a most important aspect of human life. Therefore, the head can be referred to from various perspectives: reasoning can be directly focused upon (3.1.), the head can be seen as the control centre of our body, thus mainly referring to the mind (3.2.), or it can generally be seen as a container (3.3.).[21] It should furthermore be obvious that the examples quoted do not cover the whole range of possible *head* expressions in English, but that they nevertheless indicate the main usages of the term.

Comparing the corpora of *head* expressions and of *heart* expressions, we find a commonality in the fact that in both cultural models the container metaphor, a very general if not universal metaphor, is active in the conceptualization of human mental faculties. If we have traditionally conceptualized our various body parts as being the locus of something mental, this something must be in a container-containment relationship to the body part itself. Taking a closer look at the two analyses, it becomes obvious that, in line with Western dualism, quite different "somethings" are seen as being contained in the two body parts in question. The heart is mainly responsible for feelings, especially emotions, which only marginally enter our basic conceptualization of the head.

The corpus analyzed contained just two instances of *head* expressions which might be classified as related to "emotions": *heady*, meaning "extremely exciting" or "rash, impetuous", and *headstrong*, meaning "obstinate". Neither of these expressions refers to prototypical, basic feelings but to more peripheral ones. In *heady*, something goes to one's head, weakening or even eliminating the control centre, and in *headstrong*, the person thinks his/her arguments are good enough not to give in to some other person's arguments, and in that way, ratio is involved, if in a distorted way. Therefore, those two emotionally tinted expressions do not belong to the same set of emotion expressions that we found for *heart*. We can still argue, therefore, that the two cultural models discussed above show an almost complete division of labour between the two body parts: the head as the centre of rational judgment and the heart as the centre of emotions.

This division of labour is, as discussed before, due to the change in philosophical thought that gained the upper hand with Descartes' mathematically based rationalism and his belief in the dualism of mind and body, inherited from the Judeo-Christian traditions and Greek philosophy. The previously prevalent ethnomedical belief in the four humours, which conceptualized the heart as also being the locus of rationality – we find a similar concept in Chinese (cf. Yu This volume) – was replaced by the idea that the head/brain/mind complex is where rational thinking takes place and that the heart is the place where our emotions are located. To Descartes, who – at least in the stereotypical view that we have of him – saw the mind as much more important than the body, the ideal thinking was logical thinking, without emotional intervention. We know today that this view is untenable, as there are complex connections between the body and the mind, neither of which can exist or operate at all without the other. Nonetheless, present-day English still conceptualizes it as such. Seitz (1999: 24) expresses this idea very clearly when he claims that "thinking is an embodied activity (...) the body structures thought as much as cognition shapes bodily experiences". But although we can agree with Seitz and with Damasio that the "body" pre-exists "thought" and is predominantly involved in all brain activities, this view has not yet entered the sphere of language.

It is interesting to observe that the conceptualization of *heart* has not in any meaningful way evolved from that of a mental faculty (emotions) to a mere organ (blood pump), whereas with *head*, the mental conceptualization of "mind" has fully found its way into the language. This may be due to the enormous symbolism, deeply rooted in Western tradition, which surrounds the heart and which is nearly absent for the head. Therefore, to this day, in Western-European folk views and languages the *head* and the *heart* are conceptualized as having quite distinct and specialized functions.

Acknowledgments

I wish to thank René Dirven (University Duisburg-Essen) for his lucid comments on earlier drafts of this paper as well as Jessica Cleary-Kemp (Monash University) for checking my English. All remaining flaws are, as usual, my own responsibility.

Notes

1. The notion of "bricolage" is an analogy used by Lévi-Strauss (1962), who saw "mythical thought ... [as] a kind of intellectual bricolage" (p. 17) and maintained that the bricoleur is "creating structures by means of events" (p. 22). Such bricolage can sometimes bring forth unforeseen results on the intellectual plane.
2. The current analysis is an elaboration of the former one, i.e. it has been changed to some degree. Still, four distinctive categories have been maintained, although they are now labelled differently and explained in more depth.
3. A precursor of the heart symbol can be traced back to times before the last Ice Age, when Cro-Magnon hunters used it in cave drawings, although its meaning is unclear. The heart symbol was used in a multitude of cultures: in Ancient Egypt, it symbolized the centre of life and morality; in Ancient Greece it was seen as the centre of the soul and as the source of heat within the body; the Romans were the first to understand that the heart is the most vital organ in sustaining life. The symbol became even more widespread from the Middle Ages onwards, with a peak in the 17th century, when in Western Europe the "Sacred Heart" came to stand for the soul and love of Jesus Christ and became widely used in Christian iconography. The heart's real anatomy was only discovered in the 15th century. Leonardo da Vinci was the first artist to draw a truly accurate sketch of the organ.

   The current heart shape itself is thought to be graphically related to the medieval ideograms for fire and flight. In a medical manuscript from Cambridge (1100), the heart was associated with the element of fire (cf. www.stadt-zuerich.ch/internet/swz/home/waidsicht.ParagraphContainerList.ParagraphContainer1.ParagraphList.0015.File.pdf/waidsicht_2_04.pdf). The heart appears as a symbol in all the major cultural spheres and is not only used by Christians, but also by Jews, Buddhists, Muslims, Hindus, Taoists and, from a historical viewpoint, by Celts, Aztecs and others (cf. www.symbols.com/encyclopedia/20/2018.html). A further theory states that the heart symbol was derived from the fig symbol, which in Buddhism stands for all-encompassing love, and developed later into a symbol for erotic love, and still later was desexualized by the Catholic church (cf. matriarchat.info/symbolik/herz.html).

   Finally, the colour red implies the cultural model of the heart as blood vessel, which also comes to the fore in expressions such as my heart is bleeding.
4. These examples speak against Wierzbicka's (1992: 47f.) claim that the heart is seen as "an organ of empathy and benevolence".
5. This anthropomorphic perspective of the heart is also to be found in expressions which do not refer to the topic of love, e.g. *to one's heart's content/delight, heart's ease*, etc.
6. Interestingly enough, the word *courage* is related to the Latin word *cor* 'heart' and was borrowed into English via Old French *corage*.

7. *Heart-burning* is an instantiation of the Western conceptual metaphor NEGATIVE EMOTIONS ARE FIRE and as such indicates that the heart, i.e. the person, is consumed with anger, envy or jealousy. This kind of *heart-burning* is different from the physiological condition of *heart-burning* which acts in a metonymic way as the passage of heat from the stomach to the throat (cf. Collins English Dictionary 1999: 713: "a burning sensation beneath the breastbone caused by irritation of the oesophagus, as from regurgitation of the contents of the stomach").
8. Also the expression *to be chicken-hearted* belongs to this sub-category. In our stereotypes of animals chickens are renowned for the fact that their small hearts (as a symbol of fear, i.e. the opposite of courage) often stop beating when they are frightened or under stress. Thus, a physiological feature of these animals is first used metonymically to refer to a certain behaviour pattern and then it is metaphorically mapped onto human beings who display a similar behaviour to that of fearful chickens: they shy away even from the slightest risk and prefer to remain passive and inert.
9. This is confirmed by many papers in the present volume, all of which yield evidence for a universal trend formulated by Johnson as the image schema of container (1987: 112–138, esp. 126).
10. James' famous and frequently quoted example is that if we see a bear, we run, and because we run, we feel fear – and not vice versa, as is commonly assumed (cf. en.wikipedia.org/wiki/William_James).
11. One of the most fervently discussed questions in philosophy is the potential difference between the concepts of reason and logic, cf. Plato, Aristotle, Augustine, Heine, Peirce, Hume, Einstein and others. Even Lakoff and Johnson (1999: 3f.) have commented on this topic: "Reason includes not only our capacity for logical inference, but also our ability to conduct inquiry, to solve problems, to criticize, to deliberate about how we should act, and to reach an understanding of ourselves, other people, and the world."
12. Translation: "The heart has its reasons which are unknown to reason", *Pensées*, ed. 1670, cf. www.croixsens.net/pascal/page6.php.
13. Dirven (2002) defines this type of metonymy as "a relationship of inclusion, which at first sight seems to be similar to the relationship of 'a whole for a part': intelligence as a property of the mind is metaphorically reified as a concrete object and metonymically situated in the brain, which itself is situated in the head".
14. Examples for this meaning cluster of salience are: *The show was 2 $ per head; head count; heads and tails* (referring to a coin); *head-hunter* (relating to cannibalism), *spearhead, masthead* etc.
15. This dominant philosophy of the 17th and 18th centuries was also adhered to and developed by John Locke (1632–1704) and somewhat later by Kant (1724–1804), for example.

16. Cf. for example www.nea.org/teachexperience/braik031120.html.
17. None of my sources yielded a clear-cut picture of the origin of the "balanced mind" metaphor, but most sources pointed towards religious, especially Buddhist origins. The concept also has a prominent place in Kabalarian philosophy. Furthermore, it is mentioned in the "Kybalion", a mystical, hermetic book, first published in 1908 and containing the essence of the teachings of Hermes Trismegistus, a syncretism of the Greek god Hermes and the Egyptian god Thoth, thus a duality in itself (cf. en.wikipedia.org/wiki/Kybalion).
18. Our cultural model of senses and mind is mainly based on Thomas Aquinas' (1225–1274) famous saying "*Nihil est in intellectu quod non sit prius in sensu*" (De verit., II, 3). These ideas were taken up by a great many philosophers, among them Descartes (1596–1650) as the founder of rationalism, Locke (1632–1704) as the main proponent of empiricism, and Kant (1724–1804), who propagated a synthesis of the two approaches.
19. This cultural model has been discussed for centuries, cf. for example Rabelais' (1493–1553) educational ideal of a "*tête bien remplie*" (a well-filled head), which can be seen as "the passive side of culture", and which was contradicted by Montaigne (1533–1592), who claimed that the ideal of education was a "*tête bien faite plutôt qu'une tête bien remplie*" (a well-made, i.e. well-functioning head rather than a well-filled head"), an idea which can be connected to the active, dynamic and creative side of culture (cf. www.viepedagogique.gouv.qc.ca/numeros/121/vp121_73.pdf).
20. One example for this would be the abundance of railroad terms that are still used in present-day English, although they refer to outdated technology: e.g. *letting off steam, backtrack, fast track, derailed, to get sidetracked, one-track mind, off track, sabotage* (cf. cprr. org/Museum/RR_Words_and_Phrases.html).
21. From a historical perspective, this sequence should of course be turned around.

References

Damasio, Antonio
1993 *Descartes' Error: Emotion, Reason and the Human Brain.* New York: Avon.

Dirven, René
2002 Metonymy and metaphor: Different mental strategies of conceptualization. In *Metaphor and Metonymy in Comparison and Contrast* René Dirven and Ralf Pörings (eds.), 75–111. Berlin: Mouton de Gruyter.

Dirven, René
1993 Metonymy and metaphor: Different mental strategies of conceptualization. *Leuvense Bijdragen* 82: 1–28.
Fauconnier, Gilles and Mark Turner
2003 Metaphor, metonymy and binding. In *Metaphor and Metonymy in Comparison and Contrast*, René Dirven and Ralf Pörings (eds.), 469–487. Berlin: Mouton de Gruyter.
Geeraerts, Dirk
2003 The interaction of metaphor and metonymy in composite expressions. In *Metaphor and Metonymy in Comparison and Contrast,* René Dirven and Ralf Pörings (eds.), 435–465. Berlin: Mouton de Gruyter.
Geeraerts, Dirk and Stefan Grondelaers
1993 Looking back at anger. Cultural traditions and metaphorical patterns. In *Language and the Construal of the World*, John R. Taylor and Robert E. MacLaury (eds.), 153–180. Berlin: Mouton de Gruyter.
Gibbs, Ray W. and Gerard J. Steen (eds.)
1999 *Metaphor in Cognitive Linguistics*. Amsterdam: Benjamins.
Goossens, Louis
1994 Metaphtonymy: The interaction of metaphor and metonymy in figurative expressions for linguistic action. In *By Word of Mouth: Metaphor, Metonymy and Linguistic Action in a Cognitive Perspective*, Louis Goossens et al., 175–204. Amsterdam: Benjamins.
Holland, Dorothy and Naomi Quinn (eds.)
1987 *Cultural Models in Language and Thought*. Cambridge: Cambridge University Press.
Ibarretxe-Antuñano, Iraide
this vol. Guts, heart and liver: the conceptualisation of internal organs in Basque.
Jäkel, Olaf
1995 The metaphorical conception of mind: 'Mental activity is manipulation'. In *Language and the Construal of the World*, John R. Taylor and Robert E. MacLaury (eds.), 197–229. Berlin: Mouton de Gruyter.
Johnson, Mark
1987 *The Body in the Mind. The Bodily Basis of Meaning, Imagination, and Reason*. Chicago: Chicago University Press.
Kövecses, Zoltán
1995 Anger: Its language, conceptualisation, and physiology. In *Language and the Construal of the World*, John R. Taylor and Robert E. MacLaury (eds.), 181–197. Berlin: Mouton de Gruyter.

Lakoff, George and Mark Johnson
1999 *Philosophy in the Flesh.* New York: Basic Books.
Lévi-Strauss, Claude
1962 *The Savage Mind.* Chicago: University of Chicago Press.
Niemeier, Susanne
2000 Straight from the heart – metonymic and metaphorical explorations. In *Metaphor and Metonymy at the Crossroads. A Cognitive Perspective*, Antonio Barcelona (ed.), 195–213. Berlin: Mouton de Gruyter.
Niemeier, Susanne and René Dirven (eds.)
1996 *The Language of Emotions.* Amsterdam: Benjamins.
Palmer, Gary
1996 *Toward a Theory of Cultural Linguistics.* Austin, TX: University of Texas Press.
Quinn, Naomi
1987 The cultural model of American marriage. In *Cultural Models in Language and Thought*, Dorothy Holland and Naomi Quinn (eds.), 173–192. Cambridge: Cambridge University Press.
1991 The cultural basis of metaphor. In *Beyond Tropes: The Theory in Anthropology*, James W. Fernandez (ed.), 56–93. Stanford, CA: Stanford University Press.
Quinn, Naomi and Dorothy Holland
1987 Culture and cognition. In *Cultural Models in Language and Thought*, Dorothy Holland and Naomi Quinn (eds.), 3–40. Cambridge: Cambridge University Press.
Seitz, Jay A.
2000 The bodily basis of thought. *New Ideas in Psychology: An International Journal of Innovative Theory in Psychology*, 18 (1): 23–40.
Wierzbicka, Anna
1992 Semantics, Culture, and Cognition: Universal Human Concepts in Culture-Specific Configuration. Oxford: Oxford University Press.
Yu, Ning
this vol. The Chinese heart as the central faculty of cognition.

Internet sources

cprr.org/Museum/RR_Words_and_Phrases.html (last accessed on April 8, 2008)
en.wikipedia.org/wiki/Kybalion (last accessed on April 8, 2008)
en.wikipedia.org/wiki/Reason (last accessed on April 8, 2008)
en.wikipedia.org/wiki/William_James (last accessed on April 8, 2008)
www.croixsens.net/pascal/page6.php (last accessed on April 8, 2008)
www.etymonline.com (last accessed on April 8, 2008)

matriarchat.info/symbolik/herz.html (last accessed on April 8, 2008)
www.nea.org/teachexperience/braik031120.html (last accessed on April 8, 2008)
www.symbols.com/encyclopedia/20/2018.html (last accessed on April 8, 2008)
www.stadt-zuerich.ch/internet/swz/home/waidsicht.ParagraphContainerList.ParagraphContainer1.ParagraphList.0015.File.pdf/waidsicht_2_04.pdf (last accessed on April 8, 2008)
www.viepedagogique.gouv.qc.ca/numeros/121/vp121_73.pdf (last accessed on April 8, 2008)

Dictionaries

Cassell Dictionary of Word and Phrase Origins (ed. by Nigel Rees), 1997, London: Cassell.
Collins English Dictionary (ed. by Diana Treffry et al.), 1999, Aylesbury: Harper Collins Publishers.
DCE, 1995, London: Langenscheidt-Longman.
Oxford Collocations Dictionary, 2003, Oxford: Oxford University Press.
Roget´s Thesaurus (ed. by Betty Kirkpatrick), 1987, London: Longman.
The Cambridge Thesaurus of American English (ed. by William D. Lutz), 1994, Cambridge: University of Cambridge Press.

The heart as a source of semiosis: The case of Dutch

Ad Foolen

Abstract

This contribution presents Dutch idiomatic expressions with the word *hart* 'heart' which are used for conceptualizing aspects of inner life, in particular emotions. A comparison is made with German and English data, which shows that there are differences in detail. The descriptive part of the paper is preceded by a discussion on the question in how far such conceptualizations are universal or culture specific, concluding that an adequate theoretical model should integrate both aspects. The last part of the paper considers the history of conceptualizations of inner life in western culture, in particular the question where inner life is to be located. In Antiquity, different views were defended, some polycentric, others more holistic, some more cardiocentric and others more cerebrocentric. After the Middle Ages, a polycentric view has become dominant, with the head and the heart as the two places where inner life is to be located.

Keywords: cardiocentrism, cerebrocentrism, dualism, Dutch, embodiment, emotion, heart, reason

1. Introduction

The human body plays an important role in the conceptualization of experience. People differentiate between experience coming from the outer world ("I see a bird"), the own body ("I have a stiff leg") and the inner world or inner life ("I think that it rains", "I feel happy"). "Inner life" is used here as a catch all term for thinking, feeling, intuition, attitudes, intentions, etc. Besides providing the basis for this tripartition of experience, the body functions as a source domain for figuratively conceptualizing the other two domains, the outer world and inner life.

The present paper deals with the question how parts of the body are used to conceptualize inner life. In recent years, this topic of research has shown to be a productive one, see, among many others, the special issue of *Pragmatics & Cognition* which was devoted to 'The body in description of emotion' (Enfield and Wierzbicka eds. 2002) and the special issue of *Cog-*

nitive Linguistics (14: 2/3) on how people talk about thinking (Palmer, Goddard and Lee eds. 2003).

Like the other papers in the present volume, my paper focuses on the heart as a source of semiosis for inner life, see section 3, where I will use examples from Dutch as an illustration and make a comparison with English and German. In section 4, diachronic changes will be discussed in relation to cultural changes in western culture. But first, in section 2, I will discuss some general theoretical issues regarding the relation between the body and the conceptualization of inner life.

2. Inner life and the body

In present day western cultures, the body is strongly present in public space and discourse: there is no taboo on talking about it, showing it, enjoying bodily experience, or on asking for plastic surgery. Cognitive studies also "discovered" the body, after two thousand years of philosophical thinking about the relation between language, thought and reality, in which the body did not play any serious role (cf. Lakoff and Johnson 1999). Whether there is a relation between the present cultural and scientific popularity of the body is a question that I happily leave to the sociology of science.

2.1. A short historiography of embodiment

The observation that the human body is an important source of semiosis fits in with the more general idea of embodied cognition, which is now considered as a fundamental dimension of cognitive linguistic thinking. 1987 was an important year in this respect, with the publications of Lakoff (1987) and Johnson (1987), cf. Lakoff (1987: xiv): "Thought is *embodied*, that is, the structures used to put together our conceptual systems grow out of bodily experience and make sense in terms of it", succinctly echoed in the first sentence in Lakoff and Johnson (1999: 3): "The mind is embodied". This view has been supported in many empirical studies, for example by Heine, Claudi and Hünnemeyer (1991, chapter 5), who analyzed the original meaning of the spatial prepositions 'on', 'under', 'in', 'front' and 'back' in 125 African languages. The human body turned out to be the most important source domain, which led the authors to propose the Body Part Model (p. 126). In English, expressions like 'the foot of the mountain', 'the head of the department' and 'the heart of the city' (to mention only a few), also illustrate the idea of the body as a source of semiosis. For more examples see Goschler (2005), who shows that machines, computers and nations are "anthropomorphized" in terms of the body.

Embodiment is programmatic in modern cognitive linguistic research, but similar ideas have been proposed earlier in the twentieth century within the context of philosophy and cultural studies. First and foremost, phenomenology must be mentioned here, in particular in the version that is proposed in the work of Merleau-Ponty ([1945] 1962). The Nijmegen dissertation by de Witte (1948) was inspired by phenomenology. De Witte studied body part terms and their secondary meanings in different languages, in particular in Dutch, Bare'e (a Malayo-Polynesian language, spoken by the Toraja on Sulawesi) and Marind (a language of Papua New Guinea). De Witte was further inspired by Ernst Cassirer, cf. the following quote (1948: 9, from Cassirer 1923: 266): "Der menschliche Körper und die Unterscheidung seiner einzelnen Gliedmassen dient als eine der ersten und notwendigen Grundlagen der sprachlichen 'Orientierung' überhaupt" [The human body and the distinction of its different parts serves as one of the first and necessary foundations of linguistic orientation]. Much of the linguistic data that de Witte analyzed was taken from anthropological studies, where the study of body part terms and their extended meanings has a long tradition too.

In his overview on the "corporeal turn" in twentieth century thinking, Ruthrof (2000: 14) points out that "for some time now, the body has been a dominant feature in feminist theorizing ... The most relevant work in this respect is that of Julia Kristeva, Luce Irigaray and Hélène Cixous". Ruthrof sees certain shortcomings in phenomenological and feminist thinking, whereas he is rather enthusiastic about the cognitive linguistic work of Lakoff, Johnson and Turner.

The "corporeal turn" in cognitive linguistics had as its most direct impetus the felt need for extralinguistic 'grounding' of (formal) linguistic categories. In a first step, language was linked to conceptualization processes. This step was, in fact, the starting point of the cognitive enterprise in the 1980s. Soon it was realized that cognition is embodied, see Johnson (1987). And in the 1990s, Damasio pointed out that cognition is strongly related to emotion, cf. Damasio (1994: xii): "I began writing this book to propose that reason may not be as pure as most of us think it is or wish it were, that emotions and feelings may not be intruders in the bastion of reason at all: they may be enmeshed in its networks, for worse *and* for better".

It is interesting to note that recent studies of interaction also show interest in embodiment, see Zlatev (2002) and Linell (2005: 19): "Talk-in-interaction involves bodily gestures: participants use their voices, faces, bodily orientations and movements to signal messages. In brief, spoken language and interaction are *embodied*". It seems that Gibbs (2006: 13) is right when he states that "the time is ripe for [a] reappraisal of the body's role in human cognition. Our bodies, and our felt experiences of our bodies in action, finally take center stage in the empirical study of perception, cognition, and language and in cognitive science's theoretical accounts of human behavior".

2.2. The universalistic approach

In people's perception of their inner life and their body, the latter shows symptoms that are felt as caused by the first. In particular emotions have physiological effects: "For example, heart rate and temperature go up when people feel angry, but go down when they feel sad" (Gibbs 2006: 250). This perception can lead people to locating emotions in certain body parts, cf. Gibbs (2006: 257): "Anger in Chinese individuals is frequently located and experienced in the chest and heart, depression is often experienced in terms of something pressing into the chest, or down on the head, and grief may be experienced in terms of a kind of back pain".

In its first stage, the theory of conceptual embodiment (cf. Lakoff 1987: 12) had a universalist outlook. In this approach, the human body is considered to be 'the same' in all cultures. The body provides a 'substrate' that conceptualizations depart from, cf. Palmer (1996: 292): "There are several sources of universal constraints on language, including the universal form and functions of the human body and brain (perhaps including physiological components of a few basic affective states) …".

Against this background, one would expect that people all over the world tend to locate emotions and other aspects of inner life in the same way. This is not the case, however. As Gibbs (2006: 252) says, "people are poor perceivers of their physiological changes". This thus leaves much room for variation in conceptualization processes. Even with respect to the question whether inner life is located at all, there is variation. North Americans often describe their emotion experience as "intrapsychic feelings", without specifically locating these emotions (Gibbs 2006: 257). And if location takes place, again we might expect a certain variation.

2.3. Cultural models

1987 was not only a pivot year for the general idea of embodied cognition (Lakoff 1987; Johnson 1987), it was also the year when the volume of Holland and Quinn (eds.) appeared, in which the notion of "cultural model" was introduced, which, since then, has played an important role in cognitive linguistic theorizing.

People in different cultures conceptualize the world differently, including their "inner world". In some cultures and languages certain emotions are "hypocognized" whereas other emotions are "hypercognized". In Tahiti, for example, anger plays a prominent role in social cognition, whereas sadness and grievance are suppressed. This corresponds with the number of words and expressions that are available in the language (cf. Röttger-Rössler 2002), and probably with the strength of experiential presence of certain parts of inner life for people in that culture (cf. also Bamberg 1997). People can also focus on different symptoms belonging to the same emotion, for example on 'heat' or 'inner pressure' as different symptoms that belong to anger. Kövesces (2005: 246) calls such variation "experiential focus".

The cultural relativity also shows in the bodily localization of inner life in different cultures. De Witte (1948) reports that in Marind (Papua New Guinea), the ear, *kambét*, is the organ that occurs in expressions for thinking, knowledge and remembering, whereas the emotions are associated with *békai*, which refers to heart and lungs, the organs involved in breathing. A similar dualistic model seems to be widespread in Australian languages, see Gaby (this volume), who shows that in Kuuk Thaayorre, intellect and emotion are located in the ear and the belly, respectively. In Japanese (see Ikegami, this volume), the dualistic model seems to correspond to the western one. In comparison to this 'dualistic' cultural model, we find a more holistic model in Tsou, an Austronesian language in the highlands of Southwest Taiwan, cf. Huang (2002: 172): "The body part most intimately associated with cognition (thinking or intending) or feeling in Tsou is *koyu* 'ear', the seat of Tsou emotion and mentation". Another holistic model was found with the Ifaluk, a Micronesian people, living on the island of Ifaluk, cf. D'Andrade (1987: 142): "The model used on Ifaluk … differs from the present western model in considering the mind to be located primarily in the gut, which includes the stomach and abdominal region. Thus, thoughts, feelings, desires, hunger, pain, and sexual sensations are all experienced in the gut".

If we focus on the heart, we see variation on a more fine grained level. Šileikaitė (2004) showed that German, Lithuanian and Georgian not only differ in the number of phraseological expressions involving 'heart' for feelings, but she also showed that in a quantitative ranking positive feelings (love, sympathy) rank first in German, but second and sixth in Lithuanian and Georgian, respectively. The latter languages have negative feelings like worrying and sadness at the top of the ranking.

The examples just given make clear that an adequate study of the heart in relation to inner life needs, besides a universalistic theory of embodied cognition, the framework of 'cultural linguistics', cf. Palmer (1996: 290): "The term [cultural linguistics] invokes the anthropological tradition that culture is the accumulated knowledge of a community or society, including its stock of cognitive models, schemas, scenarios, and other forms of conventional imagery". The present volume is a piece of proof for this view. A balanced view on the universalist and the 'social constructionist' approach and a synthesizing account can be found in Kövecses, Palmer and Dirven (2002).

2.4. Metaphor, metonymy and beyond

The title of the present paper is a variation on the phrase "the body as a locus of semiosis", that was used by Ikegami (1989: 102). After introducing this phrase, Ikegami continues: "[A]s the generator of semiosis, the body works in two ways: either it may project itself metaphorically onto other entities and processes or it may extend itself metonymically to other entities and processes." Ikegami (1989: 110) perceived an interesting east-west difference in preference for the two types of semiosis: "the former type [metaphor] is dominant in the Western and the latter [metonymy] in the Eastern tradition". If Ikegami is right, one might expect that this preference also shows in western based cognitive linguistic theorizing (linguists are not immune against their cultural background). And indeed, there is a tendency in the Western literature to analyze expressions that involve body parts in relation to aspects of inner life in metaphorical terms, cf. Kövecses (2000: 168): "... the Zulu metaphor that involves the heart: ANGER IS (UNDERSTOOD AS BEING) IN THE HEART". But we should realize that the relation between bodily organs and inner life is not the same as, for example, the relation between a building as a source domain for conceptualizing scientific theorizing ('the foundations of a theory', etc.). The inner organs have a direct experiential relation to inner life. This implies that when people talk about (properties of) the heart when they mean (properties of) emotion, they use a metonymic way of conceptualizing emotions: the heart as location-for-located, effect-for-cause, part-for-whole, etc. Panther and Thornburg (2004) call such relations "conceptual metonymies". Now that cognitive linguistic theorizing realizes that in processes of conceptualization metonymy is as important as metaphor, and that the two often go together, it is time to reconsider some analyses of emotion expressions. What has been analyzed in terms of metaphor might as well, or at least partly, be analyzed in terms of metonymy.

The link between emotions and the heart, mediated by the anterior cingulate cortex, can be considered as the neurophysiological basis for the figurative, metonymic use of the heart in the description of emotions. The specific aspects of the heart that are subsequently focused on, and which lead to 'images' of the heart can be the result of additional metaphor, conceptualizing the heart as a container, construed out of certain material, with a certain color, taking a certain position in the body, changing its place and size under the influence of certain forces, and serving as an object which can be manipulated by oneself and others. Underlying the metaphor in which the heart is seen as a container (for the emotions), lies, thus, a metonymic relationship between the emotions and the heart (cf. Kövecses and Radden 1998: 61).

It is very well possible that metonymy and metaphor are not sufficient as analytic instruments to catch the whole picture of how expressions involving body parts function to characterize inner life. Take a "simple" example like *a heart of gold*. Here, such connotations of 'gold' as 'precious' or 'pure' are "blended in". In other words, the conceptualization works with different input domains at the same time, cf. Fauconnier and Turner (2002).

Finally, cultural practices in which body parts play a role might be relevant for a full understanding of the linguistic expressions. This seems to hold in particular for the heart. In different cultures, the heart has become a symbol, an emblem, for certain feelings or aspects of the inner self (romantic love, religious feelings, the personality). The heart gets "a life of its own", it becomes a cultural model, separated from the body. The "cultural heart" manifests itself in a variety of cultural practices, as is illustrated by the following examples.

- Western culture: When Pope John Paul II died in the beginning of April 2005, the Polish people expressed the wish that 'their pope' would be buried in his

homeland. In addition, they said that if the Vatican would not allow this, they would like to have at least his heart be returned. Not a hand or an ear, but his heart. In the nineties, an American woman, Claire Sylvia, got a new heart, and afterwards she developed a taste for beer and chicken nuggets and motorcycles, as she reports in her (1997) book. Later, she found out that her new heart had belonged to a young man who indeed had these preferences. Similar stories have never been reported by patients who got a new kidney or lung. In autumn 2004 and spring 2005 the Teylers Museum in Haarlem, the Netherlands, and the Museum Dr. Guislain in Gent, Belgium, had an exhibition on representations of the heart in popular and religious culture. We would not expect a similar exposition on the liver or the bladder.
- Egyptian culture: In their mummification rituals, the heart was removed and placed in its own sacred urn to signify importance. The brain was unceremoniously removed and discarded as so much trash (http://goosetheantithesis.blogspot.com, 2006).
- Meso-American culture: Fischer (1999: 482) discusses the notion of heart *k'u'x* as a key concept in Mayan culture and cosmology. In Aztec religious rituals, the heart was cut out of living bodies and shown to the gods.
- African culture: The Yaka in Congo have an initiation ritual called "mbwoolu". Part of the ritual is spitting on the heart of a wooden cult statue. The heart is considered as the seat of acquired knowledge and of morality.

Such cultural practices go beyond the strict processes of conceptualization of body and inner life. The impact of such practices on discourse is something that needs further study if cognitive linguistics wants to integrate the cultural dimension in its full consequence, see Dobrovol'skij and Piirainen (2005), Ziemke, Zlatev and Frank (eds.) (2007) and in particular its twin volume Frank, Dirven, Ziemke, and Bernárdez (eds.) (2008).

3. The heart in West Germanic languages

Niemeier (2000, this volume) shows that in English the heart is a very productive source of semiosis for conceptualizing inner life. The expressions that are listed in 3.1 show that the same holds for Dutch.

3.1. The 'heart' in Dutch

The expressions presented below, are mainly taken from the *Van Dale* Dutch dictionary of 2005. They include words and phrases with varying categorical status: adjectival derivations like *hooghartig* 'high-heartedly', noun phrases like *een hart van goud*, 'heart of gold', and verb phrases like *van zijn hart geen moordkuil maken* 'not to turn one's heart into a murderer's den', *iemand een hart onder de riem steken*, 'to stick a heart under someone's belt', meaning 'to encourage someone'. Occasionally, I will add relevant observations on other languages.

In this inventory, I will not try to categorize the expressions into metaphorical or metonymic ones. In many cases, both figurative processes work together, and, in addition, blending from other domains is involved. For a serious analysis, each expression should be analyzed in detail, for which there is no room in the present context. Taking together the multiplicity of expressions, one gets the impression that we have to do with a 'heart' discourse, which is an integral part of the cultural practice of heart symbolism, in which the heart has become a "place holder" for all inner life that goes beyond the purely intellectual dimension: feeling, involvement, intuition, personal character.

Presence, absence and involvement of the heart.
Hartelijk 'heartily', *harteloos* 'heartless'; *heb het hart eens!*, lit. 'have the heart', 'don't you dare!'; *hij heeft er hart voor* 'he has his heart in the matter'. The adjective *hartelijk*, 'heartily' implies warm and friendly feelings, *harteloos*, 'heartless', implies cold and cruel behavior.

The heart has a default location but can be dislocated. The dislocation 'stands for' particular feelings or character traits.
Default position: *het hart op de juiste plaats hebben* 'to have got one's heart in the right place': to have a good character; *zijn hart vasthouden*, lit. 'to hold one's heart'. By holding one's heart, it doesn't move to a different position. The expression is used when someone is concerned about a process, worrying about how it will end.

Marked position: *Hooghartig* lit. 'high-hearted', 'arrogant', *laaghartig* lit. low-hearted, 'mean'. The negative meaning of Dutch *hooghartig* can be explained by the EXCESS image-schema, which, in this case, overrules the positive axiological charge of UP (cf. Peña Cervel 2003: 287; see also Hampe 2005 for a relativization of the positive and negative axiological charges of the image-schemas). That deviation of the normal position of the heart is negative (at least in Dutch), can also be seen in two expressions for being afraid: *Zijn hart bonsde in zijn keel* 'his heart was beating in his throat' (upward movement) and *Zijn hart zonk 'm in de schoenen* 'his heart sank into his shoes' (downward movement). *Zijn hart sprong op van vreugde* 'his heart jumped up with joy'. This expression, which can be found frequently in the Bible, involves positive feelings, which would be in accordance with the axiological UP schema.

A few observations regarding the position of the heart in other languages: Ewe (cf. de Witte 1948: 357): the heart is in my belly = I am happy, it does not bounce in my throat). English*:* *wear your heart on your sleeve* 'to show your true feelings openly'.

<u>The heart as an object that you can win, conquer, lose, steal, give away</u>.
Je hart aan iemand verliezen 'to lose your heart to someone'; *iemands hart stelen* 'to steal someone's heart'; *hartendief* lit. 'heart thief' 'sweetheart', *de harten van de mensen veroveren* 'to win one's way into people's hearts', *hartveroverend*, 'heart conquering, nice, attractive'. *Iemand een hart onder de riem steken*, lit. 'to stick a heart under someone's belt', 'to encourage someone'.

<u>The heart is a container for feelings.</u>
Pijn in het hart, 'pain in the heart'; *het hart is vol vreugde* 'the heart is full of joy'; *je neemt een grote plaats in in m'n hart* 'you take a big space in my heart', implying positive feelings towards someone; *waar het hart van vol is, loopt de mond van over* 'if the heart is full of something, the mouth runs over'*, je hart luchten/ je hart uitstorten* 'speak out one's heart'*, uit het oog, uit het hart* lit. 'out of sight, out of heart', 'if one does not see someone anymore, one stops thinking of that person'. One can talk about more specific positions in this container: *Diep in m'n hart* 'deep in my heart', *uit de grond van m'n hart* 'from the bottom of my heart'.

<u>The heart has a certain size and this size can change.</u>
Een groot hart hebben 'to have a big heart', *een grote mond, maar een klein hart* 'a big mouth but a small heart', *ruimhartig* 'big-hearted', *mijn hart kromp ineen van medelijden* 'my heart shrank with pity', *zijn hart zwelt van trots* 'his heart swells with pride'. In Dutch, a small heart stands for fear, lack of courage. In Moroccan Arabic it stands for not being able to tolerate a joke, for being easily annoyed.

<u>The heart is made of a certain material.</u>
Een gouden hart/ hart van goud, 'a heart of gold' (cf. the song by Neil Young); *een hart van steen*, 'a heart of stone' (cf. The Rolling Stones). In

Dutch 'houten hart' is not a viable expression, but in a recent pop song (by De Poema's), it does occur, probably under English influence (cf. Elvis Presley 'Wooden Heart'). However, the Middle Dutch dictionary (Verwijs and Verdam 1894, part III, p. 388) mentions *van herten hout, hout van herten*. In Moroccan Arabic, a heart of iron stands for 'mercilessness'. The *Van Dale* dictionary also mentions 'iron heart' for Dutch, but this is not really in use. The same holds for 'heart of ice'. *Van Dale* mentions for Flemish *peperkoeken heart*, 'ginger cake heart', which means 'empathic'.

<u>The heart has weight.</u>
Licht, 'light' and *zwaar* 'heavy' are frequent collocations of 'heart'. The adjectives indicate positive and joyful emotions and negative, depressing emotions respectively.

<u>There can be pressure on the outer surface of the heart.</u>
Op iemands hart trappen 'to trample on someone's feelings', *iets op zijn hart hebben* 'to have something on one's mind'; *het is een pak van m'n hart* 'the burden is off my heart, I am relieved', *het gaat me na aan het hart*, lit. 'it comes close to my heart', 'it really affects me', *de schrik sloeg me om het hart* lit. 'the fear wrapped itself around my heart', 'my heart missed a beat'. *Iemand iets op het hart drukken*, lit. 'to press something onto someone's heart', 'to tell someone that he should not forget (to do) something'.

<u>The heart has a temperature and can melt or burn.</u>
Hartverwarmend 'heartwarming', *een koud hart* 'a cold heart'. *Mijn hart staat in vuur en vlam*, 'my heart is in fire and flame'. *Mijn hart smelt als ik hem zie*, lit.'my heart melts when I see him', 'I feel strong sympathy for him'.

<u>The heart can be damaged.</u>
Een gebroken hart 'a broken heart', *hartverscheurend* 'heartbreaking', *een bloedend hart*, 'a bleeding heart'. *Een steek in het hart*, 'a stab in the heart'.

<u>The heart can have color (but not in Dutch).</u>
In Moroccan Arabic, someone with a white heart is good, generous, someone with a black heart has a bad character. In Endo (Joost Zwarts, p.c.), someone with a white heart is generous, someone with a black heart is annoyed. In Shakespeare's Macbeth, Lady Macbeth says, in Act II, Scene II:

My hands are of your color, but I am ashamed to have a heart so white, by which she means that she is innocent.

Besides color, there are many other properties ascribed to the heart that are relevant in relation to inner life in other languages: the heart can be clean or dirty (Modern Eastern Syriac: clean heart = above suspicion, see Wolk and Yalda, this volume), wet or dry (Modern Eastern Syriac he has a dry heart = he is cruel), sweet or bitter (In Endo, a bitter heart stands for a jealous person, Joost Zwarts, p.c.).

In a cultural analysis, it would be interesting to relate the language-specific properties ascribed to the heart and the relevance of these properties in the respective cultures. Can one say, for example, that the distinction between clean and dirty is more relevant in Middle East cultural models than in western ones?

In so far as the expressions describe emotions, a first step of a more general analysis could start along the following lines. In many of the listed expressions, a property with scalar value is involved: big-small, cold-warm, heavy-light, high-low, in rest-moving, etc. Even when such a property seems absent at first sight, it might be indirectly there. Take material like iron, wood and ginger cake. These materials can be put on a hard-soft scale. The material stands metonymically for the property. The values on the scale go from neutral to extreme, and they are associated with positive and negative evaluations. These scalar properties can be metaphorically related to general properties of emotions. Emotions are positive or negative and they are more or less intensive. In the linguistic expressions listed, the properties are not related to the emotions directly, but via the heart. The properties are ascribed to the heart, whereby the question whether real hearts can have such a property, seems rather insignificant. The heart stands for emotions, on a metonymical basis. But as the heart has taken the position of a cultural symbol, this metonymical basis does not play a significant role in the genesis of new expressions anymore, with the consequence that the distinction between metonymy and metaphor seems neutralized.

3.2. Germanic comparisons: Dutch, English and German

In Dutch-English and Dutch-German bilingual dictionaries, many of the expressions involving 'heart' have a literal equivalent in the other language. For example *een gebroken hart, ein gebrochenes Herz, a broken heart.* At the same time, there are several phraseological expressions in these languages where the dictionary prefers a translation that does not involve the heart. This can be illustrated with a few examples from the Dutch → English dictionary in the *Van Dale* series: *met een gerust hart*, lit. 'with a calm heart' → *with an easy mind/conscience*; *het is niet veel, maar het komt uit een goed hart*, lit. 'it is not much, but it comes out of a good heart' → *it's not much, but the intention is good*; *iemand een goed hart toedragen*, lit. 'to bear a good heart to someone' → *to be kindly disposed towards someone*.

The German Duden, Nr. 11 (2002) (*Redewendungen*), a phraseological lexicon, specifies, under the entry *Herz*, 'heart', several expressions that can not be translated literally with a *hart*-expression into Dutch. Some examples: *sich ein Herz fassen/nehmen*, lit. 'to take oneself a heart', 'to gather all your courage'; *sein Herz in die Hand nehmen*, lit. 'to take one's heart in his hand', 'to gather all one's courage'; *seinem Herzen einen Stoß geben*, lit. 'to give one's heart a kick', 'to overcome one's inner resistance and come to a decision'; *jemandem ans Herz gewachsen sein*, lit. 'to be grown to someone's heart', 'to have become dear to someone'; *jemandem wird das Herz schwer*, lit. 'someone's heart becomes heavy', 'to become sad'.

English has the expression *to eat one's heart out*, which has no counterpart in Dutch or German. Dutch has *met de hand op 't hart*, lit. 'with the hand on the heart', with a counterpart in German, *Hand aufs Herz*, but no literal counterpart in English, 'to swear to something faithfully'. Dutch has *de hand over het hart strijken*, lit. 'to brush one's hand over one's heart', 'to give in', which has no literal counterpart in English or German, etc.

The general picture that arises from the dictionary entries is, that on the overall level, Dutch, English and German have a similar conceptualization of the heart as the organ that is associated with feelings and the self, but that, at the same time, each language has conventionalized its own range of expressions. Even if these languages share the same cultural model, we can not predict the exact inventory of expressions in each language. The language-specific inventory of heart-expressions is the result of more or less idiosyncratic conventionalization processes, within the boundary of the shared western cultural model.

4. Diachrony and dualism

When we consult (etymological) dictionaries of Dutch (EWN, Verwijs and Verdam, WNT), we find that under the lemma for 'hart' several mediaeval expressions are listed that have since then disappeared. They typically pertain to the rational part of inner life, as is also noted by de Witte (1948: 437).

This diachronic shift, involving replacement of heart expressions by head expressions, which has also been noted for English (see Niemeier, this volume), is generally linked to the development in Western medicine and philosophy in the 16th and 17th century, when the 'real' function of the heart was discovered. The anatomist Andreas Vesalius (1514–1564) laid the foundations in his *De Humani Corporis Fabrica* ('The Fabric of the Human Body'), published in 1543, in which he gave precise descriptions with exquisite drawings of the internal organs. The next step was taken by William Harvey (1578–1657), who in 1628 in his *Exercitatio Anatomica de Motu Cordis et Sanguinis in Animalibus* ('An Anatomical Exercise on the Motion of the Heart and Blood in Animals') showed what the real function of the heart is, namely, to pump blood in the process of blood circulation.

A short excursion into the medical-philosophical literature of the Middle Ages and Antiquity shows a mix of the theory of the four humors (cf. Geeraerts and Grondelaers 1995) and a vivid discussion on the question where the powers of inner life are to be located. In the present context, we can leave out the theory of humors and concentrate on the discussion regarding the location. Two positions can be distinguished: a more monocentric one, adopting one centre for inner life, and a more pluricentric one. Within the monocentric view, there were two main competing positions: cerebrocentrism (or the encephalocentric theory, as Crivellato and Ribatti 2007 call it) and cardiocentrism, cf. Clarke and O'Malley (1968: 1):

> Although earlier Western civilizations such as those of the ancient Egyptians, the Mesopotamians, and the Hebrews had selected the heart as the central organ, the Greeks were divided on this matter. Thus Empedocles, Democritus, Aristotle, Diocles, Praxagoras, the Stoics, and the Epicureans favored the heart, whereas Alcmaeon, Pythagoras, one of the Hippocratic Writers, Plato, Herophilus, Erasistratus, Rufus, and Galen chose the brain.

Two points of comment are in order here. Firstly, the writings that are attributed to Hippocrates (460–377 B.C.) are a corpus of different authors (the 'Hippocratic Corpus', cf. Phillips 1973: 38ff.), which do not take unequivocally the same position. In one of the books from the Corpus, *The Sacred Disease*, a cerebrocentrist position is defended, cf. the passage quoted in Clarke and O'Malley (1968: 4):

> One ought to know that on the one hand pleasure, joy, laughter, and games, and on the other grief, sorrow, discontent, and dissatisfaction arise only from [the brain]. It is especially by it that we think, comprehend, see, and hear (...).

Secondly, the position of Plato (c. 429–347 B.C.) is a combination of pluricentrism and monocentrism, cf. Clarke and O'Malley (1968: 6):

> The soul was made up of three parts, and the most important part, the rational, was assigned to the head and presumably to the brain; the other parts were in the heart and in the upper abdomen.

It is surprising, that Aristotle (384–322 B.C.), in contrast to Hippocrates and Plato, defended a cardiocentric position, cf. Clarke and O'Malley (1968: 8):

> Aristotle argued that the heart was the chief organ of the body and the center for thought and appreciation of sensation. The brain, an important structure, was secondary to it and functioned only as a means of cooling the heart's heat.

Aristotle's position was criticized by Galen (A.D. 129–199) in a lively treatise (see Galen 1981), in which he defends the Hippocratic and Platonic position. This did not result, however, in a final triumph of cerebrocentrism, cf. once again Clarke and O'Malley (1968: 16–17):

> One cannot avoid admiring the way in which Galen attacked his opponent and (...) it is difficult to understand why the Aristotelian theory survived the onslaught. Its authority in the medieval period and later was probably owing to the greater availability of Aristotle's works.

After the Middle Ages, Western culture fragmented. Religious and scientific views diverged. In the context of the Catholic Church, the devotion of the Sacred Heart, which had started much earlier, became strong in the seventeenth century, cf. the Catholic Encyclopedia on the history of the 'Sacred Heart' cult:

> The image of the Heart of Jesus was everywhere in evidence, which fact was largely due to the Franciscan devotion to the Five Wounds [of the dying Jesus on the cross] and to the habit formed by the Jesuits of placing the image on their title-page of their books and the walls of their churches.

In the (Catholic) Counter-Reformation, it was stressed that religious belief is a feeling, which one should not try to understand. Placing faith in God in the heart, fitted, of course, the dualistic pressure from philosophy and science, it helps to make faith 'immune' for rational critique. At the same time, this religious cultural model might very well have had a conserving effect on expressions which relate the non-rational part of inner life to the heart.

Descartes (1596–1650) is the prototypical representative of the shift to dualism (cf. Damasio 1994). But we should not forget that it was especially the emotions that caused a problem for Descartes. He devoted a separate publication to the problem of the passions: *Les Passions de l'Ame* ('The Passions of the Soul', 1645–46), dedicated to Elizabeth, Princess of Bohemia (1618–1680), with whom he corresponded about the passions. Unavoidably, Descartes' solution gets a dualistic twist, cf. Gaukroger (1995: 401): "Passions are the perceptions which have their origin in the body". This means that emotions are considered to be in the body, but at the same time, they are perceived by the mind, duplicated intellectually, and the duplicate is called "passions". Blaise Pascal (1623–1662) had a discussion with Descartes on September 23 and 24, 1647. This discussion has not been written down, but Pascal might have protested against the idea that the emotions are perfectly transparent to the mind, cf. his well-known saying "that the heart has its reason that reason knows not of ... Do you love by reason?".

In the history of Dutch, some traces of the shift from cardiocentrism to dualism can be found: expressions in which intellectual functions like memory and wisdom were associated with the heart were gradually replaced by expressions in which the heart did not play any role. Today, memory is associated with the head: *uit het hoofd leren* 'to learn by heart', but the Middle Dutch dictionary mentions *bi der herten* 'by heart', cf. present-day French *par coeur* and English *to learn by heart*.

In the Middle Dutch dictionary, we find many expressions that conceptualize the heart as the seat of intellect and thought (p. 391): *van herten onbekent* 'with restricted intellect'; *Hi docht int hert sere* 'he thought strongly'. The WNT (volume 6: 19) states that in the Bible the heart is the seat of wisdom: *Alle vrouwen die wijs van herten waren* 'all women who were wise of hearts' (Exodus 35, 25); *Ick hebbe u een wijs, ende verstandigh herte gegeven* 'I have given you a wise and sensible heart' (1 Kings 3, 12); *Dat wij een wijs herte bekomen*, 'that we get a wise heart' (Psalms 90, 12). In the official Dutch Bible translation of 1637, called

Statenvertaling, the collocation of heart and wisdom is still frequent, but in later translations the heart is left out. For example, in 1 Kings 10,24, it is said that people went to Salomo to hear the wisdom that God had given to his heart, as the *Statenvertaling* says (*om zijn wijsheid te horen, die God in zijn hart gegeven had*), but in modern translations we read "…*om te luisteren naar de wijsheid waarmee God hem vervuld had*", 'to listen to the wisdom with which God had filled him'. A computer search in the NBV (*New Bible Translation*) of 2004, shows that there are 13 passages left in which heart and wisdom co-occur; remarkably, a search on collocations of 'heart' and feelings like 'joy' or 'sorrow' gives more results in the modern translation than in the older one.

The English expression *to learn by heart*, to memorize, shows that at least in the English language the emotional-intellectual dualism is not reflected in an absolute way in the language. Norman French *par coeur* might have played a role here: abstract domains got French labels ('spirit' instead of 'ghost') and this might have protected such expressions from change, even when they were 'calques'. Niemeier (this volume, section 2.3) comments on another expression, *to take it to heart* as follows: "If one takes something to heart, one thinks about it very seriously. This expression may well be reminiscent of the Old or Middle English concept of the heart as the seat of knowledge …". Although Dutch has given up the equivalent of '*to learn by heart*, there are, just like in English, some expressions left that we would associate, at least partly, with intellectual functions such as *Diep in mijn hart denk ik dat hij gelijk heeft* 'Deep in my heart I think that he is right' or *ter harte nemen* 'to take to heart'. Thoughts and knowledge that are strongly associated with the 'self' are still in the heart, scientific, encyclopedic and everyday knowledge are in the head. We can conclude that the brain is seen as an information processor, and the heart as the 'self', where, besides feelings, personal thoughts, convictions and intuition are located.

5. Conclusion

Expressions involving the heart for the conceptualization of inner life clearly fit the view that cognition is embodied. At the same time, we have seen that the heart has a cultural value, which makes it necessary to take the cultural context into consideration when analyzing such expressions.

The picture that arises for the Dutch data is radically different from the one that is sketched by Yu (this volume) for Chinese. In Chinese, the heart is the ruler, it supervises the emotional and intellectual functions that are located in different parts of the body. The Chinese picture is at the same time a distributed and holistic one. The Dutch data fit into the Western dualistic model, in which the heart is not the supervisor of the other organs. Heart and head may even play conflicting roles in inner life as e.g. in *mijn hart zegt ja maar mijn verstand zegt nee* 'my heart says yes but my head says no'.

The strong similarities between Dutch, German and English do not exclude micro-variation between them. A cultural model does not predict the lexical and phraseological details of a specific language. We can conclude that further research should thus continue to explore three levels of analysis: the language-specific details, the culture-specific generalities ('Western cultures', 'Chinese culture', etc.) and the universal constraints that play a role in the embodied conceptualization of inner life.

References

Bamberg, Michael
1997 Language, concepts and emotions. The role of language in the construction of emotions. *Language Sciences* 19: 309–340.

Cassirer, Ernst
1923 *Philosophie der symbolischen Formen. Erster Teil: Die Sprache.* Berlin: Bruno Cassirer. [Translation: *Philosophy of symbolic forms. Part One: Language.* New Haven: Yale University Press].

Clarke, E. and C. D. O'Malley
1968 *The Human Brain and Spinal Cord. A Historical Study Illustrated by Writings from Antiquity to the Twentieth Century.* Berkeley and Los Angeles: University of California Press.

Crivellato, Enrico and Domenico Ribatti
2007 Soul, mind, brain: Greek philosophy and the birth of neuroscience. *Brain Research Bulletin* 71, 327–336.

D'Andrade, Roy
1987 A folk model of the mind. In D. Holland and N. Quinn (eds.), 112–148.

Damasio, Antonio
1994 *Descartes' Error. Emotion, Reason, and the Human Brain.* New York: Putnam.

Dobrovol'skij, Dimitrij and Elisabeth Piirainen
2005 *Figurative Language: Cross-cultural and Cross-linguistic Perspectives.* Amsterdam, etc.: Elsevier.

Duden 11
2002 *Redewendungen*. Mannheim: Dudenverlag.

Enfield, Nick and Anna Wierzbicka
2002 Introduction. The body in description of emotion. *Pragmatics & Cognition* 10: 1–25.

EWN
2003 *Etymologisch Woordenboek van het Nederlands.* Amsterdam: Amsterdam University Press.

Fauconnier, Gilles and Mark Turner
2002 *The Way We Think. Conceptual Blending and the Mind's Hidden Complexities*. New York: Basic Books.

Fischer, Edward F.
1999 Cultural logic and Mayan identity. *Current Anthropology* 40: 473–499.

Frank, Roslyn M., René Dirven, Tom Ziemke, and Enrique Bernárdez (eds.)
2008 *Body, Language and Mind. Vol. 2: Cultural Situatedness.* Berlin: Mouton de Gruyter.

Gaby, Alice
this vol. Gut feelings: Locating intellect, emotion and life force in the Thaayorre body.

Galen
1981 *On the Doctrines of Hippocrates and Plato*. Edition, Translation and Commentary by Phillip de Lacy. Berlin: Akademie-Verlag.

Gaukroger, Stephen
1995 *Descartes. An Intellectual Biography.* Oxford: Clarendon Press.

Geeraerts, Dirk and Stefan Grondelaers
1995 Looking back at anger. Cultural traditions and metaphorical patterns. In *Language and the Construal of the World*, John Taylor and Robert E. MacLaury (eds.), 153–180. Berlin: Mouton de Gruyter.

Gibbs, Raymond W.
2006 *Embodiment and Cognitive Science.* Cambridge: Cambridge University Press.

Goschler, Juliana
2005 Embodiment and body metaphors. *www.metaphorik.de*.

Hampe, Beate
2005 When *down* is not bad, and *up* is not good enough: A usage-based assessment of the plus-minus parameter in image-schema theory. *Cognitive Linguistics* 16: 81–112.

Harvey, William
1628 *Exercitatio Anatomica de Motu Cordis et Sanguinis in Animalibus.* Frankfurt.

Heine, Bernd, Ulrike Claudi, and Frederike Hünnemeyer
1991 *Grammaticalization. A Conceptual Framework.* Chicago: The University of Chicago Press.
Holland, Dorothy and Naomi Quinn (eds.)
1987 *Cultural Models in Language and Thought.* Cambridge: Cambridge University Press.
Huang, Shuanfan
2002 Tsou is different: A cognitive perspective on language, emotion, and body. *Cognitive Linguistics* 13: 167–186.
Ikegami, Yoshihiko
1989 The East and the West from a viewpoint of cultural semiotics. In *Evolution of Culture*, W. Koch (ed.), 100–116. Bochum: Brockmeyer.
this vol. The heart: What it means to Japanese speakers.
Johnson, Mark
1987 *The Body in the Mind: The Bodily Basis of Meaning, Imagination, and Reason.* Chicago: University Press of Chicago.
Kövecses, Zoltán
2000 *Metaphor and Emotion. Language, Culture and Body in Human Feeling.* Cambridge: Cambridge UP.
2005 *Metaphor in Culture. Universality and Variation.* Cambridge: Cambridge University Press.
Kövecses, Zoltán, Gary Palmer, and René Dirven
2002 Language and emotion: The interplay of conceptualization with physiology and culture. In *Metaphor and Metonymy in Comparison and Contrast*, René Dirven and Ralf Pörings (eds.), 133–159. Berlin: Mouton de Gruyter.
Kövecses, Zoltán and Günter Radden
1998 Metonymy: Developing a cognitive linguistic view. *Cognitive Linguistics* 9 (1): 37–77.
Lakoff, George
1987 *Women, Fire, and Dangerous Things. What Categories Reveal about the Mind.* Chicago: The University of Chicago Press.
Lakoff, George, and Mark Johnson
1999 *Philosophy in the Flesh. The Embodied Mind and its Challenge to Western Thought.* New York: Basic books.
Linell, Per
2005 *The Written Language Bias in Linguistics. Its Nature, Origins and Transformations.* London: Routledge.

Merleau-Ponty, Maurice
1945 *Phénoménology de la perception*. Paris: Gallimard. Translation: *Phenomenology of perception*. London: Routledge and Kegan Paul, 1962.

Niemeier, Susanne
2000 Straight from the heart – metonymic and metaphorical explorations. In *Metaphor and Metonymy at the Crossroads*, Antonio Barcelona (ed.), 195–213. Berlin: Mouton de Gruyter.
this vol. To be in control: kind-hearted and cool-headed: The head-heart dichotomy in English.

Palmer, Gary
1996 *Toward a Theory of Cultural Linguistics*. Austin: The University of Texas Press.

Palmer, Gary, Cliff Goddard, and Penny Lee
2003 Special issue: Talking about thinking across languages. *Cognitive Linguistics* 14 (2/3).

Panther, Klaus-Uwe and Linda L. Thornburg
2004 The role of conceptual metonymy in meaning construction. *www.metaphorik.de.*

Peña Cervel, M. Sandra
2003 *Topology and Cognition: What Image-schemas Reveal about the Metaphorical Language of Emotions*. München: Lincom.

Phillips, E.D.
1973 *Greek Medicine*. London and Southampton: Thames and Hudson.

Röttger-Rössler, Birgitt
2002 Emotion und Kultur: Einige Grundfragen. *Zeitschrift für Ethnologie* 127: 147–162.

Ruthrof, Horst
2000 *The Body in Language*. London: Cassell.

Šileikaitė, Diana
2004 Der Somatismus 'Herz' als phraseologisches Weltbildelement im Deutschen, Litauischen und Georgischen. *Kalbotyra* 54 (3): 1–10.

Sylvia, Claire
1997 *A Change of Heart. The Extraordinary Story of a Man's Heart in a Woman's Body*. Warner Books.

Van Dale
2005 *Van Dale Groot Woordenboek van de Nederlandse Taal*. 14th edition, Ton den Boon and Dirk Geeraerts. Utrecht and Antwerpen: Van Dale Lexicografie.

Verwijs, E. and J. Verdam (eds.)
1894 *Middelnederlandsch Woordenboek*. Den Haag: Nijhoff.

Witte, J.J. de
1948 *De betekeniswereld van het lichaam. Taalpsychologische, taalvergelijkende studie* [The meaning world of the body. Language psychological and language comparative study]. PhD Diss., Nijmegen.

WNT
1912 *Woordenboek van de Nederlandsche Taal*, deel 6. Den Haag: Nijhoff and Sijthoff

Vesalius, Andreas
1543 *De Humani Corporis Fabrica*. [Translated by Daniel Garrison and Malcolm Hast, *On the Fabric of the Human Body*. Evanston, Il: Northwestern University, 2003].

Wolk, Daniel and Oraham Yalda
this vol. Cognitive cultural models of the heart (*Libba*) in Modern Eastern Syriac.

Yu, Ning
this vol. The Chinese heart as the central faculty of cognition.

Ziemke, Tom, Jordan Zlatev, and Roslyn Frank (eds.)
2007 *Body, Language and Mind, Vol. 1: Embodiment*. Berlin: Mouton de Gruyter.

Zlatev, Jordan
2002 Mimesis: The 'missing link' between signals and symbols in phylogeny and ontogeny? In *Mimesis, Sign and Language Evolution*, Anneli Pajunen (ed.), 93–122. Publications in General Linguistics 3, University of Turku, Finland.

http://goosetheantithesis.blogspot.com, March 14, 2006, "Christians gotta have heart"

The heart and cultural embodiment in Tunisian Arabic

Zouhair Maalej

Abstract

The Muslim concept of the heart as an instrument of understanding is evident in the teachings of the Koran (the Holy Book of Muslims). The heart in present-day T. Arabic, however, is almost exclusively the seat of emotions and cultural values, but hardly ever the instrument of thought and understanding, which are relegated to the *3aql* 'intellect'. Compared to other parts of the body, the heart is one of the most productive source domains for cultural conceptualizations in present-day T. Arabic. It is a CONTAINER for emotions, people, and objects that can enter it and leave it (IN-OUT schema). It is also capable of movement as in UP-DOWN (fear, panic, and worry), and WIDE-NARROW (worry, anxiety) image schemas. The heart provides metaphoric conceptualizations for love and sadness, and is also involved in conceptualizations of cultural values such as compassion, cruelty, courage, encouragement, generosity, hard work, kindness, laziness, meanness, (in)tolerance, conscience, remembrance, and so forth. The metonymic model of the heart's Idealized Cognitive Model describes the HEART FOR PERSON metonym, where the heart stands for the person. Compared to the conceptualizations of the heart in English, the heart in present-day T. Arabic is fairly restricted in scope. Indeed, while the heart in English describes a wider range of emotions, mental faculties, and cultural conceptualizations – equating this organ with the mind, thinking, understanding, etc. – in present-day T. Arabic the *qalb* 'heart' is largely dissociated from the mind, thinking, and understanding.

Keywords: heart, metaphor, metonymy, cultural model, cultural values, emotions, Tunisian Arabic.

1. Introduction

In the West, three models of the relation between body and mind can be isolated: (i) the humors model, which originated in the Greek culture and medicine and dominated Western thought up to the middle ages, with remnants still felt in some language use nowadays (Geeraerts and Grondelaars 1995); (ii) the body-mind split model, which was staunchly defended by Greek philosophers such as Plato and Aristotle and systematized by Descartes, and which dominated Western philosophy for many centuries (Lakoff and Johnson 1999); and (iii) embodied thought model, which called for a body-mind conflation (Johnson 1987; Lakoff and Johnson 1999). In particular, the embodied model recently questioned the other two models, especially the body-mind split model.

The embodiment thesis is the backbone on which cognitive linguistics rests, and, perhaps, also on which its future will greatly depend. Drawing on the foundational work by Lakoff and Johnson (1980), Johnson (1987) and Lakoff (1987), the literature on embodiment is growing bigger with the addition of specific volumes (Gibbs 2005; Ziemke et al., in press) and papers in journals (Sinha and Jensen de López 2000; Ziemke, 2003; Maalej 2004, 2007; Rohrer in press). Treatments of embodiment range from "embodiment as the physical substrate" (Rohrer in press) to structural coupling, historical embodiment, physical embodiment, 'organismoid' embodiment, organismic embodiment, and social embodiment (Ziemke 2003).

Lakoff and Johnson (1999: 102–4) investigated embodiment by distinguishing it into neural, phenomenological, and cognitive unconscious levels, which are all useful biological, experiential, and psycho-philosophical dimensions to embodiment. However, why do we have, for instance, to move in the direction of abstract forms of embodiment such as some of the types distinguished by Ziemke (2003) while rudimentary physiological and cultural forms of embodiment are out there awaiting treatment and recognition? It may seem a proliferation of terminology to add new concepts to the already existing ones (developed by Lakoff and Johnson 1999, Rohrer 1998 2001, Wilson 2002, Ziemke 2003). But cultural embodiment (Sinha and Jensen de López 2000; Maalej 2004), as another addition to the list, may turn out to be a promising alley into research on embodiment.

The conception of "cultural embodiment" offered here can be contrasted with the more physiological kind of embodiment. For instance, emotions are known in brain studies to be regulated by the hypothalamus brain structure, and controlled by the limbic system, which inhibits and excites them. Lakoff and Johnson (1999) call this kind of correlation between emotions and their correlative brain structures "neural embodiment." Cultural embodiment occurs when physiological embodiment is departed from in significant ways, thus constructing a culturally-situated form of embodiment. Cultural embodiment in the sense used here is when the neural synchronization between emotions, on the one hand, and the hypothalamus and the limbic system, on the other, is ignored in a given culture, and when the control of experience and its conceptualization is permeated by culture. An important type of cultural embodiment is when the physiology, function, and neural basis of body parts are imaginatively exploited and hijacked by culture. This conception of embodiment runs counter to "Lakoff & Johnson's companion formulation that the physiological body grounds cultural thought and never the reverse" (Maalej 2007: 91).

Bodily organs used in emotions and endearment are commonplace in present-day T. Arabic. For instance, in the conceptualization of anger, body parts such as stomach, brain, nerve, bone, and testicle represent different degrees of anger such as *fqa3-l-i ma3id-ti* (He burst open my stomach), *Haraq-l-i muxx-i* (He caused my brain to burn), *Haraq-l-i 3Saab-i* (He burnt my nerves), *digdig-l-i 3Daam-i* (He broke my bones into small bits), and *nfaxx-im-l-i* (He inflated my testicles) (Maalej 2004). On the other hand, organs such as the heart, liver, and eye are commonly used in endearment in present-day T. Arabic, offering conceptualizations such as *ya ruH qalb-i* (hey, soul of my heart), *ya kibd-i* (hey, my liver) or *ya mamm-u 3ayn-i* (hey, pupil of my eye). Such conceptualizations show how the dearness of children is motivated by and correlated in present-day T. Arabic with the centrality of organs such as the heart, liver, and eye to perception.

The eye, in particular, is involved in a panoply of cultural conceptualizations such as love (e.g. *flaan fi 3ainayya*: X is in my eyes: I love X so much), perseverance (e.g. *3mill l-milH fi 3ainay-h w bana daar*: He put salt into his eyes and built a house: He worked hard/persevered and managed to build a house), ambition (e.g. *3ain-ha kbiira/waas3a*: her eye is big/ spacious: She is very ambitious), over-ambition (e.g. *l-3ain ma yimlaa-ha kaan d-dud w traab*: only worms and sand can fill the eye: Her/His ambition is outrageous), desire of coveting (e.g. *3ain-u 3ali-ha:* His eye is on her/it: He covets her/it), with the object of the desire of coveting being a woman, a car, a house, a plot of land, etc. The eye can also be found in memory-related conceptualizations as in *xalli-ha bain 3ainay-k* (Keep it between your eyes: Don't forget it) or *jaat bain 3ainayya* (It came between my eyes: I remembered it). Remembrance is also conceptualized with the ear as in *xalli l-Hkaaya xirS fi wiDnik* (Keep that story as an earring: Never forget that story). However, except for eye and ear, bodily organs used as bearers of mental faculties are rare in present-day T. Arabic.

Each of the body parts used in the conceptualization of emotions, endearment, and cognitive faculties does not, however, constitute a cognitive/ cultural model as complex and elaborate as that of the heart in present-day T. Arabic. The heart as a source domain has been attested in the expression of emotions in many cultures, including Chinese (Yu, 1995, 1998, 2003), English (Niemeier 1997, 2000), Hungarian (Kövecses, 2000, 2002), and Persian (Sharifian this volume), and so forth. Building on Maalej (2004,

2007), the present chapter shows the heart to be one of the most productive source domains in present-day T. Arabic for the conceptualization of emotions and feelings, describing a complex Idealized Cognitive Model (Lakoff 1982, 1987). Its image schematic structure shows the heart both as a static container for emotions, people, and objects, and a dynamic entity moving in space. Its metaphoric model establishes the heart as a repository for emotions and cultural values while its metonymic model describes the HEART FOR PERSON metonym, where the heart stands for the person.

The data on which this chapter is based comes from conventionalized expressions that take the heart as a target domain in present-day T. Arabic. It is based on the author's intuition as a native speaker, and it is crossed-checked informally by other natives of T. Arabic to minimize the risk of interference from English expressions. There exist for T. Arabic no sizeable written documents such as dictionaries, nor even oral corpora recorded or transcribed. Since T. Arabic is only a spoken language variety of Arabic, few documents[1] are found in written form. The only book that discusses Tunisian culture and traditions in the author's sub-dialect is Zouari and Charfi (1998), which has very little to offer about the heart in present-day T. Arabic.

The structure of the present chapter is as follows. Section 2 discusses conceptions of the heart in the Arab-Islamic culture, isolating a religion-based model and a secular one, arguing that present-day T. Arabic has opted for the latter as the basis of an Idealized Cognitive Model. Section 3 addresses the conceptualizations of the heart in present-day T. Arabic in light of the humoral doctrine, which is shown to offer very little toward accounting for the Idealized Cognitive Model of the heart in present-day T. Arabic. Section 4 deals with the image schematic conceptualizations of the heart in present-day T. Arabic, especially the CONTAINER schema with regard to anger, fear, and love emotions and some cultural values. Section 5 is devoted to the metaphoric conceptualizations of the heart in emotions, cultural values, and mental faculties. Section 6 discusses the metonymy of the HEART FOR PERSON.

2. The heart in the Arab-Islamic culture

In the Koran, the holy book of Muslims, the heart is a bearer of the mental faculty of understanding, though it is not used to think with as, for instance, is the heart in Chinese (Yu this volume). Consider the following examples from the Koran:

(1) a. Many are the Ginns and men we have made for Hell: They have hearts wherewith they understand not, eyes wherewith they see not, and ears wherewith they hear not.[2]

b. Do they not travel through the land, so that their hearts (and minds) may thus learn wisdom.[3]

c. Verily in this is a Message for any that has a heart and understanding or who gives ear and earnestly witnesses (the truth).[4]

In (1a), it is clear through the use of the instrumental conjunction (wherewith) that the heart is for understanding. Emphasis is added between brackets "(and minds)" by the translator in (1b) to the translation of *qulubun* (hearts), though it does not have a counterpart in the source text. The intention of the translator is to make sure that through this added (and conjoined) parallel to "their hearts" the heart in translation is interpreted by the reader as "their minds." But the translator could have done away with the addition as understanding the heart as a mental faculty is very obvious in the cognitive verb "learn" predicated of hearts. The translator did the same in (1c), whereby he translated the word *qalb* (heart) by "heart and understanding." This extra care on the part of the translator to render "heart" as "heart and understanding" in the translated text is taken to ensure that "heart" triggers the right translation equivalent intended in the Koran.

The heart-as-instrument-of-understanding metaphor is brought out more clearly when the heart is said to be veiled/locked/covered, thus precluding it from performing its main function of understanding, signaled in many places in the Koran:

(2) a. But we have thrown veils on their hearts, so they understand it not.[5]

b. Do they not then earnestly seek to understand the Qur'an, or are their hearts locked up by them?[6]

c. And We put coverings over their hearts (and minds) lest they should understand the Qur'an, and deafness into their ears.[7]

There is, therefore, a relation of causality between lack of understanding and the heart being veiled, locked up, and covered, whereby the heart is clearly held responsible for the understanding task.

In (2c), the heart also shares the hearing function with the ear as an instrument of understanding. This is clearly understood if we add the necessary material to the elliptical construction used to avoid repetition: "… and [put] deafness into their ears" [lest they should understand the Koran]. Further confirmation of the involvement of audition and sight with the heart in the Koran, can be seen in the following contexts:

(3) a. We could punish them (too) for their sins, and seal up their hearts so that they could not hear.[8]

b. Truly it is not their eyes that are blind, but their hearts which are in their breasts.[9]

The heart in the Koran, therefore, does part of the mind's work as in Western culture. This can be inferred from the conceptual metonymy CONCRETE FOR ABSTRACT, which subsumes the specific conceptual metaphor, UNDERSTANDING IS SEEING AND HEARING, with the heart as the organ of understanding. This perceptive metaphor for understanding has been attested for Indo-European languages by Sweetser (1990: 28), who studied it under the generic Mind-as-Body metaphor.

Alongside the conceptual metaphor CONCRETE FOR ABSTRACT, where the heart qualifies as an organ of understanding, the Koran also makes use of the more universal heart-as-container metaphor. The container is signaled by the preposition *fi* (in) as in:

(4) "And God knows (all) that is in your hearts: And God is All-Knowing, Most Forbearing."[10]

Obviously, the logic of the container presupposes contents filling it, which can be positive or negative. Man's heart, thus, may be filled with faith (Suras XLIX, LVIII), hypocrisy (Sura IX), ignorance (Sura XLVIII), indignation (Sura IX), peacefulness (Suras XLVIII), perversity (Sura III), rancor (Suras LIX), regret (Sura III), sickness (Suras V, VIII, IX, X, XXIV, XLVII, LXXIV), terror (Suras III, VIII, XXXIII, LIV). In particular, the heart being filled with sickness is very frequent in the Koran, whereby sickness is metaphorically conceptualized. As will be seen later on in the chapter, Tunisian culture conceives of the heart as being filled with different contents than does the Koran.

The prevalent religious system in Islam (which it shares with Christianity and Judaism) can be captured in the soul-body dichotomy, where the soul is heavenly, immaterial, and eternal while the body is earthly, material, and ephemeral:

(5) Body Soul
Religious system

Believers should aspire to shun things bodily in favor of things heavenly. As seen in the Koran, the heart and the intellect overlap at the bodily member of the dichotomy since both heart and intellect have biological foundations. However, compared to the Western body-mind dichotomy, Islam not only does not dichotomize the heart and the intellect, it explicitly states that both heart and intellect are bodily-based. As stated by al-Jawziyya (1998: 81), "according to Islamic medicine, the primary connection of the spirit (rûh) with the body is by the heart from which the spirit arises and is sent forth into the parts of the body" (quoted in MacPhee 2003: 66):

(6)	*ruH*	*qalb*	*jasad*
	(soul)	(heart)	(body)

Thus, under this religious model the heart acts like a mediator between the body and the soul.

The Arab culture has certainly been impacted by the teachings of Islam, which explains the existence of the heart-as-intellect metaphor in olden times. Indeed, the heart among Arabs had come to stand for the intellect as attested in *The Tongue of Arabs* (Ibn ManDur 1994), one of the famous books that documents linguistic practices in Arabic history. Reporting from authorities, Ibn ManDur (1994 Book I: 687) mentioned that "it is permissible in Arabic to say: 'You have no heart' and 'Your heart is not with you', i.e., 'Your intellect/reason/mind is not with you,' and 'Where did your heart go?', i.e. 'Where did your intellect/reason/mind go?' Others said: He who has a heart, i.e., understanding and reflection."[11] There are certainly relics of ancient conceptions surviving in deeply religious communities, as documented by MacPhee (2003: 57) for the rural province of Errachidia, Morocco, where the heart is seen as "an organ and symbol that links spiritual, emotional, and physical experience." However, the religious model of body versus soul and the medical body-spirit-heart model have been abandoned in present-day T. Arabic, thus giving rise to a secular system where heart and intellect find themselves dichotomized. The heart and intellect have gradually been separated in the Arab culture probably under the influence of western philosophy, especially ancient Greek philosophy and rationalist-oriented Enlightenment in Europe, which was overwhelmingly translated and adopted into Islamic philosophy and Arab culture.

The religious conceptualization of the heart-as-understanding that existed in Arab-Muslim culture, therefore, seems to have been short-lived in many of the dialects of Arabic. As documented by MacPhee (2003: 65) for Morocco, "the growing influence of secularism and capitalist sentiments in the Sahara" are responsible for social disunity. Probably for similar political reasons, the religious component was lost in Tunisian culture, giving way to the secular heart-intellect dichotomy. The heart and the intellect in present-day T. Arabic have come apart so completely that those who think with their hearts are deemed irrational as in:

(7)	*flaan*	*yxammim*	*b-qalb-u*
	X	think-IMPERF	with heart his

X thinks with his heart.
'X is quite irrational.'

As a result, the heart has kept a very low profile in the conceptualization of mental faculties in present-day T. Arabic, but occupies a preponderant place in the conceptualization of emotions and cultural values, describing an Idealized Cognitive Model of the heart image schematically, metaphorically, or metonymically.

Curiously enough, thinking to the self is achieved with the help of the soul or between the self and the soul as in the following examples:

(8) a.	*xammimt*	*bain-i*	*w bain*	*ruH-i*
	[I] think-PERFECT	between myself	and between	soul my

'I thought between myself and my soul.'
I thought to myself.

b.	*xammimt*	*m3a*	*ruH-i*
	[I] think-PERFECT	with	soul my

'I thought with my soul.'
I thought to myself.

The "with" in (8b) expresses the notion of instrumentality rather than that of accompaniment, where the soul is conceptualized as carrying some of the burden of thinking with the self. Such a soul-as-understanding metaphor seems to be the only religious survivor in present-day T. Arabic.

3. The heart in Islamic medicine

Burnett (2004), a professor of the History of Islamic Influences in Europe, presents Avicenna's medicine as built on the four humors:

> The basis of this medicine was 'humoral pathology', i.e. an understanding that the human body consisted of four humours: blood, yellow bile (choler), phlegm, and black bile (melancholy), which were related to other `quaterneries': the elements air, fire, earth and water; the seasons spring, summer, autumn and winter; the ages of man, childhood, youth, middle age and old age; and the triplicities of the signs of the zodiac. Good health depended on the four humours being well-balanced in respect to each other (the Greek term is 'eukrasia' – a 'good mixing', 'temperament' or 'complexion').

Winsvold (2005) argues that "the success of the Arabian medicine can be credited for their endorsement of Greek and Roman medicine at a time when this knowledge was lost to the West." Winsvold (2005) adds that medicine was practiced according to the:

> Six Necessities': *Air* (including climate, soil etc.), *Food, Bodily rest and movement, Sleep, Emotional rest, Excretion and retention.* Any of these were believed to influence the temperament. This meant that people living in one climate would have a different temperament than others. Food or even substances surrounding a man, such as wood, brick or metal, would influence his health, and thus both be potential explanations for disease and form the basis of a cure.

However, the question that must be raised is whether the theory of the four humors can be traced back in cultural conceptualizations of the heart in T. Arabic.

As presented by Geeraerts and Grondelaers (1995: 156) and Yu (this volume), the humoral tradition includes physiological, psychological, and medical aspects. Physiologically, the four humors (yellow bile, black bile, phlegm, and blood) regulate the body's well-being. Psychologically, they generate corresponding temperaments (angry, fearful, phlegmatic, and sanguine), depending on the dominant humor available to the body. Medically, the humoral theory identifies diseases and their symptoms and prescribes therapies to regulate the body's imbalance. Arguing against the conceptualization of anger as the HEAT OF A FLUID IN A CONTAINER, Geeraerts and Grondelaers (1995: 174) claim that anger "has undergone the influence of the humoral doctrine, but that the original set of humoral expressions has been subjected to a process of reinterpretation and obsolescence," whereby the humoral expressions have been, under scientific influences, corrected as physiological ones.

To address the heart in present-day T. Arabic, the system of humoral correspondences is reproduced from Geeraerts and Grondelaers (1995: 158) in the following table:

Table 1. The system of humoral correspondences

	Phlegm	**Black bile**	**Yellow bile**	**Blood**
Characteristic	Cold and moist	Cold and dry	Warm and dry	Warm and moist
Element	Water	Earth	Fire	Air
Temperament	Phlegmatic	Melancholic	Choleric	Sanguine
Organ	Brain/bladder	Spleen	Liver/stomach	Heart
Color	White	Black	Yellow	Red
Taste	Salty	Sour	Bitter	Sweet
Season	Winter	Autumn	Summer	Spring
Wind	North	West	South	East
Planet	Moon	Saturn	Mars	Jupiter
Animal	Turtle	Sparrow	Lion	Goat

Originating in translations of western concepts, names for the four humors do exist in Modern Standard Arabic as *?an-nafsu l-balRamiyyatu* (phlegmatic self), *?an-nafsu is-Safraawiyyatu* (choleric self), *?an-nafsu s-sawdaawiyyatu* (melancholic self), and *?an-nafsu d-damawiyyatu* (sanguine self). However, unlike English and French, present-day T. Arabic does not seem to include the nominal concepts that stand for the humors (except, of course, blood), nor any of the adjectives derived from the four humors that designate the temperaments in the theory. Practically, the humoral theory no longer has an existence in the mind of speakers of T. Arabic.

Present-day T. Arabic does include expressions such as *damm-u sxun* (His blood is hot/He is hot-blooded), to mean that he is easily irritable, and *damm-u baarid* (His blood is cold: He is not irritable). Such expressions might be argued to evoke the humoral doctrine. However, a close scrutiny reveals a skewing of correspondences. For instance, in *damm-u sxun*, although blood is hot not just warm, the person having that kind of disposition is not sanguine but choleric in present-day T. Arabic. In *damm-u baarid*, there is a problem fitting the expression either under phlegm or black bile since both of them are cold (and it does not fit under blood as blood does not associate with coldness). Granting that it can fit under either one, there is still a problem of correspondence between blood, on the one hand, and phlegm or black bile, on the other. Blood associates with the heart while phlegm as coldness associates with the brain/bladder and black bile with spleen. If one believes, as do Geeraerts and Grondelaers (1995: 163), that blood here stands for yellow bile and phlegm because blood carries all humors, then it can be argued that these expressions have come into being under the influence of the humoral theory, and have subsequently been reinterpreted physiologically later on in history. The same problem persists with expressions using the heart as a landmark as in the following expressions describing various degrees of heartburn: *3and-i n-naar 3ala qalb-i* (I have fire on my heart), *qalb-i yaHraq ki n-naar* (My heart is burning like fire), *3and-i sihraaja naar 3ala qalb-i* (I have a flame of fire on my heart). Fire (and its derivatives burning and flames) associates with liver and stomach in the theory while in these expressions in present-day T. Arabic it is in/on to the heart.

There also exists in present-day T. Arabic the expression *qalb-u baarid* (His heart is cold: He is a lazy person) discussed earlier in connection with laziness/idleness. The laziness-as-coldness metaphor does not evoke phlegmatic placidity, which is often contrasted to the choleric temper associated with yellow bile. Rather, coldness here is contrasted to dynamism and activity. This may be attributed to the metaphor INTENSITY IS HEAT (LACK OF INTENSITY IS COLDNESS), as discussed in Kövecses (2005: 27). Coldness acts as a source domain for many experiential domains in present-day T. Arabic such as in *riiq baarid* (cold saliva: silly talk), *janab-ha baarid minn-u* (Her side is cold about him: She does not trust him), *ydai-h baarda* (His hands are cold: He does not have the courage to do anything), *wja33 baarid* (cold labor: intermittent pain during labor), *3iiša baarda* (cold life: spice-free food), *wTaa baarda* (cold land: fairly infertile land), *s-suq baarda* (cold market: bear market), *liqma baarda* (cold mouthful: food that one does not work or tire for), *s-tiqbaal baarid* (cold welcome), and so forth. (Zouari and Charfi 1998: 33–4). Presumably, such expressions do not suggest phlegm or any synonym of it, especially those expressions that talk about food, life, labor, market, and land. The lack of correlation between coldness in these conceptualizations and phlegm suggests cultural mappings between coldness and expected qualities of food, welcome, life, labor, market, and land that are experienced and judged to be below a certain level of normalcy in the culture. As a result, the humoral theory does not seem to motivate or predict emotions and cultural values addressed in the current chapter.

Such a cultural explanation of these well-established correlations between the experiential domain of coldness and various experiences points to cultural specificities. In other words, if it were the case that the four humors governed conceptualizations, all cultures would have had the very same conceptualizations of experience, and we know this is true of only a restricted number of experiences across cultures. For instance, English-speaking people conceptualize the present-day T. Arabic "cold talk" as stupid or silly talk, "cold side" as distrust, "cold land" as fairly infertile land, "cold labor" as having little labor pain, "cold market" as bear market, and so on. These conceptualizations are evidence that our cognition is fundamentally "built for encultured variation" (Levinson 1996: 177). However, this does not mean that cultures do not share some conceptualizations as in *s-tiqbaal baarid* and *s-tiqbaal Haarr* in present-day T. Arabic and "cool welcome" and "warm welcome" in English.

The following sections present cultural conceptualizations of the heart in present-day T. Arabic following Lakoff's (1982, 1987) theory of categorization known as Idealized Cognitive Model.

4. The image-schematic model of the heart in T. Arabic

Lakoff and Johnson (1980) and Lakoff and Turner (1989) isolated four semantically autonomous domains that serve as metaphoric grounding for semantically non-autonomous concepts. Such cognitive domains, which seem to be pervasive in many cultures, include spaces, substances, objects, and containers. In many (if not all) cultures, the heart is not only associated with emotions but receives them as its contents, which makes it a CONTAINER for them. The preposition *fi* (in) in present-day T. Arabic shows the heart's containment dimension in:

(9) a. *qult* *fi qalb-i*
[I] say-PERFECT in heart my
'I said in my heart.'
I said to myself.

b. *yxalli/yxabbi/yaxzin* *fi qalb-u*
[he] keep/hide/store-IMPERF in heart his
'He keeps/hides/stores everything in his heart.'
He is too secretive.

c. *HaaTiT/3aamil* *fi qalb-u*
[he] keep/do-IMPERF in heart his
'He keeps things in his heart.'
He is anxious/worried.

In (9a), things that are said to the self are kept in the heart as a secret chamber. In (9b-c), the heart is a storehouse or hiding place, tightly preserving positive and negative emotions and information.

The logic of containers predicts that the heart can be full or empty of some contents as in:

(10) a. *qalb-u* *m3ibbi* *3aliy-ya*
heart his full on me
'His heart is full about me.'
He strongly picks at me.

b. *qalb-u* *ma fi-š* *raHma*
heart his no in not compassion/mercy
'There is no compassion in his heart.'
He is/is not compassionate/merciful.

In (10a), the heart is full of bias against the speaker while in (10b) the heart is empty of mercy.

In Maalej (2004: 59–60), it was shown that in present-day T. Arabic the heart, alongside the body, is a container for anger. Because anger is considered a liquid filling the body or the heart, anger as a liquid in a container may slosh, fill up, or may explode as a consequence of incapacity to take it in, etc, as in:

(11) a. *qalb-i* *TafTaf*
heart my slosh-PERFECT
'My heart was sloshing with anger.'

b. *qalb-I* *t3abba minn-u*
my heart was full from him
'I had had enough of him.'

c. *qalb-I* *taršaq*
heart my explode-PERFECT
'My heart exploded.'

The logic of containers assumes that if the heart can be filled, it can also be relieved of its contents by emptying one's anger into someone else as in:

(12) *farraRt* *fii-h* *qalb-i*
[I] empty-PERFECT in him heart my
'I emptied my heart's contents into him.'

This expression is not equivalent to the English "to pour one's heart out;" its meaning has to do with retribution, that is someone angry finds an opportunity to take revenge on someone else. Emptying one's heart in this way seems to have a relieving/cleaning effect on the heart's contents.

Congruent with the CONTAINER image schema, the heart can function according to an IN-OUT schema, whereby beloved people and desired objects enter it and leave it as in:

(13) a. *daxl-u* *fi qalb-i* *kiif niktit l-3asal*
[they] enter-PERFECT in heart my like drop the honey
'They entered in my heart like a drop of honey does.'
I like them so much.

b. *xarj-u* *min qalb-i*
[they] leave-PERFECT from heart my
'They left my heart.'
I was sad to lose/give/sell them.

c. *sallit-l-i* *qalb-i*
[she] pull-PERFECT to me heart my
'She pulled out my heart.'
Her departure made me sad indeed.

Together with the preposition *fi*, the deictic verbs *txall* (enter) in (13a) and *xrajj* (leave) in (13b) provide an excellent illustration of the heart as a bounded space having an entrance and an exit, which people and objects can enter and leave. However, as Niemeier (1997: 94) rightly suggests, "the way from feelings to heart is no one-way street but works both ways, as feelings may go in and go out of one's heart." Even though this also takes place in T. Arabic, the way in is happy and difficult and the way out is painful and easier. It takes a lot of time and energy to get in someone's heart, but it takes very little to leave it or be washed out of it.

In (13c), however, separation between two people profiles the heart as an entity that can leave its place by being pulled out of the body. For the target domain of separation, T. Arabic metaphorically conceptualizes the heart as a MOVABLE OBJECT. Present-day T. Arabic culturally imagines the physical separation of two people as a separation between the body and the heart, in which the latter is pulled out from the former by exerting force-dynamics. In English, such a state occasions the breaking of the heart; in present-day T. Arabic the heart does not seem to be breakable as love is in the mind or intellect. In other words, love affects the mind not the heart in present-day T. Arabic. As such, love, like anger, tends to interfere with accurate perception as in *mahbul 3ali-ha* ([He] is mad/crazy about her), *xarjit-tu min 3aql-u* ([She] made him leave his intellect: She made him crazy about her), and so forth.

Beside the container or the three-dimensional in-out schemas, the heart in present-day T. Arabic admits conceptualizations using the one-dimensional UP-DOWN verticality schema as in:

(14) a. *habbaT-l-i* *id-dimmar* *fi qalb-i*
[he] lower-PERFECT to me the destruction in heart my
'He lowered destruction in my heart.'
He managed to dishearten me.

b. *qalb-i* *TaaH*
heart my fall-PERFECT
'My heart fell.'
Fear took hold of me.

c. *hazz-l-i* *qalb-i*
[he] lift-PERFECT to me heart my
'He lifted my heart for me.'
He encouraged me/He gave me a lot of encouragement.

d. *qalb-i xrajj* *min blast-u*
heart my go out-PERFECT from place its
'My heart left its usual place.'
Panic took hold of me.

Sadness in (14a) is lowered into the bottom of the heart as is clear from the verb *habbaT* (to lower). In (14b), the heart is conceptualized as dropping out of fear. Someone who needs encouragement needs their heart to be lifted as in (14c). It, thus, seems that these downward movements of the heart are motivated by the conceptual metaphor DOWN IS BAD. In (14d), the place for the heart in the body is spatially ambiguous. As Niemeier (2000: 200) notes for English, the heart does not settle in one place or one position, but is imagined as a "MOVABLE OBJECT."

As noted above, the heart in present-day T. Arabic is filled with emotions different than those mentioned in the Koran such as faith, hypocrisy, ignorance, indignation, peacefulness, perversity, rancor, regret, sickness, terror. Some of the emotions addressed image-schematically in present-day T. Arabic include compassion, fear, panic, sorrow, and others. The image-schema of Idealized Cognitive Model of the heart in present-day T. Arabic includes three sub-models: CONTAINER, IN-OUT, and UP-DOWN. The container presupposes contents, which are mostly emotions such as anger, mercy, spite, and so on. The IN-OUT schema suggests that ENTERING THE HEART IS GOOD (or IN IS GOOD) but LEAVING THE HEART IS BAD (or OUT IS BAD) while the UP-DOWN schema suggests UP IS GOOD and DOWN IS BAD. In most cases, cultural conceptualizations of the heart as motivated by embodied schemas describe it as a MOVING AND MOVABLE OBJECT.

5. The metaphoric model of the heart in T. Arabic

In Western culture, the body-mind dichotomy has been understood as a hindrance to making sense of the self, meaning, reason, and communication. In order to do so, it was urged to bring body and mind together (Johnson 1987: xxxvi; Varela et al 1991: 28). However, in the Arab-Islamic culture, as has been shown earlier on in this chapter, two dichotomies seem to be active together. In religious matters, a Muslim is an earthly body contrasted to an eternal soul. In earthly, experiential matters, the self is seen as divided between a *3aql* (an intellect) and a *qalb* (a heart). The two are thought to be hard to reconcile, because the former is rational while the latter is sentimental.

This section will look into the following target domains: (i) the metaphoric conceptualizations of the heart in emotions, and (ii) the conceptualizations of the heart in cultural values. In almost all the metaphoric conceptualizations of the heart in present-day T. Arabic, the heart could be seen as HEART FOR THE PERSON. And based on this metonymy, specific metaphoric conceptualizations will be spelled out below.

5.1. The heart in the target domain of emotions

A lot of different things can metaphorically happen to the heart in present-day T. Arabic depending on what emotion is involved. Such happenings may range from leaving one's heart behind, feeling pain in it, witnessing various mechanical transformations to it, to having it cooked as illustrated in the following examples:

(15) a. *xallayt* *qalb-i* *3and-u/-ha*
[I] leave-PERFECT heart my with him/her
'I left my heart with him/her.'
It was so painful for me to leave him behind.

b. *Haal-u* *ywijja3* *l-qalb*
state his cause pain-IMPERF the heart
'His condition is causing pain to my heart.'
I feel pain in my heart for him.

c. *qaTTa3-l-i* *qalb-i*
[he] cut-PERFECT to me heart my
'He cut up my heart.'

d. *Haal-u* *yfitfit* *l-qalb.*
state his cause to fall into pieces-IMPERF the heart.
'His condition is causing the heart to fall into pieces.'

e. *qalb-i* *walla* *Hinna*
heart my become-PERFECT henna
'My heart turned henna.'

f.	*Haal-u yDawwib*	*l-qalb*	
	state his melt-IMPERFECT	the heart	
	'His condition makes the heart melt.'		

g.	*qalb-i*	*miswi (w miqli)*	*3alii-ha*
	heart my	roast and fry-PERFECT-PASSIVE	on her
	'My heart is roasted and fried about her.'		
	I am very sad about her condition.		

Caring for others in present-day T. Arabic is imagined to occasion a split between the heart and the body or the self as in (15a), whose conceptual metaphor is CARING FOR OTHERS IS LEAVING ONE'S HEART BEHIND FOR THEIR SAKE. Based on the metonymy HEART FOR THE PERSON, this conceptualization creates a Divided-Person metaphor (Lakoff 1996: 103).

In present-day T. Arabic, the mildest form of compassion for others is represented by the conceptual metaphor, HAVING PITY IS FEELING PAIN IN ONE'S HEART, whose linguistic counterpart is in (15b). Having pity for someone is not something that one does for them but it is something that happens to one's heart as a result of its being affected by people's states or situations. Various other ways to express one's pity/compassion towards people may include the conceptualizations in (15c-f) above. No English equivalents are provided for them because these conceptualizations are highly culturally-constructed. It takes a lot of imagination to conceptualize pity in the way it is done in present-day T. Arabic. In (15c), the kind of pain felt as a result of pity is the result of a knife cutting the heart into pieces, which presupposes the letting of blood that is at the origin of pain. In (15d), the pain comes as a result of the heart turning into small crumbs, which suggests a painful disintegration of the heart. In (15e), the heart undergoes a further transformation turning it into henna powder – a fine form of grinding, suggesting extreme pain and concern. In (15f), the pain to the heart caused by pity simply makes it melt. Clearly, these expressions reflect the metaphor, PITY CAUSES CHEMICAL/MECHANICAL TRANSFORMATIONS TO THE HEART, where the degree of pain felt by the speaker is proportional to the degree of sufferance the affected person endures.

Another transformation of the heart occurs to it in sadness/sorrow in (15g), whereby it is profiled as undergoing a mutation that usually occurs to food while it is cooking. Interestingly, roasting and frying presuppose heat of fire in the heart, which is a way of imagining one's heart as cooking by roasting and/or frying. Such roasting and frying of the heart may evoke the conceptual metaphor, BEING SAD IS HAVING ONE'S HEART COOKED, not in the sense that we eat it as food, but in the sense that what happens to it in sorrow is similar to what happens to food as it is being prepared. As a structuring domain, food has been attested in many areas of experience such as sex and lust (Emanatian 1995), and women in many cultures (Maalej 2001), and so forth.

While the image-schematic model of the heart in present-day T. Arabic involves containment and movement in space, the metaphoric model with emotions mainly describes transformations, which occasions changes of state to the heart ranging from being cooked by roasting or frying, being cut into pieces, or ground into powder, to completely melting away.

5.2. The heart in the target domain of cultural values

Apart from featuring as an important source domain in the conceptualization of emotions in present-day T. Arabic, the heart is at the center of the conceptualization of many cultural values, which manipulate the heart according to various pairs of antonyms such as soft/tough, strong/weak, small/big, black/white, cold/warm, live/dead, and so forth. Niemeier (2000: 205) captures the transformations that occur to the heart in English in the conceptual metaphor, THE HEART IS A MANIPULABLE OBJECT.

Before dealing with these pairs, it is useful to address the polysemy of the heart in present-day T. Arabic as in the following metaphor:

(16) *ma 3and-him-š qalb*
no with them not heart
'They have no heart.'
They are heartless.

Contextualized, heartlessness represents either laziness or indifference. Laziness is invoked in the following proverbs:

(17) a.	*l-qalb*	*ma ySiir*	*kaan l-ir-rHa*
	the heart	no become-IMPERF	only to the grinder
	'The heart can only be made to the hand-operated grinder.'		

b. *illi ma fii-h* *qalb ymut* *smiin m3aš3aš*
who no in him heart die-IMPERF fat fatten-PASSIVE
'He who does not have a heart will die so fat.'

The moral of the proverb in (17a) is that love for work can only be self-generated. In (17b), an idle person is teased about his fatness, which is a metonymy for lack of physical exercise.

However, to profile other moral/cultural values, the heart is manipulated in various ways. Compassion, for instance, is conceptualized as softness of the heart while cruelty is conceptualized as hardness or toughness:

(18) a. *qalb-u rqiiq*
heart his affectionate
'His heart is affectionate.'
He is kind.

b. *qalb-u SHiiH*
heart his hard
'His heart is hard.'
He is unfeeling/cruel.

Compassion and affection are, thus, correlated with softness of the heart in present-day T. Arabic, and cruelty correlates with different degrees of toughness. In (18b), the degree of toughness is mild, giving rise to a moderate kind of cruelty. But cruelty can be conceptualized as having *qalb kaasaH* (a tough heart) or *qalb Hjarr* (a heart of stone), which are shared by English's *hardness of heart*, *heart of marble*, *heart of iron*, *heart of stone*, and so forth. (Niemeier 2000: 201) and Hungarian's hardness or toughness of the heart (Kövecses personal communication). The experiential mapping of soft/tough things (such as foods and non-foods) in the socio-physical environment onto affection/cruelty is captured in the conceptual metaphors, SOFT IS GOOD and TOUGH IS BAD.

Apart from softness and toughness, the heart is conceptualized as changeable in degree of strength or weakness as in the following cases:

(19) a. *qalb-u qwiyy*
heart his strong
'His heart is strong.'
He is courageous/He has a lot of courage.

b. *qalb-u* *D3iif*
heart his weak
'His heart is weak.'
He lacks courage.

If the material the heart is made of is strong, a person is said to have courage to confront hardships as in (19a). Weakness in the heart profiles lack of courage as in (19b).

The heart in present-day T. Arabic is not only conceptualized as changeable in texture (toughness/strength, softness/weakness), but also in size as in the following cases:

(20) a. *qalb-u kbiir*
heart his big
'His heart is big.'
He is generous.

b. *qalb-u SRiir*
heart his small
'His heart is small.'
He is mean.

Bigness of the heart as in (20a) is conceptualized as generosity, and smallness as meanness as in (20b). This correlation between bigness and generosity and smallness and meanness owes its existence to the conceptual metaphors, BIG IS GOOD and SMALL IS BAD. The heart changing in size was also noted for English by Niemeier (2000: 200), who captures it in CHANGEABLE IN SIZE.

Beside the change of texture, strength, and size, the heart in present-day T. Arabic is also conceptualized as changing in color as in:

(21) a. *qalb-u abyaD*
heart his white
'His heart is white.'
He is tolerant.

b. *qalb-u akHal*
heart his black
'His heart is black.'
He is spiteful (Maalej 1999).

In T. Arabic, just like in some Western cultures, white is associated with tolerance as in (21a) and black is profiled as intolerance as in (21b). This correlation of whiteness with tolerance and blackness with intolerance comes from the conceptual metaphors, WHITE IS GOOD and BLACK IS BAD.

Black hearts, which are presupposed in (22a-b) as lacking in purity, should undergo a process of purification to attain or approximate whiteness as in:

(22) a. *Saffi* *qalb-ik*
purify-IMPERATIVE heart your
'Purify your heart.'
Be more lenient/tolerant.

b. *qalb-u Saafi*
heart his pure
'His heart is pure.'
He is kind.

The heart can only be pure if its contents are purified, which can be captured as HEART IS A CONTAINER, where, by extension, THE CONTENTS OF THE HEART ARE SUBSTANCES. The process of purification that Tunisians map onto tolerance/kindness and intolerance is picked up from the purification of substances in their experience. Indeed, both liquid and solid substances undergo this purification process. A liquid such as milk is purified from its skin using a filter. Substances such as corn or olives are also purified, removing pebbles, seeds, and leaves with a sieve. Thus, in this case, if someone is intolerant, their heart needs to be purified to become tolerant and kind. Tolerance as purification of the heart is echoed in the present-day T. Arabic proverb:

(23) *xiyaar* *l-mumin* *qalb-u Saafi*
the best the believer heart his pure
'The best believer is one whose heart is pure.'

An interesting metonymy for purity of the heart is *flaan qalb-u Saafi laban* (X's heart is milk pure/white), where the whiteness of milk is a metonymy for its purity.

Along with change of texture, strength, size, and color, the heart in present-day T. Arabic is also conceptualized as changing in temperature as in:

(24) a. *qalb-u baarid*
heart his cold
'His heart is cold.'
He is idle/lazy.

b. *qalb-u* *yaHraq* *3ala* *xidmt-u*
heart his burn-IMPERFECT on work his
'His heart is burning for his job.'
He is a conscientious worker.

In (24a), coldness is a metaphor for lack of life or death, which stands for laziness. The phenomenological, felt sense of the physiological warmth of the heart is turned into a cultural metaphor for laziness as coldness. The laziness-as-coldness-of-the-heart metonymy-motivated metaphor dispenses with the heart's warmth as a physiological necessity, thus profiling lack of warmth in the heart as a typically cultural value. As Fauconnier and Turner (2002: 300) showed convincingly, HEAT OF THE BODY is not physiological, but a metaphor transferring heat to the human body.

In (24b), however, the conceptualization of industriousness takes advantage of the phenomenological, felt sense of physiological warmth of the body, turning it into fire burning in the heart. The industriousness-as-fire-in-the-heart augments, so to speak, the heart's warmth as a physiological necessity, thus profiling it as a typically cultural value. Burning as industriousness, however, is sometimes profiled ironically as in:

(25) *qalb-u* *yaHraq* *barša*
heart his burn-IMPERFECT a lot
'His heart is burning a lot.'
He could not care less about work.

The ironical use of *yaHraq barša* (burning a lot) in fact suggests its total opposite – that the heart is not burning with industriousness. Fire that metaphorically is supposed to power a person's heart can be extinguished, extinguishing with it the cultural value of industriousness.

In all the aforementioned conceptualizations where the heart's texture, strength, size, color, and temperature are manipulated, the heart is conceptualized as effecting a change of state, which profiles it culturally as a passive participant. However, the heart can also assume a more active role, thus controlling the self as in:

(26) a. *qalb-i* *kalaan-i*
heart my eat-PERFECT me
'My heart ate me.'
I had a pang of conscience.

b. *qalb-i* *ma* *kalaan-i-š*
heart my no eat-PERFECT me not
'My heart did not eat me.'
I had no pang of conscience.

In (26a), the physiological experience of eating is mapped onto the more mental experience of conscientiousness, where the heart becomes the eater and the body the object of the eating process. Inferentially, the phenomenological, felt sense of the physiological rest (i.e. the non-eating state) of the heart is when the self is doing things conscientiously. However, when the self starts doing things non-conscientiously, the heart finds itself in the physiological necessity of eating the self, profiling this physiological eating as a typically cultural value, which can be captured in the conceptual metaphor CONSCIENCE IS WHEN THE HEART EATS THE SELF or CONSCIENCE IS WHEN THE SELF BECOMES EDIBLE TO THE HEART. The counterpart of conscientiousness is rendered via the negative as in (26b), where the self does not think that it is being lazy.

Another conceptualization of laziness versus conscientiousness/industriousness is profiled through the death/life of the heart, where the heart stands for the person having that heart, as in:

(27) a. *qalb-u mayyit*
heart his dead
'His heart is dead.'
He is idle/lazy.

b. *qalb-u Hayy*
heart his alive
'His heart is alive.'
He is industrious.

Abiding by the metonymy of HEART FOR PERSON, the conceptual metaphor in (27a) could be LACK OF ENERGY/INDUSTRIOUSNESS IN AN INDIVIDUAL IS BEING DEAD and in (27b) PRESENCE OF ENERGY/ INDUSTRIOUSNESS IN AN INDIVIDUAL IS BEING ALIVE.

To sum up this section, as Niemeier (2000: 200) captured regarding size in CHANGEABLE IN SIZE for English, a characteristic of the Idealized Cognitive Model of the heart in present-day T. Arabic is its manipulability in terms of degree of texture, strength, size, color, temperature, status, and vitality. Owing to the gradable nature of the linguistic categories used to conceptualize cultural values metaphorically in present-day T. Arabic, the values themselves should not be seen as absolute pairs with two poles, the negative and the positive. Rather, the values are graded so that they constitute a cline, and are modified by quantifiers such as *barša* (a lot) and *šwayya* (a little bit). For instance, laziness can be talked about in terms of degree as in *qalb-u baarid* šwayya (He is a bit lazy), *qalb-u baarid* (He is lazy), *qalb-u baarid barša* (He is very lazy). Sometimes, the quantifier *barša* (a lot) is reduplicated to create another degree of laziness such as in *qalb-u baarid barša barša* (He is extremely lazy).

6. The metonymic model of the heart Idealized Cognitive Model in T. Arabic

As seen so far, the picture of the Idealized Cognitive Model drawn by the heart in present-day T. Arabic profiles it image schematically not only as a CONTAINER, but also as MOVING and MOVABLE in the bodily space, capable of IN-OUT and UP-DOWN movements. This dynamism is captured via directional metaphors that extend the IN-OUT and UP-DOWN image schemas metaphorically. Metaphorically, the heart's characteristic property is CHANGEABILITY and MANIPULABILITY. This kind of understanding is termed "indirect understanding via metaphor" by Lakoff and Johnson (1980: 178), who argue that directional structure is imposed via directional metaphors as projections of image schemas such as the ones dealt with in Section 4 above, whereas the structure of experience is imposed via structural metaphors as in Section 5 above.

To complement the discussion of the MOVABILITY and the CHANGEABILITY or MANIPULABILITY of the heart Idealized Cognitive Model in present-day T. Arabic, we need to address its inherent metonymic model. But before doing that, some reflection about metonymy as a conceptual phenomenon is needed. Metonymy is not simply a matter of reference or words substituting for others in the lexicon as traditional accounts had it. Langacker (2000: 199) argues that "a metonymic expression serves as a reference point affording mental access to the desired target (i.e. the entity actually being referred to)." Kövecses and Radden (1998: 39) define metonymy as "a cognitive process in which one conceptual entity, the vehicle, provides mental access to another conceptual entity, the target, within the same domain, or ICM." Talking about the cognitive and communicative function of metonymy, Langacker (2000: 199) argues that "metonymy allows an efficient reconciliation of two conflicting factors: the need to be accurate, i.e. of being sure that the addressee's attention is directed to the intended target; and our natural inclination to think and talk explicitly about those entities that have the greatest cognitive salience for us."

Thus, MENTAL ACCESSIBILITY realized via metonymy can be added to the picture. Indeed, the most prevalent metonymy is the HEART FOR PERSON, where the heart provides this mental accessibility to the person possessing the heart as in:

(28) a. *l-qalb* *3al l-qalb*
the heart on the heart.
'The heart is on the heart.'
Two hearts that beat as one.

b. *qalb-u* *baahi/Tayyib*
heart his good/fine
'His heart is good/fine.'
He is kind.

c. *qalb-u xaayib*
heart his bad/ugly
'His heart is bad/ugly.'
He is unkind/cruel.

The HEART FOR PERSON metonymy is motivated by the fact that the heart in (28a) is "salient and easily coded" (Langacker 2000: 199) in the sense that the heart inhabits an individual, which it can evoke. In (28b–c), the question is not about the metaphoricity of the heart as good or bad, but the salience between the heart and its possessor.

In the context of emotional conceptualization, an important candidate for the metonymy of HEART FOR PERSON is love in present-day T. Arabic as in:

(29)	*3and-i*	*qalb illi*	*yHibb-ha*
	With me	heart that	love-IMPERF her
	'I have a heart that loves her.'		
	I love her so much.		

It is socially motivated to avoid talking directly about one's love in public in the Tunisian culture. Using a metonymy violates "the need to be accurate" (Langacker, 2000: 199), and allows speakers to hide behind it, knowing that the heart has "the greatest cognitive salience" with the person (Langacker 2000: 199). Conversely, present-day T. Arabic does not have the metaphor 'breaking someone's heart," although Tunisians talk of "broken heart" in cases of disappointment such as in the proverb, *?in-naaSri l-maxSur w l-qalb l-miksur* (lost money and broken heart), which means that money wasted brings disappointment.

Very often the relation between metaphor and metonymy is indeterminate (Riemer 2002: 386), that is the demarcation line between the two may be fuzzy. Metonymy and metaphor may co-occur, creating MENTAL ACCESSIBILITY and MANIPULABILITY of the heart in present-day T. Arabic as in the conceptualization of greed and gratification (sexual or other):

(30) a. *qalb-u jii3aan*
heart his hungry
'His heart is hungry.'
He is greedy/He is insatiable.

b. *qalb-u šib3aan*
heart his satiated
'His heart is satiated.'
He is satiated.

MENTAL ACCESSIBILITY is captured through the metonymy, HEART FOR PERSON. Metonymically, the heart's hunger and gratification stand for the hungry or gratified person. However, when the person is substituted for the heart, this creates metaphors such as, *The person is hungry*. The linguistic metaphor can be captured in the conceptual metaphor, GREED IS HUNGER or the more generic metaphor DESIRE IS HUNGER. Conversely, social gratification is profiled in hunger terms as in (30a). The negative evaluation of a socially greedy person is captured in the T. Arabic proverb, *xuð-ha min yidd šib3aan iða jaa3 w ma taaxiðhaaš min yidd jii3aan iða šbi33* (Take it from the hand of a satisfied person if he gets hungry and don't take it from a hungry person if he becomes satisfied). Inferentially, the heart here becomes the stomach for greed and satisfaction.

One dimension of this indeterminacy has been interpreted as metonymic motivation for metaphoric mappings (Barcelona 2000). Forgetfulness is a form of malfunction of the mind, leading up to a defective memory, but in present-day T. Arabic one forgets with one's heart as in:

(31) *qalb-i ?a3ma*
heart my blind
'My heart is blind.'
I am forgetful.

It should be noted that in the metaphoric interpretation the blindness of the heart in present-day T. Arabic is forgetfulness, which presupposes that one of the heart's function is its capacity to remember. This can be captured in the conceptual metaphor, REMEMBERING IS SEEING, which is related to the more generic KNOWING IS SEEING; if my heart is blind, I do not remember things. In the metonymic interpretation, the heart is interpreted as contiguous with the person as in HEART FOR PERSON. This seems to be consistent with the conception of blindness as encoded in the Koran in (3b) above, which is repeated here for the sake of convenience:

(32) Truly it is not their eyes that are blind, but their hearts which are in their breasts.[12]

However, blindness of the heart in the Koranic conception is incapacity to understand while this Tunisian conception has to do with forgetfulness. Niemeier (2000: 208) expressed fascination as to how memory, which is normally associated with the head/brain comes to be associated with the heart in expressions like "learn something by heart." If this unscientific match between heart and memory means something, it serves to suggest that this model of the heart in English and present-day T. Arabic has a cultural basis.

In sum, the metonymic model of the heart is realized via simple metonymy having potential metaphoric counterparts, or via a metaphor-metonymy continuum. The correlative work that these do is ACCESSIBILITY via the HEART FOR PERSON metonymy or metonymic ACCESSIBILITY in conjunction with metaphoric CHANGEABILITY.

7. Conclusion

The chapter has tried to demonstrate that the heart in present-day T. Arabic is a productive source domain, profiling most emotions and culture-specific values. The cognitive/cultural model of the heart reflects these cultural conceptualizations exploiting image schema, metaphor, and metonymy. Logically, since the heart is seen as the locus of emotions and values, it is culturally affected by these emotions and values.

The cultural model of the heart, therefore, qualifies as an Idealized Cognitive Model (Lakoff 1982, 1987). Its propositional structure shows the heart to be an affected participant in interpersonal relations and an affecting participant across the image schemas, metaphors, and metonymies that profile it. As seen in the body of the chapter, the heart is conceptualized image schematically as a CONTAINER and in UP-DOWN and IN-OUT schemas. This has been captured in the directional metaphor, THE HEART IS MOVING/MOVABLE. It also offers various metaphoric conceptualizations that profile emotions in terms of CHANGEABILITY or MANIPULABILITY of the heart. Apart from these conceptualizations, the heart yields a HEART FOR PERSON metonymy, providing ACCESSIBILITY of the body/self through the heart.

As a source domain for emotions and cultural values, the heart is an excellent illustration of the cultural embodiment of the mind. If Johnson (1987) is right in claiming that the embodied mind yields embodied meaning, imagination, and reasoning, the present chapter extends embodiment to culture, thus suggesting that cognition is not just embodied but culturally embodied. Imaginative structures such as image schemas, metaphor, and metonymy contribute to cultural imagination, which motivates what Maalej (2004, 2007) called "cultural embodiment," which is a kind of embodiment mediated and motivated by cultural imagination.

To further corroborate the import of culture in conceptualization, reference has to be made to Palmer (1996: 36), who considers cultural linguistics as "primarily concerned not with how people talk about some objective reality, but with how they talk about the world that they themselves imagine." The present-day T. Arabic culture has imagined generosity to be bigness of the heart, meanness smallness, tolerance whiteness, spite blackness, compassion/mercy softness, indifference toughness, and so forth. But cultures do not randomly talk about the world as they imagine it; each culture organizes itself according to "its own priorities of grouping and differentiation" (Palmer 1996: 227).

Acknowledgements

I am thankful and grateful to the following friends and colleagues: René Dirven for generous comments on the first and second drafts of the current chapter, which made an impact on my analysis; Zoltán Kövecses for being generous with his time and knowledge of metaphor and metonymy, which has greatly improved the quality of my analyses; Farzad Sharifian for his help, patience, and encouragement; an anonymous reviewer for insightful comments on the chapter. I owe also special thanks to Jessica Cleary-Kemp (Monash University, Australia) for very useful comments that greatly improved the quality of the language of the chapter. However, responsibility for the contents remains mine.

Notes

1. As far as I know, the only written documents in T. Arabic that exist are a few plays by Taoufik Jebali (*Night's Talk*, 1997), Mohamed Idriss (*Ismail Pacha*, 1997), Fadhel Jaidi (*Familia*, 1997), etc.
2. Sura VII (A'raaf, or Heights), translated by Abdullah Yusuf Ali (1938), pp. 396–97.
3. Sura XXII (Hajj, or The Pilgrimage), translated by Abdullah Yusuf Ali (1938), p. 863.
4. Sura L (qaf, or The Matter has been decreed), translated by Abdullah Yusuf Ali (1938), p. 1417.
5. Sura VI (An'am, or Cattle), translated by Abdullah Yusuf Ali (1938), p. 289.
6. Sura XLVII (Muhammad, or The Prophet), translated by Abdullah Yusuf Ali (1938), p. 1385.
7. Sura XVII (Bani Isra'il, or The Children of Israel), translated by Abdullah Yusuf Ali (1938), p. 707.
8. Sura VII (A'raaf, or Heights), translated by Abdullah Yusuf Ali (1938), p. 371.
9. Sura XXII (Hajj, or The Pilgrimage), translated by Abdullah Yusuf Ali (1938), p. 864.

10. Sura XXXIII (Ahzab, or The Confederates), translated by Abdullah Yusuf Ali (1938), p. 1123.
11. Translation mine.
12. Sura XXII (Hajj, or The Pilgrimage), translated by Abdullah Yusuf Ali (1938), p. 864.

References

Al-Jawziyya, Ibn Q.
1998 *Medicine of the Prophet.* Cambridge: The Islamic Texts Society.

Barcelona, Antonio
2000 On the plausibility of claiming a metonymic motivation for conceptual metaphor. In *Metaphor and Metonymy at the Crossroads: A Cognitive Perspective*, A. Barcelona (ed.), 31–58. Berlin/New York: Mouton de Gruyter.

Burnett, C.
2004 *Arabic Medicine in the Mediterranean.* Retrieved 26-03-06, from http://www.muslimheritage.com/topics/default.cfm?ArticleID=459.

Emanatian, Michele
1995 Metaphor and the expression of emotion: The value of cross-cultural perspectives. *Metaphor and Symbolic Activity,* 10 (3): 163–182.

Geeraerts, Dirk & Stefan Grondelaers
1995 Looking back at anger: Cultural traditions and metaphorical patterns. In *Language and the Cognitive Construal of the World*, J. R. Taylor and R. E. MacLaury (eds.), 153–179. Berlin: Mouton de Gruyter.

Gibbs, Raymond W. Jr.
2005 *Embodiment and Cognitive Science.* Cambridge: Cambridge University Press.

Ibn ManDur, M.
1994 *Lisaanu l-3arabi.* 3rd ed. (*The Tongue of Arabs*). Beirut: Dar Saadir.

Johnson, Mark
1987 *The Body in the Mind: The Bodily Basis of Meaning, Imagination, and Reason.* Chicago/London: The University of Chicago Press.

Kövecses, Zoltán
2000 *Metaphor and Emotion: Language, Culture and Body in Human Feeling.* Cambridge: Cambridge University Press.
2002 *Metaphor: A Practical Introduction.* Oxford: Oxford University Press.

2005 *Metaphor in Culture: Universality and Variation*. Cambridge: Cambridge University Press.

Kövecses, Zoltan & Gunther Radden
1998 Metonymy: Developing a cognitive linguistic view. *Cognitive Linguistics*, 9 (1): 37–77.

Lakoff, George
1982 Categories: An essay in cognitive linguistics. In *Linguistics in the Morning Calm*, The Linguistic Society of Korea (ed.), 139–193. Seoul: Hanshin Publishing Company.
1987 *Women, Fire, and Dangerous Things: What Categories Reveal about the Mind*. Chicago/London: The University of Chicago Press.
1996 Sorry, I'm not myself today: The metaphor system for conceptualizing the self. In *Spaces, Worlds and Grammar*, G. Fauconnier and E. Sweetser (eds.), 91–123. Chicago/London: The University of Chicago Press.

Lakoff, George & Mark Johnson
1980 *Metaphors We Live By*. Chicago/London: The University of Chicago Press.
1999 *Philosophy in the Flesh: The Embodied Mind and its Challenge to Western Thought*. New York: Basic Books.

Lakoff, George & Mark Turner
1989 *More than Cool Reason: A Field Guide to Poetic Metaphor*. Chicago/London: The University of Chicago Press.

Langacker, Ronald W.
2000 *Grammar and Conceptualization*. Berlin/New York: Mouton de Gruyter.

Levinson, Stephen C.
1996 Relativity in spatial conception and description. In *Rethinking Linguistic Relativity*, J. J. Gumperz & S. C. Levinson (eds.), 177–202. Cambridge: Cambridge University Press.

Maalej, Zouhair
1999 Metaphoric discourse in the age of cognitive linguistics, with special reference to Tunisian Arabic. *Journal of Literary Semantics,* 28 (3): 189–206.
2001 Of animals, foods, objects, and plants, or how women are conceptualized: A cross-cultural perspective. Paper presented at the 4th International Conference on Researching and Applying Metaphor (RAAM IV): Metaphor, cognition, and culture, University of Manouba, Tunisia.
2004 Figurative language in anger expressions in Tunisian Arabic: An extended view of embodiment. *Metaphor and Symbol*, 19 (1): 51–75.

2007 The embodiment of fear expressions in Tunisian Arabic: Theoretical and practical implications. In *Applied Cultural Linguistics: Second Language Teaching/Learning and Intercultural Communication*, F. Sharifian and G. B. Palmer (eds.), 87–104. Amsterdam/Philadelphia: John Benjamins.

MacPhee, M.
2003 Medicine for the heart: The embodiment of faith in Morocco. *Medical Anthropology,* 22: 53–83.

Niemeier, Susanne
1997 To have one's heart in the right place – metaphorical and metonymic evidence for the folk model of the heart as the site of emotions in English. In *Human Contact through Language and Linguistics*, B. Smieja and M. Tasch (eds.), 87–106. Frankfurt: Peter Lang.
2000 Straight from the heart – Metonymic and metaphorical explorations. In *Metaphor and Metonymy at the Crossroads: A cognitive perspective*, A. Barcelona (ed.), 195–213. Berlin/New York: Mouton de Gryuter.

Palmer, Gary B.
1996 *Toward a Theory of Cultural Linguistics*. Austin: University of Texas Press.

Riemer, Nick
2002 Remetonymizing metaphor: Hypercategories in semantic extension. *Cognitive Linguistics,* 12 (4): 379–401.

Rohrer, Tim
1998 When metaphors bewitch, analogies illustrate, and logic fails: Controversies over the use of metaphoric reasoning in philosophy and science. Ph.D. dissertation: Department of Philosophy and the Graduate School of the University of Oregon.
2001 Pragmatism, ideology and embodiment: William James and the philosophical foundations of cognitive linguistics. In *Language and Ideology: Cognitive Theoretical Approaches,* R. Dirven, B. Hawkins and E. Sandikcioglu (eds.), 49–82. Amsterdam: John Benjamins.
In press The body in space: Dimensions of embodiment. In *Body, Language and Mind (Vol. 1): Embodiment*, T. Ziemke, J. Zlatev and R. M. Frank (eds.). Berlin: Mouton de Gruyter.

Sharifian, Farzad
2003 On cultural conceptualisations. *Journal of Cognition and Culture,* 3 (3): 187–207.

Sinha, Chris and Kristine Jensen de López
2000 Language, culture and the embodiment of spatial cognition. *Cognitive Linguistics*, 11 (½): 17–41.

Sweetser, Eve
1990 *From Etymology to Pragmatics: Metaphorical and Cultural Aspects of Semantic Structure.* Cambridge: Cambridge University Press.

Varela, Francisco J., Evan Thompson, and Eleanor Rosch
1991 *The Embodied Mind: Cognitive Science and Human Experience.* Cambridge, Mass.: The MIT Press.

Wilson, M.
2002 Six views of embodied cognition. *Psychological Bulletin and Review,* 9 (4): 625–636.

Winsvold, B. S.
2005 Arabian medicine in the Middle Ages: The Arabian world's contribution to science and medicine. Retrieved 26-03-06, from http://www.globalcomment.com/science&technology/article_14.asp.

Yu, Ning
1995 Metaphorical expression of anger and happiness in English and Chinese. *Metaphor and Symbolic Activity,* 10 (2): 59–92.
1998 *The Contemporary Theory of Metaphor: A Perspective from Chinese.* Amsterdam/Philadelphia: John Benjamins Publishing Company.
2003 Metaphor, body, and culture: The Chinese understanding of gallbladder and courage. *Metaphor and Symbol,* 18 (1): 13–31.

Yusuf Ali, Abdallah
1938 *The Holy Qur'an* (3rd edition). Syria: Dar Al-MuSHaf.

Ziemke, Tom
2003 What is that thing called embodiment? In *Proceedings of the 25th Annual Meeting of the Cognitive Science Society.* Retrieved 10-02-05, from http://www.cogsci.rpi.edu/CSJarchive/Proceedings/2003/pdfs/244.pdf.

Ziemke, Tom, Jordan Zlatev, and Roselyn M. Frank (eds.)
In press *Body, Language and Mind (Vol. 1): Embodiment.* Berlin/New York: Mouton de Gruyter.

Zouari, A. & Charfi, Y.
1998 *Mu3jamu l-kalimaati wa t-takaaliidi iš-ša3biyyati bi Safaaqusa* (*Dictionary of Words and Popular Traditions in Sfax*). Sfax: Société générale d'imprimerie et de cartonnage (SOGIC).

Index

Bei Fragen zur Produktsicherheit wenden Sie sich bitte an:
If you have any questions regarding product safety,
please contact:

Walter de Gruyter GmbH
Genthiner Straße 13
10785 Berlin
productsafety@degruyterbrill.com